HANDBOOK OF PERCEPTION

Volume IX

Perceptual Processing

This is Volume IX of

HANDBOOK OF PERCEPTION

EDITORS: *Edward C. Carterette and Morton P. Friedman*

Contents of the other books in this series appear at the end of this volume.

HANDBOOK OF PERCEPTION

VOLUME IX

Perceptual Processing

EDITED BY

Edward C. Carterette and Morton P. Friedman

Department of Psychology
University of California, Los Angeles
Los Angeles, California

ACADEMIC PRESS New York San Francisco London 1978

A Subsidiary of Harcourt Brace Jovanovich, Publishers

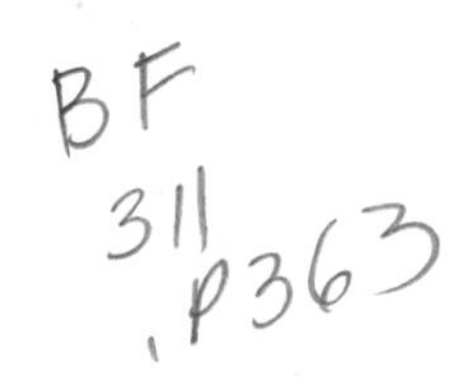

ACADEMIC PRESS, INC.
111 Fifth Avenue, New York, New York 10003

United Kingdom Edition published by
ACADEMIC PRESS, INC. (LONDON) LTD.
24/28 Oval Road, London NW1 7DX

Library of Congress Cataloging in Publication Data
Main entry under title:

Perceptual processing.

(Handbook of perception ; v. 9)
Includes bibliographies and index,
1. Perception--Addresses, essays, lectures.
2. Human information processing--Addresses, essays, lectures. 3. Visual perception--Addresses, essays, lectures. I. Carterette, Edward C. II. Friedman, Morton P.
BF311.P363 153.7 78-13533
ISBN 0-12-161909-5

PRINTED IN THE UNITED STATES OF AMERICA

CONTENTS

Chapter 3. Sorting, Categorization, and Visual Search

Patrick Rabbitt

PART II. PATTERN PROCESSING

Chapter 4. Schemes and Theories of Pattern Recognition

Stephen K. Reed

Chapter 5. Perceptual Processing in Letter Recognition and Reading

W. K. Estes

Chapter 6. Eye Movements and Visual Perception

Geoff D. Cumming

Chapter 7. Perceptual Learning

Richard D. Walk

Chapter 8. Size, Distance, and Depth Perception

Walter C. Gogel

PART III. ILLUSIONS AND DISORDERS

Chapter 9. Illusions and Hallucinations

Richard L. Gregory

Chapter 10. Disorders of Perceptual Processing

Francis J. Pirozzolo

LIST OF CONTRIBUTORS

Numbers in parentheses indicate the pages on which the authors' contributions begin.

DAVID E. CLEMENT (49), Department of Psychology, University of South Florida, Tampa, Florida 33620

GEOFF D. CUMMING (221), Department of Psychology, La Trobe University, Bundoora, Victoria, Australia 3083

W. K. ESTES (163), The Rockefeller University, New York, New York 10021

WALTER C. GOGEL (299), Department of Psychology, University of California, Santa Barbara, Santa Barbara, California 93106

RICHARD L. GREGORY (337), Brain and Perception Laboratory, University of Bristol, Bristol, England

STEVEN W. KEELE (3), Department of Psychology, University of Oregon, Eugene, Oregon 97403

W. TRAMMELL NEILL (3), Department of Psychology, University of Oregon, Eugene, Oregon 97403

FRANCIS J. PIROZZOLO (359), Department of Neurology, Minneapolis Veterans Administration Hospital, Minneapolis, Minnesota 55417

PATRICK RABBITT (85), Department of Experimental Psychology, University of Oxford, Oxford OX1 3UD, England

STEPHEN K. REED (137), Department of Psychology, Case Western Reserve University, Cleveland, Ohio 44106

RICHARD D. WALK (257), Department of Psychology, The George Washington University, Washington, D. C. 20052

FOREWORD

The problem of perception is one of understanding the way in which the organism transforms, organizes, and structures information arising from the world in sense data or memory. With this definition of perception in mind, the aims of this treatise are to bring together essential aspects of the very large, diverse, and widely scattered literature on human perception and to give a précis of the state of knowledge in every area of perception. It is aimed at the psychologist in particular and at the natural scientist in general. A given topic is covered in a comprehensive survey in which fundamental facts and concepts are presented and important leads to journals and monographs of the specialized literature are provided. Perception is considered in its broadest sense. Therefore, the work will treat a wide range of experimental and theoretical work.

This ten-volume treatise is divided into two sections. Section One deals with the fundamentals of perceptual systems. It is comprised of six volumes covering (1) historical and philosophical roots of perception, (2) psychophysical judgment and measurement, (3) the biology of perceptual systems, (4) hearing, (5) seeing, and (6) which is divided into two books (A) tasting and smelling and (B) feeling and hurting.

Section Two, comprising four volumes, covers the perceiving organism, taking up the wider view and generally ignoring specialty boundaries. The major areas include (7) language and speech, (8) perceptual coding of space, time, and objects, including sensory memory systems and the relations between verbal and perceptual codes, (9) perceptual processing mechanisms, such as attention, search, selection, pattern recognition, and perceptual learning, (10) perceptual ecology, which considers the perceiving organism in cultural context, and so includes aesthetics, art, music, architecture, cinema, gastronomy, perfumery, and the special perceptual worlds of the blind and of the deaf.

The "Handbook of Perception" should serve as a basic source and reference work for all in the arts or sciences, indeed for all who are interested in human perception.

EDWARD C. CARTERETTE
MORTON P. FRIEDMAN

PREFACE

But if we want details, I think that the experimentalists may justly point to three principal achievements: the complete recasting of the doctrine of memory and association, the creation of a psychology of individual differences, and the discovery of attention.

EDWARD BRADFORD TITCHNENER,
Lectures on the Elementary Psychology of Feeling and Attention, 1908

Attention is the very core of cognitive psychology. What limits processing and what controls the flow of information from input to output? These are central problems of any theory of attention. Apparently, a stimulus evokes a wide range of memories within and between such codes as auditory, visual, semantic, or abstract. Activation so widespread requires a limiting control process for selecting and integrating codes and memories, one which is guided by goals and by task demands. So, Volume IX, **Perceptual Processing**, begins in a sensible way with Keele and Neill's chapter on *Mechanisms of Attention*, emphasizing the *how* and *why* of the processing. The next chapter, *Perceptual Structure and Selection* by Clement, shifts to an emphasis on *what* is processed, taking the view that perception is an active process of selecting—"the organism perceives, rather than receives information." By structure is meant the correlation among stimuli, among the elements of stimuli, among the representations of stimuli, and among the representations of all of these in the neural processes of the perceiver. Many experimental results on selection and categorization in visual search, together with models of possible control processes, are taken up in Rabbitt's Chapter 3 on *Sorting, Categorization, and Visual Search*. A moral of this review is that a subject *can* use any of a large number of categorizing schemes but the one he *does* use may not be the one the experimenter *wishes* used.

"It's a bird! It's a plane! It's *Superman*!" How does one describe and recognize patterns? Many ways are possible and many theories have been proposed: template, structural, topological, and feature. And, given a description that is psychologically effective, how does one identify a particular pattern? This latter requires detailed process models. In Chapter 4, *Schemes and Theories of Pattern Recognition*, Reed deals mainly with the central problem of representing patterns. The broader cognitive aspects are reviewed in his book on psychological processes in pattern recognition (Reed, 1973).

Owing to advances in theory and method a great part of what is known about perceptual processes in reading has come only in the past 12 years. The plan of W. K. Estes' Chapter 5, *Perceptual Processing in Letter Recognition and Reading*, is to review how individual letters are processed and then ask how far this knowledge will go toward accounting for the perceptual properties of arrays of letters, possibly involving rules of interaction. Then, with adequate theoretical preparation, it is hoped to clarify such focal issues as the perceptual units of reading, limits of capacity, whether processing is parallel or serial, categorial perception, and linguistic factors in the recognizing of letters.

Eye movements may betray the focus of attention, and from them inferences are made about mental life. New methods and theories have recently proved the value of studying eye movements in perception and cognition, as Cumming points out in Chapter 6 where he surveys *Eye Movements and Visual Perception*.

Perceptual learning refers roughly to the discovery of those aspects of stimuli that govern discriminative acts. It is hard to separate perceptual learning from cognitive learning for the second embraces the first and more. Playing the French horn well is a highly learned perceptual motor act. An able conductor likewise has learned a complex motor skill. But he has learned more than keeping the tempo, adjusting the dynamics of the orchestra, and cueing first chairs. He has learned also cognitive skills, such as program balance, interpretation, management, and planning. Walk's Chapter 7 on *Perceptual Learning* reviews data and theory, in which visual research figures large but new auditory research and the interaction and integration of modalities is stressed.

From my seventh floor window I see a young woman on the street about 25 meters away. She subtends the same angle as a U. S. dime held at arm's length. I should say that she is quite attractive, about 1.75 meters tall and weighing about 55 kilograms. Notice that my perception of size contradicts the proximal stimulus at my retina. Failure of the proximal stimulus to account completely for my percept is called *stimulus ambiguity*. The review of *Size, Distance, and Depth Perception* by Gogel in Chapter 8 indicates the conditions for and possible explanations of stimulus ambiguity.

The volume closes with two chapters on perceptual anomalies, distortions and disorders. Pirozzolo's succinct outline in Chapter 10, *Disorders of Perceptual Processing*, seen from an orthodox neuropsychological view, is preceded by Chapter 9 in which Gregory treats *Illusions and Hallucinations*. Illusions are perceptual departures from physical reality, such as distortions (Müller–Lyer figure), ambiguities (Necker cube), or paradoxes. The position taken is that any proposition is a hypothesis, whether about the physical or perceptual world. Departures from physical facts are *errors*, perceptual departures are *illusions*, and so errors and illusions are similar. Assumptions about the nature of *signal* errors and *data* errors in perceptual processing lead to a distinction between illusions that are physiological and those that are cognitive. A classification of illusions as physiological or cognitive is attempted with alternative accounts within each.

Financial support has come in part from the National Institute of Mental Health (Grant MH-07809), The Ford Motor Company, and The Regents of the University of California.

Editors of Academic Press both in New York and San Francisco have been enormously helpful in smoothing our way.

Part I

Attention and Selection

Chapter 1

MECHANISMS OF ATTENTION

STEVEN W. KEELE AND W. TRAMMELL NEILL

I. INTRODUCTION

The concept of attention lies at the very core of cognitive psychology. Indeed some people consider this concept to be the primary feature distinguishing the cognitive school from classical behavior theory. According to some classic theories of learning, any conditionable stimulus that reliably precedes a conditionable response by a short amount of time will come to elicit that response. Such theories regard people and animals as passive receivers and transformers of information from the environment. It is now well established even for animals (e.g., Reynolds, 1961), however, that only one of two simultaneous stimuli might be conditioned, implying active selection of available information. The selected stimulus varies between animals and with context. In general, and not only in

ISBN 0-12-161909-5

conditioning, the nervous systems of animals and people alike are severely limited in processing information. Only some information is perceived, only some is responded to, only some is remembered.

The study of attention is concerned with the nature of these limitations and with the selective processes that deal with them. Where in the sequence of information processes do limitations occur? What options do people have about which information is selected? What happens to nonselected information? What are the costs and benefits of selection? Although many different approaches to the problem of attention exist, the present chapter concentrates on the analysis of attention in perceptual tasks by humans.

A study by Klein and Posner (1974) vividly illustrates the limited nature of processing, as well as illustrating some types of mental operations that draw on processing capacity. A target light was moved horizontally across an oscilloscope screen, reversing directions at one, three, or five points. The actual reversal points varied on different trials, and after each pattern the subject tried to reproduce the time and positions of the reversals by moving a lever. In one test, subjects visually observed the pattern before reproducing it. In a second test, they not only observed the pattern but also tracked it by moving the lever to follow the target with a cursor. In the former condition the subject has available only a visual input with which to remember the pattern, but in the latter condition both visual and kinesthetic inputs are available. Which is the easier condition to reproduce? Many people expect the dual code to be better; in fact, people were more accurate when only visual input was available.

The Klein and Posner results appear paradoxical in their indication that two inputs are worse than one. What theory could explain the paradox? Two possibilities are (*a*) that the process of correcting tracking errors requires attention, competing with the attention demands of storing the information in memory; and (*b*) that attempts to store two codes draw on a limited-storage mechanism and result in mutual interference—that is, only one code can be rehearsed at a time. Both hypotheses appear partly correct. Klein and Posner show that corrections sometimes interfere with reaction time to other signals, implying attention demands of the corrective process in tracking. They also show that when the arm is passively, rather than actively, moved in tracking, memorization of both codes also results in interference, even though corrections in tracking are eliminated.

Many people closely identify attention with being able to remember what was attended. Clearly, in view of the Klein and Posner study, such a link should be avoided. Instances exist in which apparently greater investment of attention, as in tracking, actually impairs memory. While memorization may require attention, the converse is not true. The analysis of attention depends on observations of interference between tasks, as well as on observations of the recall of information. As a general working definition, two tasks that interfere with each other are said to require attention. Of course, interference for physical reasons, such as the requirement to simultaneously move the finger in opposite directions, is not indicative of attention. Also, interference because of peripheral masking is not indicative of attention.

II. TWO KINDS OF PROCESSING LIMITATIONS

All brain processes take time, and this constitutes one limitation on processing. Moreover, the brain exhibits space limitations in terms of the number of things that can be done at once. In the 1950s and early 1960s a widely held view of attention theorized a strong relation between the two types of limitation. This *single-channel theory* (Welford, 1960) and the closely allied *filter theory* (Broadbent, 1958) were loosely built around a technical concept of information as developed by Shannon and Weaver (1949). The translation of a signal from a sensory code to a memory or response code was thought to involve a channel of limited capacity. Since only one channel existed, the translation of one signal to another code precluded translating yet another. The greater the information content of the first signal, the longer the time it would occupy the channel, and the greater the delay in processing another signal. Under certain versions of the theory, two signals may be processed at the same time. If the information load of both together is less than the capacity available, capacity may be divided between them with no resulting interference. But if the total demand exceeds the available capacity, both signals will suffer.

Evidence for single-channel theory developed on two fronts. On one front, classic experiments by Hick (1952) and Hyman (1953) showed that the reaction time to translate light signals to either key-press responses or verbal responses was related to the information processed: Reaction time increased linearly with the logarithm of the number of potential stimuli and responses for a given situation and decreased with the logarithm of stimulus probability when the total number of stimuli was held constant. When errors were made, the information transmitted from signal to response was reduced and the reaction time correspondingly decreased in a linear fashion with the drop in information transmitted. These results from the Hick and Hyman studies all suggested a channel limited to about 150–200 msec per bit of information processed.

However, the time to process each bit of information depends heavily on variables other than the amount of information. Chief among these is stimulus–response (S–R) compatibility (Fitts & Seeger, 1953): The time to translate a signal to a response depends not only on the number and probabilities of stimuli and responses, but also on the relationship between them. Translation is quicker when stimuli and responses are spatially similar, and sometimes reaction time may hardly increase with information. Leonard (1959) found this to be the case when people responded by pressing down with a finger that was touched, and Fitts, Peterson, and Wolpe (1963) found it to be the case when people pointed at lights as they were illuminated. In addition, the translation time is greatly reduced with high degrees of practice (Fitts & Seeger, 1953; Mowbray & Rhoades, 1959), again indicating that the time required to translate a signal to a response is not strictly limited by the amount of information.

On the second front that attempted to link time and space, it was shown that when two signals were presented about the same time, processing of one or the other or both was slowed. The interference appeared related to the time demands of individual signals. Welford (1959) presented two successive signals, one light followed by another. When the second signal occurred before the first response was emitted, the reaction time to the second signal was delayed. The second-signal delay was actually greater than the remaining processing time of the first signal, suggesting that only one signal could translate to a response at one time. In addition, monitoring the feedback from the first response further preempted the processing channel. Karlin and Kestenbaum (1968) and Smith (1969) increased reaction time to the first signal by increasing the number of stimulus alternatives. Broadbent and Gregory (1967) increased reaction time by decreasing S–R compatibility. Increased reaction times to the first signal were closely matched by further delays in processing the second signal. The results are exactly what would be expected were single-channel theory basically correct: Processing the second signal cannot begin until the first is finished.

Paradigms using simultaneous auditory messages led to a similar conclusion. Cherry (1953) found that when two prose passages were spoken by the same voice and came through the same speaker, selectively repeating one message and ignoring the other was very difficult. If, however, the messages came through separate earphones, the task was relatively easy. When successfully shadowing the message to one ear, subjects were unaware of the semantic input in the ignored ear. They rarely noticed switches from English to German or to English played backward. Results of this sort suggested that only one message at a time could be translated from the sensory input to the semantic code. Other variations on the dichotic listening paradigm suggested that the translation process was limited in an information sense. Broadbent (1956), for example, found the disrupting effect of a buzzer on understanding a verbal message to be larger for a two-choice buzzer decision than a one-choice buzzer decision.

A diagram of the theory growing out of these sorts of study is shown in Fig. 1, based on Broadbent (1958). When more than one stimulus or message occurs at a time, they are entered in parallel to a sensory buffer. Barring peripheral masking, no interference occurs at this stage. One message is then allowed through a filter; the other message is held in the buffer for later processing. The filter prevents overloading a limited-capacity mechanism that translates the stimulus to some other code stored in long-term memory.

III. ILLUSTRATIVE DIFFICULTIES FOR SINGLE-CHANNEL THEORY

The central assumption in single-channel theory is a processing limitation in translating information in the sensory buffer to long-term memory. The information in long-term memory may be the name of the stimulus, the meaning of the stimulus, or some other information, such as the response to be made to the stimulus. If it can be shown that an ignored stimulus actually activates information in long-term memory at the same time another stimulus is being normally processed, a problem is posed for single-channel theory. One study that used this approach was Treisman's (1964a) shadowing study with French–English bilinguals.

The subjects listened to and repeated a prose passage from Orwell's "England, Your England" played in one ear in either French or English.

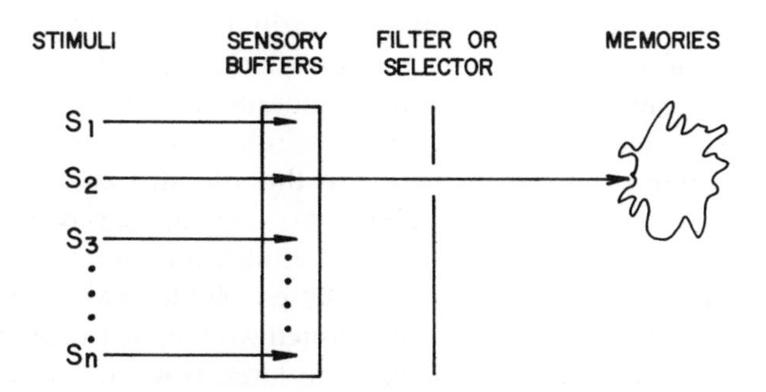

Fig. 1. A schematic representation of single-channel or filter theory. Only one signal at a time gains access to the memory system. [Adapted from Broadbent (1958).]

At the same time, and not known to the subjects, the same passage was played in the opposite ear, but in the other language. Initially the two passages were staggered in time, but gradually the gap was reduced. When the gap was small, and particularly when the shadowed message led, many of the subjects spontaneously noticed that the two messages were the same, though in different languages. Apparently the ignored message was not entirely blocked from long-term memory.

Since Treisman's study used prose passages, the messages had considerable redundancy; the meanings of early parts of a passage prepare one for later meanings. This redundancy could be important to the effect described by Treisman, and, indeed, it played a part in Treisman's explanation.

Lewis (1970) avoided redundancy by presenting a list of unrelated words to the ear to be shadowed. The other ear received other words synchronized with the attended items. The reaction time to shadow a word was slightly delayed when the unattended word was a synonym of the simultaneous word in the attended ear. The subjects, however, reported no awareness of the content of the ignored words. Apparently the nonshadowed ear receives semantic analysis and is not entirely blocked prior to long-term memory. Otherwise the semantic nature of the nonshadowed word should have no effect.

Treisman, Squire, and Green (1974) replicated some features of Lewis's experiment. They found that the semantic relationship of the nonshadowed to the shadowed word had a modest effect on reaction time to shadowed words early in a list, but the effect disappeared by the seventh item shadowed. They suggested that more effective blockage prior to long-term memory occurs as subjects go further into the list. Nevertheless, the results are damaging to single-channel theory because of the fact that, at least early in the list, more than one item may activate long-term memory at one time. Moreover, as Treisman *et al.* point out, a failure to get the synonymic effect does not necessarily imply blockage. Nevertheless, because of the Treisman *et al.* results, a modification by von Wright, Anderson, and Stenman (1975) becomes more important.

They presented a long list of dichotic words, in which the Finnish word meaning *suitable* occurred six times, to the attended ear. Two of those occasions were followed by an electric shock. Reappearance of the word produced a galvanic skin response (GSR). Following conditioning, another long list of dichotic word pairs was presented. Subjects shadowed one ear and ignored the other. When the conditioned word appeared in the ignored list, it evoked a GSR change, though not as large as when it occurred in the shadowed ear. More important, when either a synonym or a homonym of the conditioned word appeared in the ignored ear, the change in GSR, though smaller than for *suitable*, was as large in the unattended as in the attended ear. Again it appears that the ignored ear is not blocked prior to semantic analysis in long-term memory.*

These studies pose a problem for classical single-channel theory. At least some translation of one signal to memory can occur while another is being translated. To handle such results, theoretical departures from single-channel theory took two major directions.

One direction, proposed by Treisman (1964a), basically maintained the idea of a limit between sensory and memorial processes. Treisman suggested that an ignored message, rather than being perfectly filtered, is only attenuated so that some leakage occurs in the memory system. When previous content sensitizes appropriate memory units, or when particular memory units (e.g., one's name) are more permanently sensitized, the leakage is great enough to trigger a response.

The second theoretical answer to problems with single-channel theory was to suggest that more than one signal had unimpaired, parallel access to the memory system and that selectivity, aided by cues such as message location, occurred at the memory level rather than the sensory level. Deutsch and Deutsch (1963) and Norman (1968) were early proponents of this view.

The general question that emerges at this point, therefore, is whether more than one signal has simultaneous and unimpaired access to signal-related information stored in memory. It is quite evident that when two signals must be processed to the stage of separate responses, interference usually occurs; there is a capacity limitation. But is that limitation in the access to information in long-term memory, or is it in the mental operations such as rehearsal, response execution, and conscious perception that are applied to information already activated in long-term memory? The answer to this seemingly simple question has been exceedingly elusive. A number of studies quite convincingly argue that memory retrieval, while time-consuming, is nonattentive. These studies will be discussed in the next section. Other studies just as convincingly raise problems for the unlimited view. Some of those problems are described and some speculations are made on possible resolutions of the conflict.

IV. PROCESSING MULTIPLE SIGNALS

A general strategy in assessing the source of limitations is to present two signals at about the same time and vary the response requirements.

* Recently Wardlaw and Kroll (1976) attempted to replicate the demonstration that a conditioned item in an unattended ear evokes a GSR change. They were unable to replicate the phenomenon. The reason for the discrepancy is unknown.

One signal may require a response and the other may not. Both signals may converge on the same response. Both signals may require separate responses. The difficulty of the two stimulus–response mappings may be independently varied. By examining the patterns of interference or facilitation under different conditions, clues as to the sources of interference may be obtained.

This approach constitutes a microstudy of attention that looks in detail at short lasting processes that normally take less than a second. It differs from the macrostudy of attention in more complex, continuous tasks. For example, Martin, Marston, and Kelly (1973) presented a probe signal during a memory task. Reaction time to the probe indicated large attention demands during all phases of memory—encoding, retention, and retrieval. Each of these phases, however, involves several unanalyzed mental operations. For example, the process of encoding could involve the activation of the name or meaning of a presented item, conscious recognition, and elaborative rehearsal. All these are grouped under the heading of encoding. The attention demand at a macrolevel presumably reflects the mix of microprocesses, their rate of occurrence, and whether they can be delayed for split seconds while the probe task is handled. Kerr (1973) has surveyed the literature looking at macrodemands. In this review, the emphasis is on discrete trial settings in which microprocesses are examined. These include studies in which two signals converge on one response, in which two signals conflict regarding a response, and in which two different responses are required.

A. Processing Redundant Information

Consider a situation in which two signals occur simultaneously and both indicate the same response. According to single-channel theory, reaction time to the redundant stimulus should be no faster than the fastest of the two signals when presented alone. Some studies, however, have shown that the redundant stimulus is actually responded to faster than either component alone. Morton (1969a) had people sort cards on the basis of numerals 1 to 6, or numerosity of × marks ranging from one × to six ×s, or redundant numerals and numerosity in which the numeral 1 occurred once, 2 twice, 3 three times, etc. The redundant method resulted in slightly faster sorting than either component alone. Moreover, this result was true for individual subjects and was therefore not an artifact of averaging over each subject's best dimension. Biederman and Checkosky (1970) obtained similar results for size and brightness judgments.

Figure 2 shows data from a study by Ellis and Chase (1971) that make a

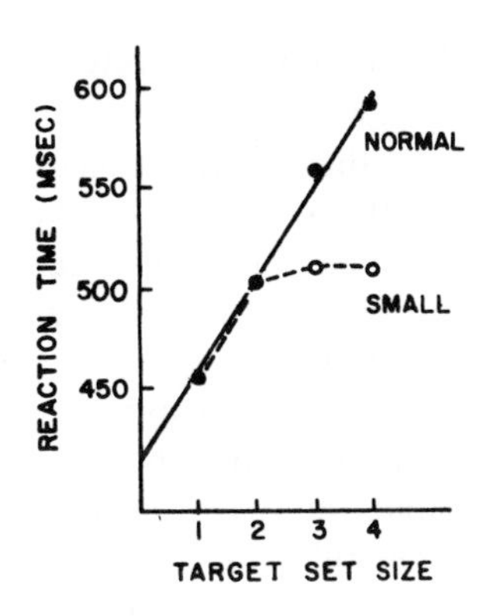

FIG. 2. Reaction time to indicate a stimulus is not a member of a target set as a function of target set size and whether the stimulus is small or normal in size. [Adapted from Ellis & Chase (1971).]

similar point. Subjects were shown a set of one, two, three, or four letters, followed by a test letter. If the test was a member of the preceding set, subjects were to press one key; if not a member, they were to press another key. On half the negative trials (those in which the test was not a member of the set) the test letter was smaller than normal. Subjects were told that whenever they saw a small letter they could immediately press the negative key.

As seen in Fig. 2, if the memory set is three or four, subjects are faster when the test item is small as well as a nonmember of the memory set. In contrast, if the memory set is only one or two, a small test item has no advantage. Similar results were also found when the negative stimulus sometimes had a distinctive color rather than size.

Ellis and Chase's data suggest that people judge size (or color) at the same time they search the memory set. Whichever process is first completed terminates the decision. Since size decisions take longer than searching one or two memory items, small size does not help until the search set consists of three or four items. It could be argued that when memory sets are of three or four items, subjects try size first and then, if that fails, search the memory set. Such a serial strategy would predict normal size negative responses to be slower than in a control condition in which only normal size letters were used. In fact, a control condition with only normal size letters exhibited the same reaction times as normal size letters in the context of small letters on some trials.

The reduction in reaction time to redundant signals is easily explained by parallel access to memory. When a signal occurs, memories associated with the stimulus begin to accrue information. If two different signals feed into the same memory, accrual will occur at a faster rate, leading to either reduced reaction time, reduced error, or both.

B. The Fate of Irrelevant Information

Studies of redundancy suggest that two signals can be processed in parallel when it is advantageous to do so. An even stronger case for automatic parallel access to memory can be made if it is shown that a second signal is processed to memory, even when the result is detrimental.

Suppose two signals occur at the same time, and that one is relevant to a response and the other irrelevant. When the signals are quite discriminable, the irrelevant one usually has no discernible effect on reaction time to the relevant one (Morgan & Alluisi, 1967; Well, 1971). At what stage of processing is the analysis of the irrelevant signal blocked?

The Stroop effect, a notable exception to the blocking of irrelevant information, casts light on the problem. When a person is asked to respond to the color of a stimulus and ignore its form, he has an unusually difficult time doing so if the form spells out a conflicting color word. For example, when the word *green* is printed in red ink, people are slow in responding to the ink's color. One interpretation of the Stroop effect, akin to single-channel theory, suggests that words often capture access to the channel before color, thereby delaying color processing. This seems unlikely, since irrelevant forms in general do not cause interference, suggesting that form has no privileged access to the translation channel (Archer, 1954; Schroeder, 1976). A second interpretation is that both the color and the word gain access of their names in parallel. When two color names are activated in memory, conflict ensues and additional time or process is required for resolution.

Keele (1972) differentiated between these two interpretations by comparing reaction times to (*a*) a colored meaningless form repeated in a string; (*b*) colored words that name another color; and (*c*) colored neutral words. If the single-channel view is correct and words are sometimes processed first, then conditions (*c*) and (*b*) should be more difficult than (*a*), since in both cases the word would be read before the color could be named. If, however, forms access memory in parallel with color, then only condition (*b*) should result in interference, since it is the only one that generates conflict. The reaction times to the three conditions respectively were 554, 604, and 559 msec and the error rates were .08, .09, and .07, supporting parallel access.

It could be argued that only the color words have access to the limited channel. This seems unlikely, since to know that a word is or is not a color word requires prior access to memory. Perhaps, however, the recognition units are more sensitive to color words due to the color context. If this were true, then on occasions where the word agreed with the color itself (e.g., *blue* written in blue ink), interference should again be observed, since the word would be processed before the color. However, Hintzman, Carre, Eskridge, Owens, Shaff, and Sparks (1972) found facilitation, compared to a control with constant meaningless forms, rather than interference on the occasional congruent word and color, indicating that the two stimuli gain simultaneous access to memory. Clark and Brownell (1975) observed a similar phenomenon with arrows and spatial position. Response to an up-pointing arrow was facilitated when the arrow was high on the display and slowed when low on the display. Moreover, Klein (1964) showed that noncolor words having strong color connotations, such as *banana* or *grass*, also interfere strongly with color naming. This observation supports the contention that words and colors are analyzed in parallel, as opposed to the hypothesis that only color words are sensitized and processed.

A potential interpretive problem with the Stroop studies regards the control condition to which other conditions are compared. The control itself involves colored forms, usually a repeated nonsense form. It is difficult in fact to present a color control without some form to bear the color. Is it possible that even meaningless forms are processed before color processing begins, invalidating their use as a control? Some evidence against this view is provided in Keele's study. The neutral-word condition involves more different forms than does the constant-form control, and yet it takes no longer than does the control. Keele also examined varying nonsense forms and found the same result. When several forms are relevant for response, variation in form increases reaction time (Beller, 1970). Since variation in form does not influence reaction time in the Stroop setting in which form is irrelevant, it is difficult to argue that form is being processed before color is started.

Perhaps more convincing evidence comes from studies such as that by Egeth, Jonides, and Wall (1972). In one of their experiments, subjects pressed a key whenever any digit was present in an array of distractor letters. No increase in reaction time to the digit occurred with an increase from zero to five in the number of distracting letters. Because any digit could occur when in fact a digit was present, it appears very unlikely that digits were being identified by selecting a unique feature not present in letters. Instead, it would appear that all the items activate either the concept of letter or the concept of digit in memory; if any one digit activates memory, it is selected. This result suggests, as do studies of the Stroop effect, that irrelevant material is being processed to a memorial stage along with the relevant, but the irrelevant material only causes a decrement in reaction time when it creates confusion about target identity.

Studies on the fate of irrelevant information suggest, therefore, that more than one stimulus has parallel access to memory and that selection occurs after memory activation. Irrelevant information appears not to be completely filtered early in processing, if at all, even when it is deleterious in its effects.

C. The Problem of Stimulus Integrality

Garner (1974a,b), Kahneman (1973), and Lockhead (1972) suggest that filtering of irrelevant information and redundancy gain depends on the nature of the stimuli. Only integral stimuli yield redundancy gain and produce difficulty in filtering. Integrality can be defined by experimental results: Stimulus dimensions that when combined redundantly result in redundancy gain and when combined orthogonally result in interference are said to be integral. To avoid circularity, however, integral dimensions can be defined as those that belong to the same object. Thus size and color of the same object are integral but size and color of different objects are separable.

When Garner and Felfoldy (1970) and Felfoldy and Garner (1971) redundantly combined value and chroma of Munsell colors on the same chip and had cards bearing these chips sorted into piles, sorting time was faster than when both dimensions varied. But when cards with two color chips were sorted, with value varying on one chip and chroma on the other, no redundancy gain was observed. The failure of redundancy gain for separated dimensions occurred whether subjects were explicitly told about the redundancy or not. Likewise, when the dimensions were combined orthogonally, variation on the irrelevant dimension interfered with card sorting on the relevant dimension only when both dimensions varied on the same color chips.

Garner and Felfoldy's results could suggest parallel processing only for integral stimuli. For nonintegral stimuli, single-channel theory may be correct. Some of the earlier studies of redundancy gain and filtering that implied parallel processing used integral stimuli. In studies of the Stroop effect, word color is normally integral with the word. In the Biederman and Checkosky study and in the Ellis and Chase study, the dimensions are clearly integral. Whether numerosity and numerals in Morton's study are integral is more problematic. The Egeth *et al.* (1972) study involving the search for a digit among letters poses a greater problem for the proposition that only dimensions of an integral stimulus are processed in parallel, for obviously the several forms are not integral.

Other problems also exist for the proposition that only integral dimensions simultaneously contact the memory system. Dyer (1973) presented a color patch on one side of a fixation point and a black color word on the other side of fixation—clearly separated dimensions. Nonetheless, form interfered in making color judgments. In a study by Morton (1969b), subjects sorted cards into piles on the basis of number of figures on the card while attempting to ignore auditorially presented numbers. Obviously, visual figures and auditory names are not integral, but nevertheless the conflicting numbers slowed card sorting. Moreover, some studies to be described later present evidence for parallel processing of visual and auditory signals.

Stimulus integrality clearly influences redundancy gain and filtering. But if nonintegral stimuli are highly related at the memory level, parallel processing is observed. Failure to observe either redundancy gain or interference effects implies only that the stimuli do not interact at either the sensory level or the memory level; it does not imply that they are processed serially.

D. Multiple Activations in Memory by One Stimulus

In Keele's study of the Stroop effect, subjects respond to colors by pressing keys. Neutral words have no preexisting associations with particular keys, and that appears to be the primary reason why neutral words do not lead to conflict. When people name the colors rather than press keys, even neutral words lead to a conflicting tendency to verbalize (Klein, 1964). This fact allows the Stroop effect to further index what has been activated in memory. Warren (1972) presented word triads, such as *oak, maple, fir,* to be recalled a few seconds later. Just before recall a colored word was presented, and subjects named the color. If the colored word was a member of the preceding triad, people were slow in naming the color. Moreover, a word associated with the preceding triad (e.g., *tree*) also received a slower response. Reaction time to the color, therefore, indexes prior activation of word meaning.

Conrad (1974) used the Warren technique to assess the activation of ambiguous word meanings in sentence contexts. For example, following the sentence, *The beans are cooking in the pot,* she presented either the colored word *pot, kettle, marijuana,* or a control word. *Kettle* and *marijuana* are both meanings of *pot*, but in the context of the sentence only one meaning is appropriate. Reaction time to name the color of the control word was less than the reaction time for either meaning of the ambiguous word. Furthermore, the appropriate and inappropriate meanings did not differ in their interfering effect. These results suggest that all meanings of words are activated during their presentation, even in a disambiguating context, and that selection of one meaning occurs subsequently. This does not mean that people are aware of both meanings, as only the appropriate meaning may have been selected.

Conrad's results bolster the earlier conclusion that more than one memory can be activated at a time, though in this instance multiple activation flows from a single stimulus.

E. Probing Mental Operations with Secondary Tasks

When two signals require separate responses, interference with one or the other normally occurs. But what process generates the interference? The preceding studies imply that the limitation is subsequent to memory retrieval. Thus when two responses are required, a technique is needed that separates in time stimulus contact with names or other memories from later processes dealing with the encoded stimulus. A secondary probe signal can then be inserted to determine whether the ongoing process was demanding of attention.

Posner and Boies (1971) exploited this idea in a matching paradigm. A visual warning signal was followed ½ sec later by the first of two letters. One second after the first letter a second one appeared, and the subjects were to press one key if they were the same and another key if they were different. An entire second was thus available for the subject to encode the first item, with no necessity during that time for matching or responding.

At various times during the decision process a probe tone was presented; reaction time to the probe was an index of the capacity demand of any ongoing process. Figure 3 shows the probe reaction times. Reaction time to the probe decreases following the warning signal, despite the fact that the warning was for the figures, not the probe. More important, probe reaction time remains low up to 300 msec following the first letter. After that point, probe reaction time begins to rise, reaching a peak after the second letter. Similar results are found when matching is based on name as well as physical identity.

Why is probe reaction time so low right after the first letter? Perhaps during that time subjects are not encoding the first letter but instead are waiting for the second before encoding either. This possibility can be discounted, because when the second letter unexpectedly occurs early, subjects still respond faster than when they have sinultaneous presentation of the letters. Subjects do take advantage of prior exposure of one letter to encode it. Despite encoding the first letter, there was no interference with the auditory probe. These results suggest, therefore, that the encoding process, in which the signal contacts an internal representation, does not require attention either for name codes or for physical codes.

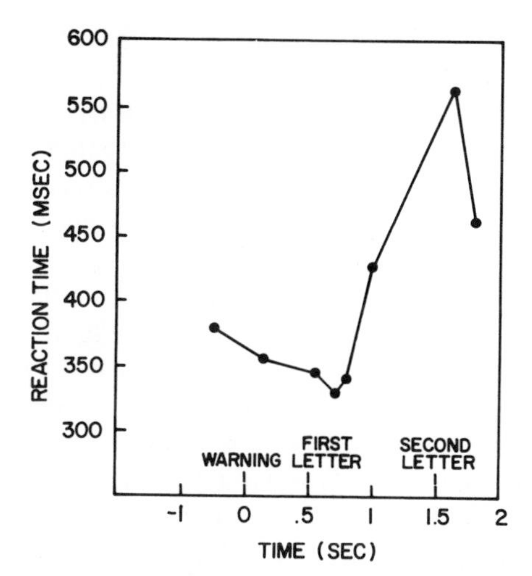

FIG. 3. Reaction time to an auditory probe signal as a function of where it occurs during the processing of letters to be matched.

The probe reaction time is high following the second item to be matched, indicating that either matching or responding or both require capacity. But probe reaction time begins to rise well before the second letter. Why? One answer, which will be elaborated later, is that encoding takes no capacity and therefore allows a limited-capacity mechanism to turn to the area of memory activated in anticipation of dealing with the next item. Some evidence for this point comes from a Posner and Klein (1973) experiment. The rise in probe reaction time is more closely related to the time remaining before a second letter than the time after a first letter. When subjects know the second letter is to be delayed, the point at which the upturn in probe reaction time begins is also delayed, beginning about ½ sec before the second letter.

F. Effects of Task Difficulty on Interference

Another method of analyzing which stage, memory retrieval, or subsequent operations, generates interference when two signals both require a response is to vary the time required by memory retrieval. Single-channel theory, which posits limited capacity for memory retrieval, would predict that increases in retrieval time for either of two signals would increase the total time required to process the signals by an amount equal to the increase in retrieval time. Models based on allocatable capacity make a similar prediction. A pair of tasks may not interfere if they do not use all the capacity; some capacity can be allocated to both tasks. But when interference does occur, it means that available capacity is already used up; no matter how capacity is allocated, any further increases in retrieval time will increase total processing time for the two signals. Parallel access to memory predicts, in contrast, that increased retrieval time for one signal may be absorbed during the retrieval time of the other, so that total processing time may not be increased by much.

The time to retrieve an appropriate response from memory can be increased by reducing stimulus discriminability, increasing the number of stimuli and responses, and decreasing the compatibility of the stimulus to response mapping. LaBerge (1973) employed the discriminability method.

Subjects were cued to expect either a 1000-Hz tone or an orange light. On most trials the expected signal (e.g., tone) occurred, but on a small number of trials the unexpected signal (e.g., light) occurred. Each signal could be made easy or difficult by requiring detection or discrimination. In the detection condition, catch trials with no signal occasionally occurred. In the discrimination condition the catch trials were a tone of 990 Hz or a yellow light. Catch signals required withholding the response; the other signals required a single button to be pressed.

When the expected stimulus was presented, discrimination took 287 msec, which is 83 msec longer than the 204 msec for detection. This result is entirely expected: Discrimination is more difficult than detection. The 83 msec difference is also nearly identical to the value obtained on control blocks of trials with only expected signals. However, when an unexpected stimulus occurred, discrimination averaged 415 msec,* only 43 msec longer than 372-msec detection and a sizable reduction in the difference score from the former value of 83 msec.

LaBerge suggested that when the expected signal failed to occur, a selection mechanism switched to the unexpected one. If discrimination did not start until after switching, then the difference between detection and discrimination should remain at about 83 msec, if not actually growing larger. The fact that the difference is actually much smaller than 83 msec suggests that discrimination is occurring in parallel with switching of the selector. By the time switching is complete, detection or discrimination is often completed. Elicitation of the response must then wait on the completion of switching and is not sensitive to the difficulty of the discrimination task. The experimental outcome, therefore, is consistent with nonattentive access to memory.

Returning to Posner and Klein's probe study, which was discussed in the preceding section, recall that probe reaction time begins to rise prior

* Reference to Table 3 in LaBerge's paper reveals a data problem with one subject in this condition that could influence the interpretation of the results.

to the expected time of appearance of the second letter to be matched. LaBerge's study yields additional insight into the behavior of probe reaction time. Apparently, the encoding of the first letter is not attention demanding. But as the time for the second letter approaches, a selector mechanism is turned to the letter in memory that has already been encoded. Once the selector mechanism is committed, occurrence of the probe requires switching and increases reaction time, even though some features of the probe may be processed during switching.

A similar explanation may be invoked to explain an apparent discrepancy with the Posner and Klein results. Comstock (1973) exposed the first of two letters to be matched for only 15 msec and followed it with a masking stimulus. Under that condition, probe reaction time began to rise much sooner than in Posner and Klein's study, suggesting to Comstock that encoding the first letter was attention demanding. Alternatively, when subjects anticipate that the results of encoding will be damaged by a mask, a selector mechanism may be turned to the first letter, resulting in an increase in probe reaction time. The encoding itself may require no attention.

There is another important aspect of LaBerge's experiment. An unexpected signal can occur in the context of expecting either a detection or a discrimination. Reaction time to an unexpected signal was 354 msec when the expected one was to have been detection, but 432 msec when the expected one was to have been discrimination. It appears to take longer to switch when the expected signal would be more difficult to perceive.

A slightly more complicated experiment by Karlin and Kestenbaum (1968) used logic similar to LaBerge's to make the same point about attention demands of memory retrieval. The logic again places the attentive mechanism on one signal and then observes what processing has occurred for a second signal of variable difficulty. In the experiment, one of two digits appeared as a first signal* and was responded to first with the left hand. At a variable interval after the first signal ranging from 90 to 1150 msec, a tone occurred for the second signal, requiring a button press with the right hand. In one condition only one tone could occur that required detection. In the other, one of two tones could occur, requiring a choice response. Task difficulty was varied differently than in LaBerge's study, but the logic was the same: Attention was directed to one signal, which in this case required a response, and then attention was switched to a second signal. Although the second response would certainly be de-

* Earlier, another portion of Karlin and Kestenbaum's study was described in which first-signal difficulty was varied. Here a portion of their data is selected dealing only with second-signal difficulty.

layed, the processing it received could be assessed by the difference in reaction times between the detection and choice situations.

Both single-channel theory and the theory of allocatable capacity lead to the same prediction, as shown in the left panel of Fig. 4. If all processing regarding choice of the second signal is delayed until the first signal is processed, choice will take longer than detection by a constant amount, yielding an additive relation between interstimulus interval and choice difficulty. The prediction is the same if some capacity is directed from the first to the second signal. The prediction for parallel access to memory is shown in the middle panel of Fig. 4. While processing the first signal, the second signal is also processed, but the selector mechanism cannot be switched to the accumulated output of signal two until the first signal has been cleared. When signal two occurs early, it will often be processed before switching and the difference between the delayed reaction times for detection and choice situations will diminish. Thus, as interstimulus interval is increased, a divergence in functions is predicted.

The actual results are shown in the third panel of Fig. 4, and they are quite consistent with parallel access to memory. However, the theory does predict complete convergence at the shortest interstimulus interval, and this failed to occur. One possible reason is that by the time the first signal is responded to, processing of the second, although in progress, has not yet been completed. A second possible reason is that the first signal is responded to more slowly at all interstimulus intervals when it is to be

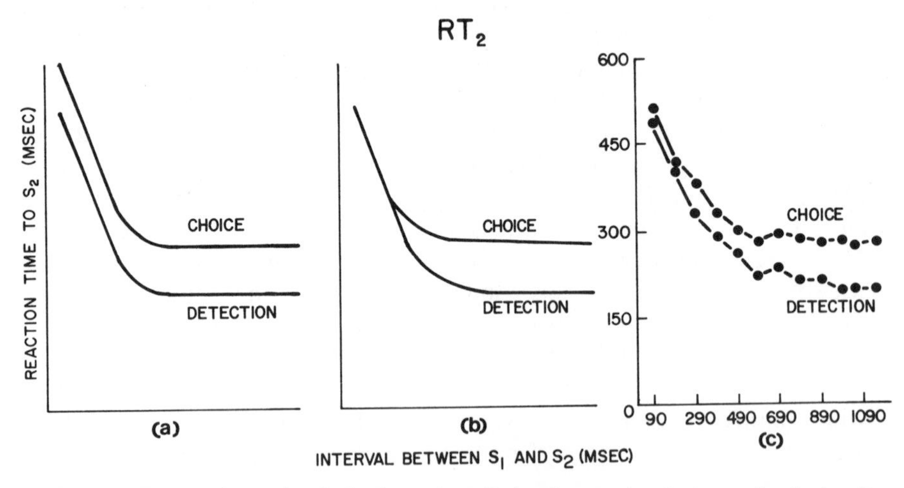

FIG. 4. Part (a) shows the single-channel prediction for reaction time to a signal when it requires detection or choice. Part (b) shows predictions for a theory of parallel access to memory followed by interference at a later stage. Part (c) shows obtained results from Karlin and Kestenbaum (1968).

followed by a more difficult second signal. Delay of the first signal further delays switching attention to the second signal. By and large, however, the choice decision of the second signal occurs while attention is diverted to the first.

At this point, it is useful to recall Karlin and Kestenbaum's manipulations of first-signal difficulty as mentioned in the introduction. Increased choice for the first signal increases first-signal reaction time and delays response to the second signal by a nearly equal amount. Observations of this sort were earlier taken to support single-channel theory. Now it can be seen that the results are equally consistent with conceptions of parallel access to memory. While the first signal is being processed, information regarding the second signal is accruing in memory. A response to the accrued information cannot be emitted, however, until a selector mechanism is switched from the first signal to the output of the second. The longer the processing time of the first signal, the longer the time before switching, and hence the greater delay in responding to the second signal.

One way of viewing the Karlin and Kestenbaum results is in terms of total processing time for two tasks performed together being less than would be predicted from reaction times were the tasks to be performed separately. Such failure of additivity was also found by Schvaneveldt (1969) when he manipulated the number of alternative stimuli and responses and compatibility of stimulus–response relations. This study is particularly important because variations in stimulus–response compatibility obviously influence the time to retrieve the response from memory. Thus if nonadditivity is found it is a powerful demonstration of the automaticity of retrieval.

A digit appeared in one of two displays in front of the subject. In the simplest situation the subject named the digit and pressed a response button on the same side as the lighted display. The total reaction time to complete both tasks was 469 msec. When digit complexity was increased by requiring the subject to add one and respond with the sum, total reaction time increased by 110 msec to a total of 579 msec. When spatial complexity was increased by requiring a button press by the hand opposite the displayed digit, total reaction time rose to 510 msec, an increase of 46 msec. Were both tasks to be increased in difficulty, single-channel theory would predict an additive increase of 156 msec (110 plus 46). The actual increase when both tasks were increased in difficulty was only 127 msec—very little more than the 110 msec increase demanded by complexity of the digit task alone. Most of the additional spatial task time was absorbed in the increased time demanded by the digit task, suggesting again that transforming input code to output code occurred in parallel for the two tasks.

Similar results were found by Schvaneveldt when the number of alternatives was varied. When both the verbal and spatial task required simple reactions, total reaction time was 356 msec. A two-bit verbal decision increased total reaction time to 607 msec. A two-bit spatial task increased total reaction time to 571 msec. But when both tasks were increased in difficulty, total reaction time was only 670 msec, considerably less than the 833 msec expected were verbal and spatial decision additive in time, as would be expected by single-channel or allocatable-capacity theories.

What justification is there for the supposition that discrimination difficulty, number of stimulus alternatives, and stimulus–response compatibility affect memory retrieval, rather than some other process such as sensory encoding? Basically, the effects of both stimulus discriminability of the LaBerge sort and number of alternatives are magnified when the mapping to response is less compatible (e.g., Brainard, Arby, Fitts, & Alluisi, 1962; Broadbent & Gregory, 1962, 1965). Thus, the manipulated times appear primarily to be the times necessary to retrieve the appropriate response. The studies by LaBerge, Karlin and Kestenbaum, and Schvaneveldt lead to the conclusion, therefore, that memory retrieval is largely nonattentive.

G. Transforming Information into Action

Numerous phenomena converge on the conclusion that more than one stimulus simultaneously activates information stored in memory. In addition to redundancy gain, filtering, probe reaction time, and retrieval-time phenomena, other phenomena, such as processing sequential information (Keele & Boies, 1973), combining signals into a single response (Keele, 1970), and the superiority of word perception over single-letter perception (Reicher, 1969) argue for the same conclusion.

The general model that emerges is portrayed in Fig. 5 and may be

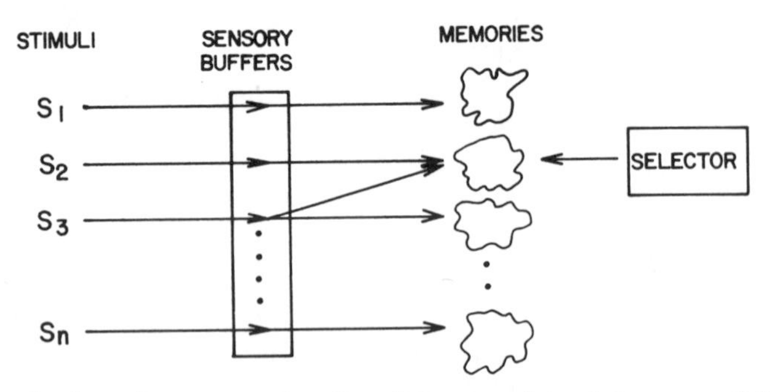

FIG. 5. A schematic representation of parallel access of signals to memory followed by a mechanism that selects one activated memory or another.

contrasted with Fig. 1. Information enters the sense organs and, barring peripheral masking, activates sensory representations. Sensory information, in turn, leads to the parallel accrual of memory information. One source of sensory information can diverge to more than one memory; more than one sensory source can converge on the same memory. However, for the information activated in memory to be converted to action, a selector mechanism must intervene, allowing subsequent operations (e.g., releasing a response, rehearsing the activated memory, matching it to some other memory). This view is basically similar to earlier ones by Deutsch and Deutsch (1963), Norman (1968), Morton (1969c), and LaBerge (1975).

One may fairly ask why, if activation of memories is automatic, are not the final responses also automatic? Why does interference occur near output? The answer may be that although it is useful to the organism for all stimulus associations to be activated, it is not useful for all possible actions to be released. Once information associated with a stimulus is available, several options exist. If the activated memories are to be stored in episodic memory for later use, they can be rehearsed; but if not, rehearsal may be avoided. If an immediate response is not desired, then it would be deleterious were a response to be reflexively activated. A selector mechanism at a point prior to final action appears necessary, therefore, for coordinating information available from the environment with information regarding goals. Together, information elicited by the stimulus and goals determines whether the action should be rehearsal, response, counting, comparing, searching for more information, or whatever. The mechanism that determines one action or another appears, therefore, to be a major source of limitation, corresponding to what we mean by attention. This view is similar to one promulgated by Shallice (1972).

V. ELABORATIONS OF ATTENTION THEORY

The evidence discussed suggests that information accrues at more than one memory location at a time and with little or no interference. But other experimental results following very similar logic fail to support parallel accrual. Attempts must be made to rectify these discrepancies. Moreover, the model evolved places selectivity only late in the processing sequence. That assumption may be wrong, and under some circumstances earlier selection may occur.

A. Problematic Results for Parallel Access to Memory

Conrad (1974), as described earlier, used a variation of the Stroop effect to show that more than one meaning of an ambiguous word is activated even in the context of a disambiguating sentence. However, Schvaneveldt, Meyer, and Becker (1976), using a different paradigm, failed to confirm this result. Subjects classified letter strings as words or nonwords by pressing one of two response keys. Earlier they had shown that when two related words occurred in succession (e.g., *bread* followed by *butter*), the time necessary to decide that the second item was a word was reduced in comparison to unrelated words. Thus, the occurrence of one word appears to activate other related words in memory. In the aforementioned study, three successive letter strings were presented in a block. Sometimes a triple such as river–bank–money was presented. The middle word is related to the first, and it shows the usual relatedness effect. The middle word itself has two meanings, however, and the second meaning is related to the third word. If both meanings of the ambiguous word are activated, as suggested by Conrad's Stroop study, then facilitation of the third word ought to be observed. But such facilitation was not found. These results appear to support a model of selective, rather than parallel, access to memory: The meaning of the first word determines which meaning of the second is accessed. The conflict between Conrad and Schvaneveldt *et al.* is not adequately resolved. Perhaps one paradigm or the other is at fault. Perhaps complicating explanations are needed. At any rate, the conflict necessitates some caution regarding conclusions reached to date.

Becker (1976) used the word–nonword (lexical) decision task in conjunction with the Karlin and Kestenbaum paradigm. The first of two signals was a letter string requiring a word or nonword response. After an elapse of either 90 or 190 msec following the letter string a tone occurred, requiring a second response. The tone task could be simple or require choice, and control conditions showed the choice time required for the task to be 141 msec slower than reaction time for a simple task. If the word decision required no attention through the encoding stage, then presumably the tone would be processed during the encoding of the word but not responded to until a selector mechanism could be switched to it following response to the word. Thus, when the tone followed a letter string, the difference between simple and choice reaction time would diminish or even vanish. This result did not obtain, however. When the tone followed 90 msec after a high frequency word, simple and choice tone reactions differed by 264 msec. When a tone followed a low-frequency word, the difference was even larger (348 msec). Both values are much larger than the 141-msec difference found in control conditions. These results dramatically differ from those of Karlin and Kestenbaum and of LaBerge (1973). Obviously, something either in the encoding of the word or tone required attention.

Becker also more exactly replicated Karlin and Kestenbaum's study, using digit identification as the first task, and again found that the difference between simple and choice tone-reaction time did not diminish when it followed the first signal.

1. Diffuse Memory Activation and Coordination of Codes

Why do Becker's results differ so sharply from expectations based on the previous theory? An important experimental difference pointed out by Becker (personal communication) is that most of the studies described earlier involved either very few alternative stimuli and responses, or large amounts of practice, or both. In contrast, Becker's subjects never saw the same letter string more than once during the course of the experiment, and the number of experimental sessions was small.

A possible resolution could take the following approach. When a stimulus appears, consistent with previous theorizing, information stored in memory is automatically activated. In many situations, particularly with words, a large amount of information is activated. A word such as *saw* may activate the concepts of *wood saw, tool, hammer, lumber,* and other meanings, such as *to have seen.* Studies by Conrad (1974) and Warren (1972) using the Stroop effect provide good evidence for such broad activation. Moreover, if this forward activating process is imprecise, *saw* may also activate other items such as *sow, paw, how,* and *sam* by virtue of physical similarity. Nonwords, to the extent that they are physically similar to real words, may also activate several memories. The great spread of diffuse activation may prevent a subject from deciding the exact identity of a stimulus unless an additional verification process is invoked. Concepts or meanings activated in memory also have other codes associated with them. For example, the concept *saw*, once activated, may in turn activate an orthographic code detailing how it is spelled. By comparing the orthographic code with the actual input, the exact stimulus can be verified.

To put the notion another way, a stimulus may activate physical codes, orthographic codes, phonemic codes, and semantic codes. Within the semantic area a large number of meanings may simultaneously be activated. An attentional process may link the different codes to each other, resulting in a precise identification of the stimulus.

This view is quite consistent with the one evolved earlier. Information may accrue simultaneously in different memory locations. The attentive mechanism is one that coordinates information of goals with what is activated in memory. It also is one that may coordinate information available in different codes.

The idea of diffuse but parallel memory activation followed by a limited process that coordinates information in memory is largely suggested by Becker's (1976) verification model, though it differs in details. It is also similar to Collins and Loftus's (1975) idea of spreading activation in semantic memory, but envisions an attentional control of the linkage between the semantic system and a dictionary system that contains phonemic and orthographic information.

How does this generalized model deal with Becker's data, on the one hand, and the earlier data, on the other? When the number of possible stimuli and responses is very limited and practice is high, the activation of a specific memory by a specific stimulus may be sufficient to precisely match stimulus with response. Under those circumstances, results such as those of Karlin and Kestenbaum may be obtained; but when practice is low or many different stimuli are used, as in Becker's study, so many memories may be activated that attention-demanding processes are needed to coordinate codes.

What useful function would be subserved by diffuse activation in memory if it requires attention-demanding operations to further specify stimuli and choose responses? First, despite the diffuseness, the number of concepts activated would be tremendously less than the totality of memory, greatly simplifying memory search. Second, most information processing occurs in contexts that further reduce the relevant memory locations. Together, context and stimulus may so greatly converge on one meaning that often little or no verification is needed.

2. Implications of Dyslexia

The general idea of widespread memory activation and subsequent code coordination has received a great deal of support and will be reviewed in greater detail later. However, investigations by Marshall and Newcombe (1973) and Shallice and Warrington (1975) of dyslexic patients provide striking support for the model. Because these studies so clarify the general proposition, it is useful to mention them at this point.

The dyslexic patients studied had all learned to read at one time and then suffered brain damage that impaired reading. Because prior reading had been established, certain errors occurred that otherwise would not have been observed. Some dyslexics mistakenly read words that were physically similar to the ones actually presented: *Chair* might be substituted for *charm, dug* for *bug, wash* for *was,* and so on. For other dyslexics, visual errors seemed particularly triggered by letters that had more than one pronunciation. For example, *incense* might yield *increase,* triggered by the ambiguity in how to pronounce the letter *c. Guest* might similarly yield *just.* The most illuminating class of dyslexics, however, yielded errors not only of physical confusion but also very often of semantic confusion when reading individual words. *Speak* might yield *talk, employ* yield *factory, found* yield *lost, hurt* yield *injure.*

What are the implications of such semantic errors? They appear to support the contention that a word leads to very diffuse activation of concepts in semantic memory. Such activation by itself is not sufficient, however, to pinpoint the precise word. The semantic dyslexic appears to have sustained damage to the mechanism that allows such pinpointing. Either visual or phonemic codes have been lost or the verification mechanism that coordinates visual, phonemic, and semantic codes is impaired. It may be that visual dyslexics suffer a similar impairment, but of lesser degree. When the impairment is partial, some verification occurs, eliminating words that are semantically related but physically very dissimilar, but concepts activated in memory that are physically similar (e.g., *dug–bug*) escape accurate verification. Under more severe impairment, it appears that little verification occurs, as practically any item activated in memory by the stimulus may be reported.

Marshall and Newcombe note that errors similar to those of a semantic dyslexic occur also for normal readers in tachistoscopic presentations, and they can sometimes be observed in reading where one word is more appropriate in a context than a word actually presented. Thus it is likely that conclusions derived from the dyslexic apply also to the normal reader.

3. Optional Filtering and Structural Interference

The general idea about attention that is emerging is that it is a mechanism that coordinates information activated in the different memory systems. Sometimes the coordination may be among different codes. Sometimes it may be between activated memories and other goals. Other controlling capabilities may also exist. In particular there may be situations in which two sources of information mutually interfere at a rather peripheral level of processing, and some control may be exercised at that level. This view, of course, is similar to Broadbent's (1958) traditional filter theory, but it differs in suggesting that the control is optional and that it may be most likely for structurally interfering signals.

Consider a situation posed by Ninio and Kahneman (1974). Subjects monitored an auditory list of words for names of animals, pressing a key when they heard one. In one situation, subjects monitored messages to both ears; in another, they ignored words in one ear and listened to the other. Which situation resulted in the fastest reaction time to an animal name? If the words in both ears were simultaneously encoded without interference, and if a selector mechanism were attuned to the "animal area" of memory, the divided-attention situation would be as fast or faster than the focused condition. In the focused condition people might be slower because they also would have to be sure that the words were from the correct ear.

In fact, Ninio and Kahneman found divided attention to be slower, not faster, by about 135 msec and to result in more missed animal names. One interpretation invokes a serial decision process at some point. Each word could be encoded in parallel and tagged for location of entry. In the divided case, the selector mechanism must then serially interrogate twice as many items activated in memory as in the focused case. Alternatively, first the word from one ear and then the word from the other ear could be encoded, as supposed by traditional filter theory. Both explanations predict that reaction-time variance should be greater in the divided situation. If a person is attending to the correct ear when an animal name appears, he will be fast; if he is attending to the incorrect ear he will be slow. The mixture of fast and slow reaction times will lead to high variability.

Variability, however, was no greater in divided than focused attention, invalidating a serial switching model. Apparently both messages were processed to the memory level in parallel in the divided situation and the selector was attuned to the animal area of memory.

In a similar study by Treisman and Fearnley (1971), subjects in one condition heard either a single nonsense syllable or a digit. In the other condition they heard a pair of items, one in each ear. If they heard a digit they were to press one key, otherwise they were to press another key. Sometimes they were cued in advance which digit to respond to, but other times any digit required a digit response. When only a single item occurred in only one ear, the unknown digit was responded to more slowly than the cued digit by about 103 msec.

What would be the expected results when the digit could occur in either ear? If the two items to the two ears are serially encoded, the unknown digit should take longer than the cued digit by about 154 msec (50% larger than the 103-msec difference when subjects focus on one ear). If the correct ear is selected first, then the cued digit can be identified 103 msec faster than the noncued digit, but if a digit does not occur in that ear, the subject must switch to the other ear and attempt identification again, doubling the total time difference. Averaging the switch and nonswitch trials together leads to a predicted 50% increase in the difference score. However, as in the Ninio and Kahneman study, reaction time increases in the divided attention situation over the focused situation, but the increase is about equal for both the cued and noncued cases, leaving the difference between them approximately the same at 92 msec. This study too leads to the conclusion that items in both ears are simultaneously encoded, but at a cost. Why the cost?

Ninio and Kahneman suggest that encoding requires processing capacity, but that the capacity can be allocated all to one ear or divided between the two ears, allowing parallel but slower processing in both. Another possible explanation is that auditory messages to the two ears tend to mask each other or merge with one another, making discrimination more time-consuming. When subjects are instructed to attend to one ear, however, the message from the other ear can be partially filtered (attenuated), reducing masking. This possibility is similar to single-channel or filter theory, but differs in two important ways. One difference is that filtering is optional and sometimes processing to memory may occur in parallel. The other is that mutual interference is closely tied to the structures doing the processing. Two inputs coming through a structurally related system, such as the two ears, may interfere more with each other than two inputs coming through different structures. These views are actually quite similar to those held by Treisman (1969).

Some evidence for optional filtering and the reduction of masking comes from a study by Hawkins, Thomas, Presson, Cozic, and Brookmire (1974). A briefly presented tone was followed by a second, masking tone that tended to bias identification of the first in the direction of the masking tone. Such masking also normally occurs when the masker is presented to the ear opposite the tone to be identified. However, when subjects know in advance that the masker will be in the opposite ear and also know its frequency, masking is markedly reduced, suggesting that the masker is filtered by ear and by frequency.

One implication of a masking-type explanation of task interference is that the decrement in divided attention should be highly sensitive to the similarity of the two inputs. Little study has been devoted to this important issue, but some evidence appears to support it. Treisman and Davies (1973) required subjects to monitor simultaneous messages for the occurrence of animal names. The simultaneous words were both visual, or both auditory, or one auditory and one visual. Detection of animal names was considerably higher when the messages were divided between the two modalities. In fact, with two modalities there was little difference between attention divided between the two message sources and attention focused on only one message.

Structural interference from similar input signals may only be apparent when signal discrimination is relatively difficult. When signals are simple or practice extensive, little decrement due to structural reasons may be apparent. Schwank (1975) presented two signals in succession, requiring a response to each. The response to the second signal was delayed—a typical effect—but the delay was no greater when both signals were colors or both were letters as opposed to one letter and one color. Greater structural interference might have been expected when both were letters or both colors, but the stimuli were simple and nonconfusable, and the major portion of processing time may have been used in retrieving the appropriate response rather than differentiating the signals.

Ostry, Moray, and Marks (1976) likewise found that by 5–10 sessions of practice subjects could monitor digits to both ears for occasional letter targets as efficiently as they could monitor a single ear. Similar results were found when monitoring for animal names mixed amongst nonanimal names. Here the simplicity of stimuli and/or a large degree of practice may overcome structural problems.

Although critical studies are few in number, leaving considerable leeway for other interpretations, a tentative conclusion places some constraints on the model developed earlier. When complex, similar stimuli come through the same modality simultaneously, the stage is set for masking, merging of signals, or signal confusability. Some protection against mutual interference is provided by filtering, or attenuation, of one message prior to memory, much as claimed by Broadbent and Treisman in early theories. However, filtering appears optional and parallel access to memory can occur, though it may add delays to processing time and increase errors. When signals are less confusable, parallel access to memory appears to incur little or no additional cost. The idea of attention as a control mechanism that can set filter location is quite consistent with the idea that an attentional mechanism coordinates codes.

B. Codes and Code Selection

The two ideas of attentional control—one of very diffuse activation in memory followed by coordination of codes, and the other of flexibility in filter location—require further development to appreciate the great deal of diversity exhibited by the processing system. A primary point to be made is that presentation of a stimulus elicits not only several items within a coding system, such as several meanings, but also different codes, and the activation of more than one code is also done in parallel.

The notion that the stimulus may elicit information in two different coding systems stems partly from observations by Posner and Mitchell

(1967). Subjects classified letter pairs as the same or different by pressing keys. When the letters had not only the same name but the same shape (e.g., *A* and *A*), classification was about 50–100 msec faster than when the letters had the same name but differed in shape (e.g., *A* and *a*). This difference in speed would not be expected were subjects selecting only on the basis of a name code; instead, either visual or name codes appear to be used for matching, and under simultaneous presentation the visual code is faster.

The Posner and Mitchell results could reflect differences in *level* of coding, with the visual code preceding and being necessary for the name code. Other observations suggest, however, that this view is incorrect. If the letters to be matched are flanked by visually similar letters, the time for a physical match is slowed but the time for a name match is not (Posner & Taylor, 1969). If the physical code preceded the name code, then slowing the physical code should also slow the name match. Conversely, if two letters have the same name, that fact does not necessarily slow the subject in responding to their physical difference. Cohen (1969) had people match strings of three unrelated letters. If any letter between the two triplets was different, subjects were to respond "different." When different letters were physically similar, matching was not slowed. Also, when different letters were similar in name, matching was not slowed. Only when items were confusable on both counts was reaction time increased. This suggests that both codes are independently processed. Finally, Corcoran and Besner (1975) showed that when two letters to be matched differed in size or in brightness, the differences affected physical matches but not name matches. Again, if the physical code were a necessary precursor of the name code, then factors that affected physical matches should have also affected the name match. A number of other examples that make the same point are cited by Posner (in press).

Results of this sort suggest that, rather than differing in level, the two codes, physical and name, are independently manipulatable and derived in parallel, with mismatches being determined by whichever code is first finished.

Studies of word perception also indicate that more than one code is available. First, it is clear from earlier cited studies (Conrad, 1974; Marshall & Newcombe, 1973; Schvaneveldt, Meyer & Becker, 1976; Shallice & Warrington, 1975; Warren, 1972) that words activate semantic codes (i.e., synonyms, associated words, and perhaps alternate meanings become activated by a word). The written word also can activate a phonetic representation, a physical representation, an orthographic representation, and perhaps a more general visual code.

Since Reicher (1969) and Wheeler (1970) it has been known that letters from single syllable words are perceived in tachistoscopic recognition as accurately, or more accurately, than isolated letters. This perceptual superiority occurs not only for real words but also for pronounceable nonwords (Baron & Thurstone, 1973). Because the pronounceable nonwords have no meaning, their superiority must be due to either the pronunciation or orthographic rules of English.

Hawkins, Reicher, Rogers, and Peterson (1976) very briefly presented an item such as *sent* followed by a word pair such as *sent* and *cent*. Subjects had to indicate which word they had seen, and because of the brief visual exposure, they often made errors. Both alternatives in the example are pronounced exactly the same (i.e., they are homophones). If the perceptual superiority of words is sometimes based on a phonetic code, even though the word is visually presented, then subjects should have more difficulty with homophonic pairs than with control items (e.g., *sold* followed by *sold* and *cold*) in which the critical letters are pronounced differently. As long as homophone test pairs were not very frequent, they indeed resulted in less accuracy than the control pairs—58.3% correct versus 72.5% correct. These results support the contention that a phonetic code is activated by the visual presentation of the word.

On the other hand, Pollatsek, Well, and Schindler (1975) demonstrated word superiority that must be attributed to an orthographic code rather than a phonetic code. They presented two letter strings and asked subjects to respond only if all letters between the two words were physically identical. Earlier, Eichelman (1970) had shown that actual words were matched faster than nonword strings, even under physical match instructions, but this could be attributed to the use of a name code in the visual condition. Pollatsek *et al.* avoided this interpretation by using letter strings of mixed case (e.g., leAF). Such a string has little visual familiarity, but it is orthographically familiar. Subjects were able to say that a pair such as leAF and lEAF were physically different more rapidly than nonword strings of the same length. If the match were being made on a name basis, interference should occur. Because facilitation occurred instead, the results support the independent existence of an orthographic code.

Familiarity may also be based on yet other codes (Henderson, 1974). People match familiar acronyms such as FBI, USSR, and IBM faster than they match nonfamiliar letter strings such as BFI, RSSU, and IMB. The familiar acronyms are neither phonetically nor orthographically regular. The basis for faster matching must therefore be either visual or semantic familiarity.

Finally, Rogers (1975) examined the activation of codes for faces. Line-drawing faces that varied in similarity were associated to names that also varied in similarity. Subjects were presented with a name followed by a face or vice versa and indicated whether they matched. When the face belonging to a presented name was similar to a subsequently presented face, judgments of difference were impaired, indicating that visual codes were being used in the match. Conversely, when the name belonging to a presented face was similar to a succeeding name, judgments of difference were again impaired. Thus, it appears that both faces and names can activate face or name codes on which subsequent matches are based. When simultaneous name–face pairs are matched, effects of both face and name confusability occur, suggesting that both codes are activated at the same time.

1. Code Selection and Integration

How are the numerous codes coordinated into a final response? Perhaps all codes converge without attention on the final response. The earlier attempt to reconcile Becker's results with parallel access to memory suggested, however, that attention is used to select and combine codes. If this is true, then the type of code used in word perception should be flexible, varying with task demands. Moreover, individual differences may exist in the use of codes.

The Hawkins *et al.* (1976) study that provided evidence for phonemic codes also showed that phonemic codes are damped when their use would be detrimental to success. When a target such as *sent* is followed by homophonic tests such as *sent* and *cent,* recognition impairment occurs only when homophonic test pairs are rather rare in the experiment. When homophonic test pairs frequently occur, the phonemic strategy would obviously be a poor one, and the subjects' code selection appears to change: Choice between homophones is no worse (66.5% correct) than choice between nonhomophonic controls (67.8% correct). Both control and homophonic cases are superior to single-letter recognition (53.8% correct), demonstrating that the codes adopted, while not phonemic, still take advantage of word familiarity.

Johnston and McClelland (1974) observed a paradoxical word-recognition effect that relates to the same point. Subjects are superior in recognizing a letter in a word as compared to an isolated letter, but only when they are not informed in advance where the letter will appear. If they are told where in a word to look for the letter, paradoxically, perception of the letter becomes worse. It appears that subjects can selectively attend either to a word code, deducing the letter from the word, or attend to the letter code. The latter appears less effective.

Evidence for code control also comes from James (1975). Subjects classified letter strings by pressing one button for words and another button for nonwords. In one experiment the nonwords were pronounceable. The words were either high, medium, or low in English-language frequency and were either concrete or abstract in meaning. Concreteness had no effect on reaction times to high-frequency words (616 msec), but low-frequency, concrete words were faster (665 msec) than abstract, low-frequency words (745 msec). Apparently, high-frequency words are so familiar at a visual level that a decision is made prior to semantic influences. Low-frequency words are less familiar visually, and the semantic aspects of concreteness versus abstractness influence reaction time.

Similar results occur when the nonwords are not only pronounceable but are also homophonic to real words, ruling out the possibility that high-frequency words were judged on phonemic familiarity rather than visual familiarity. However, when the nonwords were all made nonpronounceable, reaction time speeded considerably and the effects of word frequency and concreteness were nearly abolished. Subjects apparently shifted to a phonemic strategy for word decisions.

Although these studies by Hawkins *et al.*, Johnston and McClelland, and James argue that people have some code control, individuals probably also differ on the weighting given to different codes. Baron and Strawson (1976) asked people to rapidly read aloud lists of phonemically regular words or lists of phonemically irregular words (e.g., *tongue*). Some people read phonemically regular words faster than irregular ones. Baron and Strawson suggested that those people transformed the visual presentation to a phonemic code and then articulated it; these people were therefore called "Phonecians." Other subjects showed less difference in pronunciation time for the two types of words, appearing to transform a visual code to semantics and then to articulation, or directly from a visual code to articulation. The latter subjects named the word as they would have named an object without intervention of a phonemic code, so Baron and Strawson called them "Chinese," based on the notion that reading Chinese characters is not mediated by a phonemic code.

Support for this dichotomy came from two other tasks. On one task subjects attempted to decide whether pseudo words (e.g., *caik*) were pronounced the same as real words. On the other task, subjects were given a spelling task on words that are commonly misspelled. During the initial spelling they could not correct their answers. Then they were shown pairs with the correct spelling and an incorrect spelling and asked to choose the correct one. Some people improved their spelling more than others when they could *see* the words. Subjects good on the pseudo word task and poor on visual spelling correction were classified as Phonecians. People good on spelling correction and poor on the pseudo word task were classified as Chinese. This dichotomy was correlated with the degree to which people had difficulty in reading irregularly spelled words as opposed to regularly spelled ones. Thus, one subject type appears to rely heavily on a phonemic code; the other type relies heavily on a visual code.

Baron and McKillop (1975) have presented related evidence for phonetic versus nonphonetic individual differences.

These individual differences are quite significant as further sources of evidence for different kinds of codes and for their practical implications for understanding reading.

2. The Snyder Effect and Related Phenomena

An incoming stimulus is broken into different features or codes, including semantic codes, that develop simultaneously. A later attentive process reintegrates selected features or codes into a unified percept. This point would be strengthened if it were shown that integration is occasionally or even systematically in error. Snyder (1972) tachistoscopically presented a circle of 12 letters, all normal and in black ink except for one letter, which was either fragmented, colored red, or inverted. Following a tachistoscopic presentation of the array, subjects identified the letter that was altered. A frequent error was to misattribute the fragmentation, color, or inversion to a letter next to the one actually altered.

What is the implication of this error? When a stimulus is processed, different analyzers parse and code different features of the stimulus. Concepts such as color, inversion, size, name, and so on become activated. Only subsequently are the various codes integrated, via the fact that they come from the same spatial location. To the degree that the spatial origin of the cues are uncertain, misassignments may be made.

The Snyder effect poignantly illustrates the processing system proposed. Stimuli enter the system and, in parallel, a variety of codes and meanings within a code become activated. It is only later that they are integrated and that an attentive mechanism exerts some control in the integration.

Improper integration of codes has also been observed in other settings, sometimes in a regular way, resulting in illusions. Deutsch (1975) noted that when a high tone and a low tone alternate in one ear and the same sequence occurs one note out of phase in the other ear (so that the high note in one ear coincides with the low in the other), listeners may hear only one alternating sequence, but it may switch from ear to ear. The high tone is normally heard by right-handed persons in the right ear and the low tone in the left ear. This constitutes an illusion, because if subjects alternate ears in time with the notes they should hear a steady stream of either high or low notes and not an alternating stream. This and other musical illusions described by Deutsch indicate that messages to both ears are being analyzed for features and then reassembled.

In yet another case, Studdert-Kennedy and Shankweiler (1970) observed misassignments in phoneme perception. When a voiced consonant with frontal articulation (i.e., *b*) is played to one ear and simultaneously an unvoiced consonant with middle articulation (i.e., *t*) is played to the other ear, the voicing and place of articulation appear to be separately analyzed and then reassembled. Errors of assembly are more common than other errors, with subjects reporting *p* and *d* more often than *g* and *k*. The error of *p* involves combining the unvoiced feature from one ear with the feature of frontal articulation from the other ear, whereas errors of *g* and *k* both impart a feature not presented.

C. Costs and Benefits of Selection

A selection system that can optionally select information at an early stage reduces errors whenever complex signals in the same system mask one another. It also reduces errors when different codes lead to conflicting behaviors. However, these benefits of selectivity fail to capture one of the major aspects of a selective mechanism, even when selectivity occurs at later stages. The prior allocation of attention to a particular code or memory location within a code may greatly enhance the overall efficiency of processing, even when there is no masking or conflict.

When a signal occurs, information regarding it accrues in memory to some criterion at which time the information can be selected and related to the codes or goals. The selective system may take a measurable amount of time to operate. If the selector is uncommitted, selection may take more time than if a prior expectation for the stimulus is established. Thus, occurrence of an expected signal should result in a benefit to reaction time. Conversely, if an unexpected signal appears, the selector must switch from an already committed state, incurring a delay, or cost, to reaction time. Overall efficiency would increase whenever benefits occurred more frequently than costs. In essence, a selective attentional mechanism would help to efficiently use redundancy in any skilled setting.

Does attention switching make sense—namely, can the costs and benefits of switching be measured?

The answer is not simple, for another mechanism also exists that generates costs and benefits in reaction time. Expectation can lower the criterion for the amount of information that must be necessarily accrued before further decisions are made. If the expected signal does occur, on the average it will reach criterion sooner than it would were the criterion not lowered. However, the criterion-shift model makes a specific prediction not made by attention switching: An unexpected signal will often reach the criterion for the expected memory, resulting in a false alarm, and such false alarms will occur more often than a neutral condition with no expectation. Thus, confirmation of the switching concept requires paradigms in which criterion shifts are a less plausible interpretation.

One such paradigm by LaBerge (1973) was discussed earlier for a different purpose. LaBerge primed a person to expect either a color or a tone. When an unexpected signal occurred, reaction time was considerably increased. This cost cannot easily be explained by a criterion shift, as subjects had to make the *same* response regardless of signal. If the response criterion were lowered by the expectation, either signal should benefit. Since cost rather than benefit occurred for the unexpected signal, the effect appears due to attention switching. This conclusion is further bolstered by the observation that longer switching times occurred when the expected signals were more difficult. This is not easily explained by a criterion shift. Also, the earlier point of LaBerge's study was that information about the unexpected signal was accruing prior to switching. Again, that evidence is difficult to accommodate to a criterion-shift model, because, when the criterion is raised for an unexpected signal, it should increase reaction time more for a difficult than for an easy signal. In fact, LaBerge found the opposite to occur. So LaBerge's experiment makes the important point, in addition to automatic accrual of information in memory, that a selective device can be preset to receive the memory output from a particular signal.

Another paradigm that supports the switching concept has been developed by Posner and colleagues. Posner and Snyder (1975) had subjects match a pair of letters as being the same or different. In a high validity condition, a priming letter preceded the pair, indicating an 80% chance that the primed letter would be in the pair. Because of the high predictability, one would expect subjects to switch their attention to the area of memory called for by the prime. In another condition, the prime was valid only 20% of the time, so it presumably would not induce an attention shift. In a control condition the prime was replaced by a neutral warning signal. It is important to note that the prime in the high-validity case did not differentially predict sameness or difference, but only that a primed letter was likely to occur in a pair. Because of this, criterion-shift explanations are less attractive.

When an expected signal occurred in the high-validity condition, reaction times were about 50–80 msec faster than they were following a neutral prime. This reaction-time benefit began to develop the moment a priming signal occurred, and although the exact time course is uncertain, it reached near asymptote when the priming signal led the letter pair by somewhere around 150 msec. When an unexpected signal occurred in the high-validity condition, reaction time was slowed by about 40–50 msec relative to the neutral condition. The cost did not begin to accumulate until perhaps 150 msec or more after the priming signal.

Attention switching would suggest that benefit and cost would occur at the same time. Yet benefit begins to accrue before cost. Two different mechanisms appear to produce benefit—one begins at the time the prime occurs and the other, related to attention, begins later. The first source of benefit is called *pathway activation* by Posner and Snyder. The prime automatically activates the memory to which it feeds, even when attention is not directed toward that area. Considerable evidence for automatic activation was developed earlier. The prior activation facilitates the processing of any other signal using the same pathway.

Pathway activation was isolated in the low-validity condition, in which the prime had little predictive validity for the following letter pair. A modest reaction-time benefit of about 30 msec occurred when the succeeding letter pair contained the prime, and the benefit began to accrue the moment the prime was presented. However, the low-validity condition produced no cost, indicating that no attention was deployed to the prime.

The main features of the Posner–Snyder results have been replicated by Neely (1977) using word versus nonword decisions and priming with other words. He also separated automatic and attended components of benefit in another way. In one condition, subjects were told that if *bird* appeared as a prime, they should expect a bird (e.g., *robin*) for the lexical decision. Here both automatic and attended components should occur. In another condition, subjects were told that if primed with *building* they should expect a body part (e.g., *arm*). In this case attended benefit should occur but no automatic benefit, since *building* is not associated with particular body parts. As expected, benefit occurred at the shortest prime interval (250 msec) in the first case but not until later in the second case. In both cases cost developed after a 250-msec prime-to-word interval.

Studies of cost and benefit can be summarized as follows: When a signal occurs, information that is well associated in memory with the stimulus is automatically activated. This conclusion concurs with earlier ones and presents another line of evidence. For the activated information to be translated into action, a selector must be turned to the activated area of memory. If the selector is preset to the expected memory area, reaction time will benefit if the expected signal in fact appears; reaction time will suffer cost if an unexpected signal appears.

1. Implications of Cost–Benefit Analysis for the Study of Flexibility

Cost–benefit analysis suggests a possible way to study flexibility of attention. Though flexibility appears to be an important attribute of attention, it has received scant study, perhaps because of the lack of a theoretical structure.

The importance of flexibility has been suggested by Gopher and Kahneman (1971) and Kahneman, Ben-Ishai, and Lotan (1973) investigating Israeli pilots and bus drivers. Drivers and pilots were given a dichotic listening task in two parts. In the first part, a tone indicated which ear to attend to. Digit–word and word–word pairs occurred in the two ears and subjects were to report only the digits in the cued ear. Immediately following 16 pairs another tone occurred, directing attention to the relevant ear for part two. Part two involved 3 pairs of digits, and subjects reported the digits in the cued ear. In neither part one nor part two, when run in isolation, were errors of shadowing predictive of either pilot success or bus-driving accidents. However, when part two immediately followed part one, errors in part two were correlated at about .35 with pilot rating in two samples and bus-driving accidents in a third. These modest correlations are rather impressive considering that the pilots had been preselected on other criteria, that pilot ratings are of low validity, and that bus-driving accidents, even when corrected for fault, are extremely chance-dependent.

Why did dichotic errors in part two predict performance only when it immediately followed part one? According to Kahneman and colleagues, only in that situation were the subjects required to switch attention from one selective set to another. Flexibility of switching appears to be an important part of piloting and driving.

The Kahneman studies suggest that flexibility is a basic trait of attention in which subjects differ, but the concept has received little exploration. Perhaps flexibility in the dichotic listening paradigm is a different manifestation of reaction-time cost in the LaBerge and Posner–Snyder paradigm, and this is a promising issue for further research.

2. Does Selection Involve Inhibition?

When one message is selected, what happens to rejected messages? One possibility is that selection is analogous to setting up a gate for the selected material, thereby facilitating it and blocking all other material. Alternatively, selection of one message may sometimes involve inhibition of others. A suggestive bit of evidence was provided by Treisman (1964b). When subjects shadowed prose in the right ear, it was more difficult to ignore two other verbal messages if one was in both ears and the other only in the left ear than if both messages were in both ears or both in the left ear. Likewise, shadowing a female voice was more difficult if two competing messages in the other ear were male and female rather than both female. These results suggest that not only is the relevant message selected, but competing ones are inhibited, with inhibition being more difficult when the variability of unwanted messages is greater.

More direct evidence for inhibition, using the Stroop effect, has come from Neill (1977). In his experiment, a color (e.g., red) printed out a color word (e.g., *blue*). The subject had to respond to the color and ignore the word. In the preceding text it was suggested that both colors and words activate their memory representations in parallel and that a later process selects the color-activated memory. What happens to the word-activated memory? If, during selection, the word-activated memory is inhibited, then the inhibited concept should be less available on the succeeding trials. Neill included some occasions in which the word on one trial became the color on the next trial. For example, on trial N the color red and word *blue* required the vocal response *red*. On the next trial the color blue and the word *green* required the response *blue*. If the response *blue* had been inhibited during selection on trial N, blue should be responded to more slowly on trial $N + 1$ than would an unrelated color. Neill found reaction time on trial $N + 1$ to be 855 msec when the previously irrelevant memory became the relevant one, which is slower than the control reaction time of 823 msec. Thus, some inhibition of the irrelevant message appears to occur.

The concept of inhibition in the context of selective attention is not well developed and could use more investigation. Inhibition could, for example, be an element in explaining the discrepancy of the Conrad study and the Schvaneveldt, Meyer, and Becker study on activation of ambiguous word meanings. Perhaps both meanings become activated, and then as one is selected others are inhibited. Thus whether the paradigm finds evidence for activation or inhibition may depend on the timing of events and other details of the experimental set-up. Indeed, by slight changes in paradigm, Neill was able to alter the outcomes of his experiment so that evidence for inhibition disappeared.

VI. THE FINAL MODEL

Our analysis of attention has been rather drawn out and several side issues have been explored. Thus, a brief map of the line of inquiry and a final statement of a model of attention would be useful.

The early theories of Welford and Broadbent placed selective attention at an early point in processing, between the sensory and memory systems. Only one signal at a time was transformed to a memory code. Stimulated by these theories, a great deal of subsequent evidence, while confirming limitations of processing, tended to place the limitations further into the sequence of processing, after the accrual of information in memory. The fate of irrelevant information, observations of redundancy gain, reaction-time probes inserted during encoding, and the effects of retrieval time on interference all suggested widespread, parallel access to the memory system.

In many situations, however, parallel access to memory entails costs. When stimuli are complex and in the same modality, they may mask or merge with one another; or if only one message is relevant and other codes or messages lead to conflict, then it would be to the organism's advantage to filter prior to memory. Some evidence does accord with that idea. It appears that optional filtering is related to structural limitations, being more likely when signals are from the same modality. Thus, rather than viewing attention as always being invested at a particular place in processing, it is better viewed as a control process that can influence the flow of information. When messages interfere with one another, the control process can attenuate one source of information prior to memory, although attenuation is quite likely incomplete (otherwise phenomena such as the Stroop effect would not be observed). When the messages do not interfere, they can be processed in parallel through at least the memory stage with no cost. This last point is the primary emphasis of this review.

Not only may different memory locations within a coding system be activated in parallel, but also different codes may develop in parallel. A given word may activate physical, orthographic, phonemic, and semantic codes. This widespread and perhaps imprecise activation requires coordination of information, and it is at the coordination stage that an attentional process again exercises control. Some people suffering dyslexia appear to lack either specific codes or mechanisms for coordinating them. Cross-matching of the different codes (what Becker calls *verification*) allows the normal reader to specifically identify the input item behind semantic activation, but although error analyses indicate semantic activation with some dyslexics, they seem unable to cross-match with a visual code. Results from normal readers also indicate that the different codes do not automatically coalesce into a single decision. Instead, optional control exists over which codes are used. Thus, if many homohpones on a test make phonemic discrimination difficult, people seem able to shift emphasis to another code. In addition to differences in codes used depending on circumstances, there are also individual differences in the facility of using one code or another.

One may ask whether word recognition requires attention. The answer is that some processes of recognition do require attention and others do not. Many habitual associations and codes are automatically activated, too many for specific identification of the input. They require cross-matching for specific identification, and that process requires an attention mechanism. It is only when stimuli are few in number, practice is high, and strong context is available that recognition is nonattentive to the final percept. Even when these special features fail to narrow the field on activated memories to a large degree, automatic activation of associations would seem very functional. All the potentially relevant information regarding a stimulus will be activated and the number of activated memories will be vastly less than the totality of memory.

The attentive mechanism that controls the selection and coordination of information from different codes can be preset for information to accrue to particular memories or particular coding systems. Such presetting takes advantage of redundancy, improving overall efficiency. Whenever expectations are confirmed, processing-time benefits occur; on rarer occasions, when expectations fail, reaction-time cost occurs.

Our overall view of attention has shifted from a notion of limited capacity at particular stages to one of attention as a control process for the flow of information. The receipt of stimuli by a receptor passively activates much information stored in memory. But an active control process can modulate the flow of information to memory, sometimes attenuating an input. In other circumstances the control process allows entry to all codes, but then selects only some codes or some information to be integrated with other information. The control process can preset itself for expected information, thereby improving overall efficiency, and when it selects some information, other conflicting information can be inhibited.

Acknowledgments

Preparation of this review occurred while our research was supported by ONR Contract N0001476-C-0344, and we gratefully acknowledge that support. We wish also to thank Curt Becker and Sandy Pollatsek for their helpful comments on an earlier draft.

References

Archer, E. J. Identification of visual patterns as a function of information load. *Journal of Experimental Psychology*, 1954, **48**, 313–317.

Baron, J., & McKillop, B. J. Individual differences in speed of phonemic analysis, visual analysis, and reading. *Acta Psychologica*, 1975, **39**, 91–96.

Baron, J., & Strawson, C. Use of orthographic and word-specific knowledge in reading words aloud. *Journal of Experimental Psychology: Human Perception and Performance*, 1976, **2**, 386–393.

Baron, J., & Thurston, I. An analysis of the word-superiority effect. *Cognitive Psychology*, 1973, **4**, 207–228.

Becker, C. A. The allocation of attention during visual word recognition. *Journal of Experimental Psychology: Human Perception and Performance*, 1976, **2**, 556–566.

Beller, H. K. Parallel and serial stages in matching. *Journal of Experimental Psychology*, 1970, **84**, 213–219.

Biederman, I., & Checkosky, S. F. Processing redundant information. *Journal of Experimental Psychology*, 1970, **83**, 486–490.

Brainard, R. W., Arby, T. S., Fitts, P. M., & Alluisi, E. A. Some variables influencing the rate of gain of information. *Journal of Experimental Psychology*, 1962, **63**, 105–110.

Broadbent, D. E. Listening between and during practiced auditory distractions. *British Journal of Psychology*, 1956, **47**, 51–60.

Broadbent, D. E. *Perception and communication*. London: Pergamon, 1958.

Broadbent, D. E., & Gregory, M. Donders' B- and C-reactions and S–R compatibility. *Journal of Experimental Psychology*, 1962, **63**, 575–578.

Broadbent, D. E., & Gregory, M. On the interaction of S–R compatibility with other variables affecting reaction time. *British Journal of Psychology*, 1965, **56**, 61–67.

Broadbent, D. E., & Gregory, M. Psychological refractory period and the length of time required to make a decision. *Proceedings of the Royal Society*, B, 1967, **168**, 181–193.

Cherry, E. C. Some experiments on the recognition of speech, with one and with two ears. *Journal of the Acoustical Society of America*, 1953, **25**, 975–979.

Clark, H. H., & Brownell, H. H. Judging up and down. *Journal of Experimental Psychology: Human Perception and Performance*, 1975, **1**, 339–352.

Cohen, G. Some evidence for parallel comparisons in a letter recognition task. *Quarterly Journal of Experimental Psychology*, 1969, **21**, 272–279.

Collins, A. M., & Loftus, E. F. A spreading-activation theory of semantic processing. *Psychological Review*, 1975, **82**, 407–428.

Comstock, E. M. Processing capacity in a letter matching task. *Journal of Experimental Psychology*, 1973, **100**, 63–72.

Conrad, C. Context effects in sentence comprehension: A study of the subjective lexicon. *Memory & Cognition*, 1974, **2**, 130–138.

Corcoran, D. W. J., & Besner, D. Application of Posner technique to the study of size and brightness irrelevancies in letter pairs. In P. M. A. Rabbitt & S. Dornic (Eds.), *Attention and Performance V*. London: Academic Press, 1975.

Deutsch, D. Musical illusions. *Scientific American*, 1975, **232**, 92–100.

Deutsch, J. A., & Deutsch, D. Attention: Some theoretical considerations. *Psychological Review*, 1963, **70**, 80–90.

Dyer, F. N. Interference and facilitation for color naming with separate bilateral presentations of the word and color. *Journal of Experimental Psychology*, 1973, **99**, 314–317.

Egeth, H., Jonides, J., & Wall, S. Parallel processing of multielement displays. *Cognitive Psychology*, 1972, **3**, 674–698.

Eichelman, W. H. Familiarity effects in the simultaneous matching task. *Journal of Experimental Psychology*, 1970, **86**, 275–282.

Ellis, S. H., & Chase, W. G. Parallel processing in item recognition. *Perception & Psychophysics*, 1971, **10**, 379–384.

Felfoldy, G. L., & Garner, W. R. The effects on speeded classification of implicit and explicit instructions regarding redundant dimensions. *Perception & Psychophysics*, 1971, **9,** 289–292.

Fitts, P. M., Peterson, J. R., & Wolpe, G. Cognitive aspects of information processing: II. Adjustments to stimulus redundancy. *Journal of Experimental Psychology*, 1963, **65,** 423–432.

Fitts, P. M., & Seeger, C. M. SR compatibility: Spatial characteristics of stimulus and response codes. *Journal of Experimental Psychology*, 1953, **46,** 199–210.

Garner, W. R. Attention: The processing of multiple sources of information. In E. C. Carterette & M. P. Friedman (Eds.), *Handbook of perception* (Vol. II). New York: Academic Press, 1974. (a)

Garner, W. R. *The processing of information and structure*. Potomac, Maryland: Erlbaum, 1974. (b)

Garner, W. R., & Felfoldy, G. L. Integrality of stimulus dimensions in various types of information processing. *Cognitive Psychology*, 1970, **1,** 225–241.

Gopher, D., & Kahneman, D. Individual differences in attention and the prediction of flight criteria. *Perceptual and Motor Skills*, 1971, **33,** 1335–1342.

Hawkins, H. L., Reicher, G. M., Rogers, M., & Peterson, L. Flexible coding in word recognition. *Journal of Experimental Psychology: Human Perception and Performance*, 1976, **2,** 380–385.

Hawkins, H. L., Thomas, G. B., Presson, J. C., Cozic, A., & Brookmire, D. Precategorical selective attention and tonal specificity in auditory recognition. *Journal of Experimental Psychology*, 1974, **103,** 530–538.

Henderson, L. A word superiority effect without orthographic assistance. *Quarterly Journal of Experimental Psychology*, 1974, **26,** 301–311.

Hick, W. E. On the rate of gain of information. *Quarterly Journal of Experimental Psychology*, 1952, **4,** 11–26.

Hintzman, D. L., Carre, F. A., Eskridge, V. L., Owens, A. M., Shaff, S. S., & Sparks, M. E. "Stroop" effect: Input or output phenomenon? *Journal of Experimental Psychology*, 1972, **95,** 458–459.

Hyman, R. Stimulus information as a determinant of reaction time. *Journal of Experimental Psychology*, 1953, **45,** 188–196.

James, C. T. The role of semantic information in lexical decisions. *Journal of Experimental Psychology: Human Perception and Performance*, 1975, **1,** 130–136.

Johnston, J. C., & McClelland, J. L. Perception of letters in words: Seek not and ye shall find. *Science*, 1974, **184,** 1192–1194.

Kahneman, D. *Attention and effort*. Englewood Cliffs, New Jersey: Prentice-Hall, 1973.

Kahneman, D., Ben-Ishai, R., & Lotan, M. Relation of a test of attention to road accidents. *Journal of Applied Psychology*, 1973, **58,** 113–115.

Karlin, L., & Kestenbaum, R. Effects of number of alternatives on the psychological refractory period. *Quarterly Journal of Experimental Psychology*, 1968, **20,** 167–178.

Keele, S. W. Effects of input and output modes on decision time. *Journal of Experimental Psychology*, 1970, **85,** 157–164.

Keele, S. W. Attention demands of memory retrieval. *Journal of Experimental Psychology*, 1972, **93,** 245–248.

Keele, S. W., & Boies, S. J. Processing demands of sequential information. *Memory & Cognition*, 1973, **1,** 85–90.

Kerr, B. Processing demands during mental operations. *Memory & Cognition*, 1973, **1,** 401–412.

Klein, G. S. Semantic power measured through the interference of words with color-naming. *American Journal of Psychology*, 1964, **77,** 576–588.

Klein, R. M., & Posner, M. I. Attention to visual and kinesthetic components of skills. *Brain Research*, 1974, **71,** 401–411.

LaBerge, D. Identification of two components of the time to switch attention: A test of a serial and a parallel model of attention. In S. Kornblum (Ed.), *Attention and Performance IV*. New York: Academic Press, 1973.

LaBerge, D. Acquisition of automatic processing in perceptual and associative learning. In P. M. A. Rabbitt & S. Dornic (Eds.). *Attention and Performance V*. New York: Academic Press, 1975.

Leonard, J. A. Tactual choice reactions. *Quarterly Journal of Experimental Psychology*, 1959, **11,** 76–83.

Lewis, J. L. Semantic processing of unattended messages using dichotic listening. *Journal of Experimental Psychology*, 1970, **85,** 225–228.

Lockhead, G. R. Processing dimensional stimuli: A note. *Psychological Review*, 1972, **79,** 410–419.

Marshall, J. E., & Newcombe, F. Patterns of paralexia: A psycholinguistic approach. *Journal of Psycholinguistic Research*, 1973, **2,** 175–199.

Martin, D. W., Marston, P. T., & Kelly, R. T. Measurement of organizational processes within memory stages. *Journal of Experimental Psychology,* 1973, **98,** 387–395.

Morgan, B. B., Jr., & Alluisi, E. A. Effects of discriminability and irrelevant information on absolute judgments. *Perception & Psychophysics*, 1967, **2,** 54–58.

Morton, J. Interaction of information in word recognition. *Psychological Review*, 1969, **76,** 165–178. (a)

Morton, J. Categories of interference: Verbal mediation and conflict in card sorting. *British Journal of Psychology*, 1969, **60,** 329–346. (b)

Morton, J. The use of correlated stimulus information in card sorting. *Perception & Psychophysics*, 1969, **5,** 374–376. (c)

Mowbray, G. H., & Rhoades, M. V. On the reduction of choice reaction times with practice. *Quarterly Journal of Experimental Psychology*, 1959, **11,** 16–23.

Neely, J. H. Semantic priming and retrieval from lexical memory: Roles of inhibitionless spreading activation and limited-capacity attention. *Journal of Experimental Psychology: General*, 1977, **106,** 226–254.

Neill, W. T. Inhibitory and facilitatory processes in selective attention. *Journal of Experimental Psychology: Human Perception and Performance*, 1977, **3,** 444–450.

Ninio, A., & Kahneman, D. Reaction time in focused and in divided attention. *Journal of Experimental Psychology,* 1974, **103,** 394–399.

Norman, D. A. Toward a theory of memory and attention. *Psychological Review*, 1968, **75,** 522–536.

Ostry, D., Moray, N., & Marks, G. Attention, practice, and semantic targets. *Journal of Experimental Psychology: Human Perception and Performance*, 1976, **2,** 326–336.

Pollatsek, A., Well, A. D., & Schindler, R. M. Familiarity affects visual processing of words. *Journal of Experimental Psychology: Human Perception and Performance*, 1976, **1,** 328–338.

Posner, M. I. *Chronometric exploration of mind: An analysis of the temporal course of information flow in the human nervous system*. Hillsdale, New Jersey: Erlbaum, in press.

Posner, M. I., & Boies, S. J. Components of attention. *Psychological Review*, 1971, **78,** 391–408.

Posner, M. I., & Klein, R. M. On the functions of consciousness. In S. Kornblum (Ed.). *Attention and Performance IV*. New York: Academic Press, 1973.

Posner, M. I., & Mitchell, R. F. Chronometric analysis of classification. *Psychological Review*, 1967, **74,** 392–409.

Posner, M. I., & Snyder, C. R. R. Facilitation and inhibition in the processing of signals. In P. M. A. Rabbitt & S. Dornic (Eds.), *Attention and performance V*. London: Academic, 1975.

Posner, M. I., & Taylor, R. L. Subtractive method applied to separation of visual and name components of multiletter arrays. *Acta Psychologica*, 1969, **30,** 104–114.

Reicher, G. M. Perceptual recognition as a function of meaningfulness of stimulus material. *Journal of Experimental Psychology*, 1969, **81,** 275–280.

Reynolds, G. S. Attention in the pigeon. *Journal of the Experimental Analysis of Behavior*, 1961, **4,** 203–208.

Rogers, M. G. K. *Visual generation in the recognition task.* Unpublished doctoral dissertation, Univ. of Oregon, Eugene, Oregon, 1975.

Schroeder, R. *Information processing of color and form*. Unpublished honor's thesis, Univ. of Oregon, Eugene, Oregon, 1976.

Schvaneveldt, R. W. Effects of complexity in simultaneous reaction time tasks. *Journal of Experimental Psychology*, 1969, **81,** 289–296.

Schvaneveldt, R. W., Meyer, D. E., & Becker, C. A. Lexical ambiguity, semantic context, and visual word recognition. *Journal of Experimental Psychology: Human Perception and Performance*, 1976, **2,** 243–256.

Schwank, J. *Dichoptic and binocular viewing effects on selective attention to dual-signal inputs*. Unpublished doctoral dissertation, Univ. of Oregon, Eugene, Oregon, 1975.

Shallice, T. Dual functions of consciousness. *Psychological Review*, 1972, **79,** 383–393.

Shallice, T., & Warrington, E. K. Word recognition in a phonemic dyslexic patient. *Quarterly Journal of Experimental Psychology*, 1975, **27,** 187–199.

Shannon, C. E., & Weaver, W. *The mathematical theory of communication*. Urbana, Illinois: Univ. of Illinois Press, 1949.

Smith, M. C. The effect of varying information on the psychological refractory period. In W. G. Koster (Ed.), *Attention and Performance II.* Amsterdam: North-Holland, 1969.

Snyder, C. R. R. Selection, inspection, and naming in visual search. *Journal of Experimental Psychology*, 1972, **92,** 428–431.

Studdert-Kennedy, M., & Shankweiler, D. Hemispheric specialization for speech perception. *Journal of the Acoustical Society of America*, 1970, **48,** 579–591.

Treisman, A. M. Monitoring and storage of irrelevant messages in selective attention. *Journal of Verbal Learning and Verbal Behavior*, 1964, **3,** 449–459. (a)

Treisman, A. M. The effect of irrelevant material on the efficiency of selective listening. *American Journal of Psychology*, 1964, **77,** 533–546. (b)

Treisman, A. M. Strategies and models of selective attention. *Psychological Review*, 1969, **76,** 282–299.

Treisman, A. M., & Davies, A. Divided attention to ear and eye. In S. Kornblum (Ed.). *Attention and Performance IV*. New York: Academic Press, 1973.

Treisman, A. M., & Fearnley, S. Can simultaneous speech stimuli be classified in parallel? *Perception & Psychophysics*, 1971, **10,** 1–7.

Treisman, A., Squire, R., & Green, J. Semantic processing in dichotic listening? A replication. *Memory & Cognition*, 1974, **2,** 641–646.

von Wright, J. M., Anderson, K., & Stenman, U. Generalization of conditioned GSRs in dichotic listening. In P. M. A. Rabbitt & S. Dornic (Eds.). *Attention and Performance V*. London: Academic Press, 1975.

Wardlaw, K. A., & Kroll, N. E. A. Autonomic responses to shock-associated words in a nonattended message: A failure to replicate. *Journal of Experimental Psychology: Human Perception and Performance*, 1976, **2,** 357–360.

Warren, R. E. Stimulus encoding and memory. *Journal of Experimental Psychology*, 1972, **94,** 90–100.

Welford, A. T. Evidence of a single channel decision mechanism limiting performance in a serial reaction task. *Quarterly Journal of Experimental Psychology*, 1959, **11,** 193–210.

Welford, A. T. The measurement of sensory-motor performance: Survey and reappraisal of twelve years' progress. *Ergonomics*, 1960, **3,** 189–230.

Well, A. The influence of irrelevant information on speeded classification tasks. *Perception & Psychophysics*, 1971, **10,** 79–84.

Wheeler, D. D. Processes in word recognition. *Cognitive Psychology*, 1970, **1,** 59–85.

Chapter 2

PERCEPTUAL STRUCTURE AND SELECTION*†

DAVID E. CLEMENT

I. INTRODUCTION

The title of this chapter incorporates two important assumptions concerning perception. One is that perception deals with structure, or more specifically, with correlation among stimuli, among elements in stimuli, and among the representations of all these in the neural processes of the perceiver. The second is that perception is an active process, involving selection; that is, the organism *perceives* rather than *receives* information. The enormous panoply of sensory stimulation available to a waking or-

* Preparation of this chapter was supported in part by a Faculty Development Leave from the University of South Florida.

† This chapter was completed in February, 1975, and thus does not reflect relevant research published subsequent to that date.

ISBN 0-12-161909-5

ganism far exceeds its processing capability, at least in terms of the encoding and storage of all elements of the ambient stimulus array. As the perceiving organism necessarily processes and keeps only a portion of this array, the important questions are what is processed, how is it processed, and why is it processed. My emphasis is on the "what" question, leaving most of the speculation concerning "how" and "why" to others. Clearly this means that I am more concerned with stimuli than with neural response and overt response factors. Of course, the latter cannot be ignored, for any organism functions in toto rather than alternately as now a perceiver, now a learner, and so forth. But my focus keeps returning to the discussion of perception in terms of stimuli present before, now, and in the future; what has been perceived, what is being perceived, and what will or could be perceived.

Visualization of my concept of information processing by humans is given in Fig. 1. Actually, a more accurate representation would involve more arrows, indicating the totality of the interrelationships among the arbitrarily labeled terms. The emphasis of this chapter is upon the box labeled *organization*, which might have been labeled *perception* if the diagram included terms such as *sensation*, *learning*, and *motivation*. A consideration of perception must include extensive development of all these processes, as well as others, and other chapters in the volumes of the *Handbook of Perception* deal with these individually and in combinations.

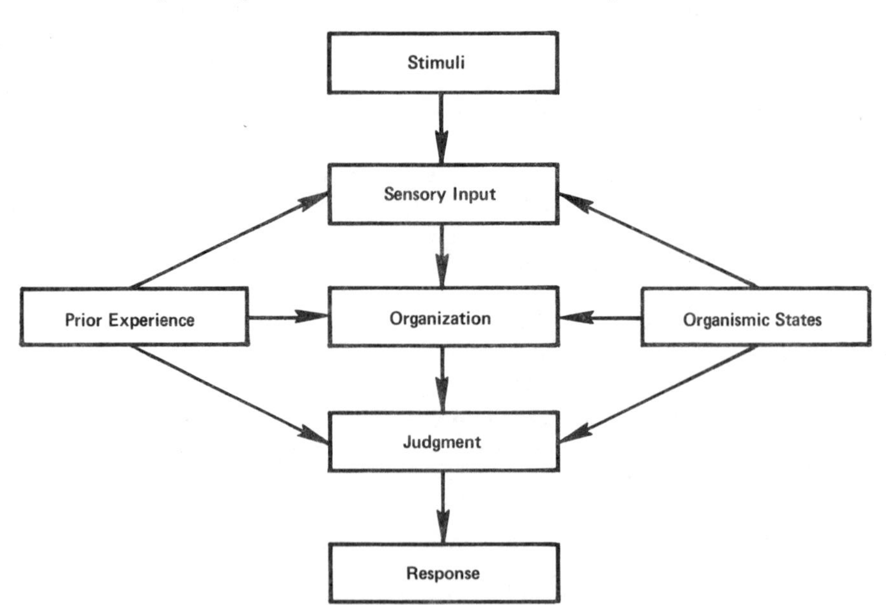

FIG. 1. A schematic diagram of information processing in humans.

I am limiting my discussion to the area of stimulus structure, both simultaneous and sequential, and how the human observer deals with this structure. Almost all examples deal with visual perception, and the discussion, for the most part, ignores questions of epistemology, neuropsychology, and motivation.

What I shall be discussing in this chapter are the basic tenets of structure and perception, including the concept of redundancy; the measurement of structure, both directly and indirectly; the development of relevant perceptual processes in humans; and the directions which further research might take in specifying and perhaps explaining the ways in which humans process information from their environment. The choice of references is selective. The reader may branch from these references to other sources of my way of thinking, but will have to rely upon references from other chapters (e.g., Dodwell, 1975; Haber, 1974) as well as from other books (e.g., Zusne, 1970) if he wishes to proceed beyond my particular frame of reference.

A. Structure and Perception

Wertheimer (1974) describes the study of perceptual structure as dealing with the way in which an essentially continuous environment is processed into discrete neural events, which then are combined in some fashion within the organism. He contrasts two views of this process—associationism and Gestalt psychology. Associationism, in a simplified sense, treats the combination of inputs into a "whole" as determined by the relations among the elemental parts. The Gestalt view, again simplified greatly, views the parts as being determined by a prepotent "whole" perception. Both these views, as adopted by various psychologists, are similar in that they emphasize the neural events as necessarily determining the resultant perceptual structure, and both start with an assumption of primary neural events (either atomistic or holistic). My own preference is to place the emphasis upon relations among all possible parts and wholes in the environment, deriving from the work of Garner (e.g., Garner, 1962, 1974). This emphasis upon environmental context does not deny the necessity for neural events, but rather assumes that perception must deal with the processing of information *from* the environment. Thus, one appropriate tactic is to study the effects of the environmental stimuli upon perception across various conditions of prior learning, motivation, and physiological state. Furthermore, the well-known effects of simultaneous context argue against a limited associationism, and the effects of sequential context argue (though less strongly) against a limited Gestalt approach.

Correlations, or structure, exist in the external stimulus world. The apparent constancy of external objects under various conditions of viewing suggests the importance of relations rather than specific stimulus values in processing by humans. For example, we identify other people by selective attention to preexisting stimulus relationships, as these relationships provide the necessary cues for appropriate discrimination. The anecdotal and actual difficulty of most persons of one race in distinguishing among persons of another race is but one indication of the kinds of simplistic strategies we test in selecting relations for person recognition. If I am white and have but one black acquaintance, there is no need for me to pay attention to any dimension other than skin color in correctly identifying her. If I have but two such acquaintances, I must pay attention to a combination of characteristics (such as color and gender) to tell her from him. If I have 27 black friends, I must use relational cues of more complexity (perhaps eye–face–hair–body relations) to make accurate identifications. I suggest that processing of a single dimension is an exception to typical perceptual processing and occurs only under extremely unlikely circumstances (e.g., that I never see any black person other than the one I happen to know). The correlational structure used to identify a person is reliable; that is, it has sequential stability or correlation. The relations among physical features for someone at this moment in time are the same relations that existed yesterday, and that will exist tomorrow. Another kind of correlational structure is that used in discriminating one entity from another when they are present simultaneously, or the specific differential relations existing at one moment in time, such as those mentioned by Rubin (1915) in consideration of figure–ground relations. His discussion of what aspects "go together" in providing the figure anticipates the present emphasis upon stimulus relations, for those parts which go together usually emanate from the same object.

The view of perceptual structure as existing in the environment, to be perceived by the observer, is an attractively simple one. It is practical, for ultimately we wish to understand and predict perceptions in a specific existing environment. It is more amenable to measurement at this stage of our knowledge than are neural events, for it is much easier to reach agreement about the way in which external stimuli are to be measured than about the way in which internal events are to be measured (indeed, if agreement is possible about the latter). Furthermore, if one views the human being as an information processor, one cannot deny the importance of the information in the real world that is available for processing. This does not suggest that the observer imposes no structure upon the environment, but rather that any structure imposed by the observer typically will be consonant with the structure existing in the real world that the observer contacts.

B. Redundancy

The quantification of the stuff of perceptual processing in terms of context and organization has been advanced by the application of information theory to psychology (e.g., Attneave, 1959; Garner, 1962; Garner & McGill, 1956). This particular method of measurement deals with sets of events, emphasizing probabilities of occurrence of events and the correlations among events. Garner (1962) prefers the use of the term *uncertainty analysis* to *information analysis,* as the former better captures the emphasis upon variability and possibility, whereas the latter seems to imply stability and certainty. Uncertainty analysis provides a nonmetric treatment for variability among events that may be stimuli, stimulus components, responses, or response components. The measurement is based upon the negatively signed logarithms of the probabilities of occurrence for events in the set under consideration, each weighted by the probability of occurrence; usually, the logarithm is taken to the base 2, leading to the unit of bits (binary digits) of information. A very frequent event would have a high probability and thus a small uncertainty value; an infrequent event would lead to a high uncertainty value. Of course, as these events contribute to total average uncertainty proportionately to their relative probabilities, the high-probability (low uncertainty) event would be weighted more heavily than the low-probability event in calculating the average uncertainty for a specified set of events. For a detailed description of the procedures, the reader is referred to Garner (1962).

Redundancy refers to reduction in uncertainty, and derives from one of two kinds of situations. The extent to which events in a set of possible events differs from being equally probable reduces uncertainty from the nominal value it would have if they were equally probable. This reduction in uncertainty, using Garner's (1962) terminology, is distributional constraint, or distributional redundancy. Correlations between events (for example, between two dimensions of a set of stimuli), which serve also to reduce the uncertainty of the total set of events, lead to correlational constraint, or correlational redundancy. Correlational redundancy refers to information or uncertainty that is contingent in nature. If a set of stimuli contains three dimensions upon which stimuli vary, contingent relations include simple redundancy (between two dimensions) and interaction redundancy (among all three dimensions). Thus, redundancy can refer to increasingly complex kinds of relations, but in all cases the relations are correlational in nature. Redundancy refers to the reduction in *amount* of uncertainty due to correlation; that is, to the amount of variability in one or more dimensions that potentially may be explained by variability in another dimension or dimensions. Ratios may be formed between contingent uncertainty and uncertainty of one dimension to give the *proportion* of uncertainty in that dimension that is predictable. This is analogous to the correlation ratio or to the square of the product–moment correlation in determining the proportion of variance predictable in one term from another (see Clement & Carson, 1961, for this usage). Uncertainty analysis also can be used to partition variance, analogous to the analysis of variance techniques (Garner & McGill, 1956), thus providing powerful analytic procedures for nonmetric data.

An example of a simple form of correlational redundancy would be a population of males in which all tall males wore beards and all short males were clean shaven; in this population, the dimensions of height and hirsuteness would be perfectly correlated or totally redundant. Knowledge of the height of a man in this population would give perfect information (total reduction in uncertainty) concerning his facial hairiness. If *most* tall males had beards, and *most* short males did not, the dimensions would be redundant, but not totally so. The use of this measure of correlation requires only discriminable categories, and is quite valuable in avoiding concerns over the correctness of interval-scale values attached to stimulus characteristics.

The use of uncertainty analysis in the 1950s approached faddism, with unrealized expectations of information theory serving as a model of how human neural processes operate. This unfortunate romance with the method alienated some potential users who might have utilized it as a tool, rather than a model. The use of uncertainty analysis since about 1960, however, has become more moderate and effective. It has served as a model of what environmental information the human information processor must process, and has suggested performance characteristics for humans insofar as the human functions parallel the procedures of uncertainty analysis. In this respect, uncertainty analysis is comparable to other heuristic procedures, such as computer simulation, in providing clues for further work, rather than as a formal model of neural functioning. Uncertainty analysis reflects the probabilistic structure of the world as revealed in imperfect correlations among events and aspects of events. Humans must learn to deal with this structure by neural processes that are fixed or probabilistic in nature.

Redundancy has been a valuable concept in attempting to quantify the insightful organizational principles of Gestalt psychology (the so-called laws of organization). It is not necessary to accept Gestalt tenets of neural functioning in order to recognize the accuracy of the examples used to illustrate the organizational principles (see Hochberg, 1974, for illustrations). However, the applications of these principles by Gestalt psychologists have been rather nonproductive, for they seem to rely upon a skilled observer rather than upon some hypothesis or procedure that could be tested in the usual empirical fashion. Specifically, saying that a good figure is one that looks well-organized provides little basis for further exploration of figural goodness. Saying that a good figure is one with a higher level of redundancy at least suggests measurement, and saying that redundancy relates to the number of similar figures that can exist in a population generated by specific rules allows testable speculations concerning the relationship between figural goodness and measurable stimulus characteristics. It allows investigation of both perceptual structure, as it relates to identifiable stimulus structure, and perceptual selection, as it relates to the stimulus structure that is chosen by the observer for processing.

II. MEASUREMENT OF STRUCTURE

Perceptual structure, as previously noted, is assumed to be dependent upon structure existing in the stimulus environment. It can be measured directly, in terms of the dimensions of stimuli, descriptions of stimulus sets, and consideration of alternative possibilities for stimuli. It can be measured indirectly in terms of performance and learning characteristics of observers presented with stimuli in different kinds of tasks. If results with varied tasks and varied sets of stimuli support the same particular hypotheses, this use of converging operations (Garner, Hake, & Eriksen, 1956) can eliminate possible alternative explanations of perceptual structure. The studies and examples cited in the following sections are drawn mostly from the work of Garner, his students, and his colleagues. Most relate to visual perceptual structure, but there is no reason to suspect that the results would differ for other perceptual modalities.

A. Structure Defined in Physical Terms

A set of stimuli is structured in two basic respects. Considering the variability possible in each of the dimensions of a stimulus, a complete theoretical stimulus population can be defined in which the dimensional variability is orthogonal. That is, the theoretical population will contain one stimulus for each possible combination of values in each dimension. The extent to which the set of stimuli is smaller than this theoretically possible population describes the distributional redundancy, or distributional constraint, of the stimulus set under consideration. Furthermore, considering this stimulus set as a defined population for experimental purposes, it may contain correlations among dimensions. Such correlation is the correlational redundancy, or correlational constraint. Subsets drawn from this set may themselves contain additional correlational redundancy. The importance of the consideration of sets and subsets of stimuli cannot be overestimated (e.g., Clement, Guenther, & Sistrunk, 1972; Garner, 1962; Levy & Kaufman, 1973). It provides the basis for identifying structure as obtaining to a single stimulus; in fact, it would be silly to speak of structure for a single stimulus if there were no set of stimuli to which it belonged. To phrase it differently, structure refers to correlation; correlation cannot exist without variability, and thus structure is dependent upon stimulus variability. A single stimulus does not contain variability in itself, but rather is an element of a variable set.

Easy specification of objective structure is possible only for very simple stimuli. For example, consider a population of five-dot patterns, where each dot is placed in one of the nine cells of an imaginary 3 × 3 matrix. Figure 2 contains examples of patterns which can be formed in this way. The total number of patterns possible is 126. If the restriction is placed upon the set of patterns that each row and each column of the imaginary

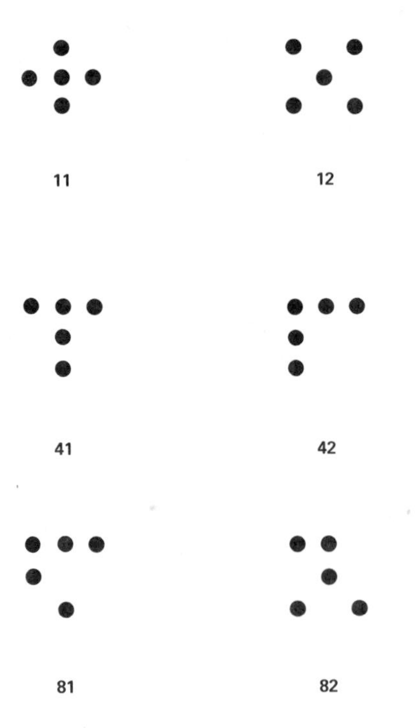

FIG. 2. Sample five-dot patterns, from three different sizes of equivalence sets.

matrix must contain at least one dot (a form of distributional constraint), the number of patterns in the total set is reduced to 90. Within this set of 90 patterns, structure may be specified further by considering equivalence sets of patterns. Assume the following kinds of transformation operations that can be performed on each stimulus: reflections about a horizontal or vertical axis, rotation by 90° increments, or combinations of these operations. Patterns which can be made to coincide by these operations will be defined as equivalent patterns and together will make up an equivalence set. Patterns 11 and 12, for example, are unique and make up two sets of size 1. Patterns 41 and 42 each come from different sets of size 4, while patterns 81 and 82 each come from different sets of size 8. This procedure leads to the identification of 17 equivalence sets of patterns among the 90 patterns making up the population. The smaller sets are more redundant with respect to the population than are the larger sets. Presumably this structure exists in the stimuli and should affect perceptual structure and performance of the subjects who use these stimuli in some kind of task. (In fact it does, as we shall see subsequently.)

A more complex example of structure is shown in Fig. 3. Assuming a theoretical population of stimuli each consisting of four letters of either upper- or lowercase, there are 7,311,616 possible patterns. The reduction to a limited number of letters in the patterns in (a) and (b) represents distributional constraint. The particular 8 patterns in each example represent correlational redundancy. In (a), only the letters *A* and *B* are used, which could lead to 256 different patterns. A subject having to learn this set of patterns would quickly determine that only two letters are used and the subjective population of stimulus alternatives would total 256. The first two positions in each pattern contain only *A* and the last two positions contain only *B*. This correlation between letter identity and position, or redundancy between letter and position, reduces the number of pat-

1. A A B B
2. A a B B
3. a A B b
4. a a B b

5. A A b B
6. A a b B
7. a A b b
8. a a b b

(a)

FIG. 3. Two sets of four-letter patterns, with different amounts and forms of redundancy.

1. A a a A
2. A B B A
3. B b b B
4. B A A B

5. C c c C
6. C D D C
7. D d d D
8. D C C D

(b)

terns possible to 16. The correlation between the case of the first and last letters reduces the possible stimulus alternatives from 16 to 8. These two simple constraints are all the subject must learn, and the amount of redundancy reflects the reduction from 256 to 8 patterns (5 bits of redundancy). Figure 3(b) also contains only 8 patterns, but these are members of a set of 4096 possible patterns (with the limitations of four letters, four positions, and upper and lower cases). Thus the amount of redundancy processed by the subject would be greater. Also, the redundancy takes a more complex form. The limitation of first and last letters to the upper case reduces the alternatives by a factor of 4; correlation between letters (*A* and *B* alone or together; *C* and *D* alone or together) reduces the alternatives by an additional factor of eight; the identity of first and last letters, and of the middle two letters, reduces the alternatives by an additional factor of four; the use of the lower case for the middle two letters only when all four letters are the same, and the upper case for the middle two letters when they are not, reduces the alternatives by an additional factor of four. Obviously, this is a more complex form of redundancy, as well as a greater amount (a reduction from 4096 to 8 patterns, or nine bits of redundancy), and a subject learning these patterns might be expected to have more difficulty in processing the perceptual structure than when given the stimulus set in Fig. 3(a). It is important to determine what the *perceived* set of alternative patterns is for a human observer in such a task. If a subject thought initially that all 26 letters of the alphabet were used in the generation of the stimulus patterns in (a) and (b), the processing of structure might proceed differently than if the number of letters used were correctly perceived. However, by the completion of a learning or discrimination task with these stimuli, all subjects should be aware of the actual letter population used (that is, would be aware of the distributional constraint). The perceived set of alternative patterns would be the only discrimination between (a) and (b) in terms of correlational redundancy; the subject must be aware of the *possible* number of stimuli given the restriction upon the alphabet—4096 in (b) while only 256 in (a), or the sets of patterns would have the same amount of redundancy. This subjective approach is still amenable to confirmation by indirect measurement in terms of relative performance characteristics when factors other than redundancy are controlled.

In summary, an observer must be aware of the stimuli which could occur, whether they do or not, and in this sense the approach is subjective. However, if care is taken in presentation of the stimuli, actual measurements of structure in terms of stimulus redundancy may be made independently of any performance on the part of the subject. This measurement rapidly becomes difficult as one proceeds from very simple two-dimensional patterns to more complex patterns (see, for example, Fig. 4), and becomes overwhelming with three-dimensional arrays. Thus, findings based upon simple patterns should not be generalized to real-world situations uncritically. It does seem reasonable, however, to assume that principles that work for simple patterns are inherent in the processing of more complex patterns, but that as potential complexity of the system increases, more and more principles will become necessary for explanation. I believe that human behavior in any meaningful task is so complex that it cannot be reduced to a small number of general rules, but rather that humans differ in terms of preferred strategies and, furthermore, have a rather large number of strategies available from which to select in a particular situation. (Garner, 1970, discusses this complexity as related to possible stimulus characteristics; on the other hand, Hock, 1973, considers discrimination between structural and analytic subjects.) The purpose underlying investigations of physical measures of structure for simple stimuli, and of performance correlated with this structure, is to

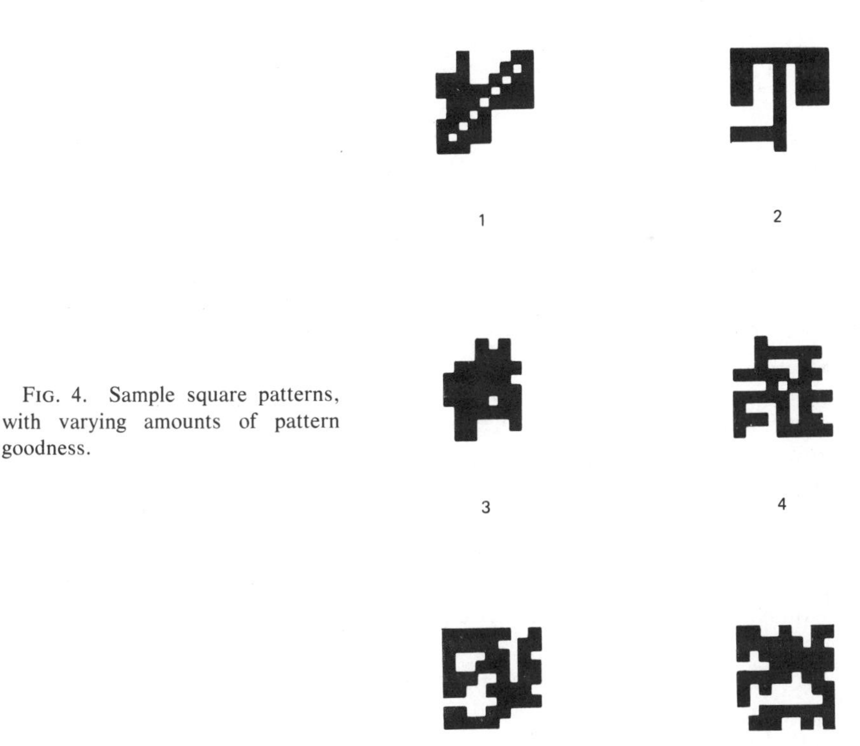

FIG. 4. Sample square patterns, with varying amounts of pattern goodness.

determine *some* of the ways in which humans perceive, not *the* way in which they perceive.

B. Structure Reflected in Performance

I shall describe a series of experiments demonstrating the efficacy of the pattern-redundancy approach as reflected in performance, focusing upon concepts coming from Garner's (1962) book as well as from earlier studies (e.g., Attneave, 1954; Hochberg & McAlister, 1953). The experiments to be described used dot patterns such as those in Fig. 2, and thus had stimuli that could be described objectively as varying in levels of redundancy. Higher redundancy accrued to patterns from smaller sets of equivalent patterns, as previously discussed. I emphasize that this redundancy was measured as a property of the stimulus, prior to investigation of possible correlations with performance.

1. Pattern Goodness and Pattern Uncertainty

Garner and I found that subjects asked to group 90 five-dot patterns into about eight groups tended to (*a*) keep patterns from the same equivalence sets together; (*b*) keep equivalence sets of the same size together; and (*c*) place patterns from smaller equivalence sets into smaller groups (Garner & Clement, 1963). Thus, subjects used some kind of clustering procedure that reflected the objective classification of stimuli into equivalence sets. This was not an artifact of the instructions, for 17 different equivalence sets and three levels of redundancy occurred in the 90 patterns, and the suggested eight groups for sorting did not in turn suggest the use of the equivalencies except on rational grounds. Also, in that study, a different group of subjects were asked to rate each of the 90 patterns for *pattern goodness,* with no experimenter-imposed definition of that term. The subjects demonstrated reliable ratings of pattern goodness, which correlated highly ($r = .84$) with the size of the group into which the other subjects had sorted the patterns. Thus, objective measures of perceptual structure in terms of redundancy correlated with subjective size of equivalence set, which in turn correlated with subjective ratings of pattern goodness. Good patterns (e.g., patterns 11 and 12 in Fig. 2) are those which come from small sets of equivalent patterns, are highest in redundancy, and represent the greatest reduction in uncertainty from the stimulus population. Glushko (1975) has confirmed the primacy of the relation between pattern goodness and redundancy using multidimensional scaling and hierarchical clustering of these dot patterns, based upon pairwise preferential judgments of goodness.

2. Latency and Uncertainty of Naming Responses

This work was extended in investigations of correlations between other kinds of performance and objective redundancy, subjective redundancy, and pattern goodness. I argued (Clement, 1964) that simpler, more redundant patterns are more easily encoded and thus could be given verbal descriptions more rapidly and more consistently. Subjects were presented with 100 patterns and asked to provide a one-word description of each. The median response latency and response uncertainty (variability) were determined for each stimulus. Stimuli included 50 dot patterns of the type shown in Fig. 2, and 50 patterns made up of blackened cells in an imaginary 9×9 matrix such as in Fig. 4 (square patterns). The objective equivalence set size could not be calculated in any simple fashion for the latter, but the assumption was made that, just as with the dot patterns, ratings of pattern goodness would reflect pattern redundancy as derived from equivalence set size. A separate set of subjects rated the square patterns for pattern goodness. (The rating of goodness of the examples in Fig. 4 decreases from pattern 1 through pattern 6.) As expected, redundant patterns were given verbal labels both faster and more consistently (in terms of response uncertainty) than less redundant patterns. For each set of patterns, the correlations between ratings of pattern goodness by one group of subjects and latency of response and uncertainty of verbal labels by the other group ranged from $r = .66$ to $r = .87$ for individual dot patterns or square patterns; from $r = .84$ to $r = .96$ for equivalence sets of dot patterns. Glanzer and his associates (Glanzer & Clark, 1963, 1964; Glanzer, Taub, & Murphy, 1968) anticipated this finding, but assumed that the verbal descriptions preceded other responses to visual patterns, and thus were the primary source of pattern structure. This seems an unnecessary reliance on an internal processing mode, whereas prediction from stimulus characteristics selected by the observer seems more straightforward. Also, their own data indicated that verbal descriptions were more sensitive to the actual stimulus set presented than to the inferred but not present subsets (see Garner, 1974, pp. 39–45 for a discussion of this). Thus, their results, but not their interpretations, are consonant with the emphasis upon inferred subset used throughout this section.

3. Paired-Associate Learning

College students were presented with 17 dot patterns (one from each of the 17 equivalence sets making up the 90 dot patterns used previously), either as stimuli or as responses in a paired-associate learning task (Clement, 1967). The other member of each pair was a double-digit number.

There were no differences in trials to criterion between patterns used as stimuli and patterns used as responses in this task. The good patterns (most redundant patterns) were learned fastest, providing further confirmation of the efficacy of ratings of pattern goodness as a measure of ease of encoding. The correlation between mean number of trials to criterion and mean rating of pattern goodness was .93. (The important finding was that the *relative* ease of learning of patterns of different redundancy was the same whether used as stimuli or responses; the equality of trials to criterion when used as stimuli or as responses almost certainly was a chance occurrence.)

4. Pattern Discriminability

Further research tested the hypothesis that a pattern is encoded first as a member of an equivalence set, and then as a specific member of that set discriminable from others within the same set. Using a card-sorting task with college students as subjects and with total sorting time as the dependent variable (Clement & Varnadoe, 1967), we found that discrimination between patterns from the same equivalence set was more difficult than discrimination between patterns from different sets. Furthermore, discriminability was greater for smaller sets (higher redundancy) than for larger sets (higher uncertainty). The tendency for processing such patterns as members of equivalent sets was so strong that specific instructions demanding the processing of only a single dot location were ineffective in preventing subjects from encoding patterns into equivalence sets (Clement & Weiman, 1970).

5. Encoding Time

The studies just cited imply that encoding is easier (takes less time and is done more consistently) when patterns are more redundant. Garner and Sutliff (1974) tested this implication directly by using good and poor patterns in a discrete reaction-time task. Each trial consisted of a two-stimulus discrimination task; that is, the subject was presented with a single stimulus and had to press one of two keys corresponding to the two possible patterns. Conditions involving good patterns (good versus good; good versus poor) produced faster reaction times than those involving only poor patterns (poor versus poor). Furthermore, in the good versus poor condition, the reaction times to the good stimulus were consistently faster than to the poor stimulus. Their conclusions were that (*a*) good patterns are encoded faster than poor patterns; (*b*) patterns of equal goodness are encoded equally fast; and (*c*) the relative discriminability between patterns (related to pattern similarity) affects the speed of encoding.

6. Nested or Mutually Exclusive Subsets

Most of the studies cited in the preceding sections found results consistent with a classification of patterns from a given population into mutually exclusive subsets. This interpretation naturally follows from the objective definition of equivalent subsets by the use of identity under reflection and rotation transformations. However, the observer is not necessarily limited to treating subsets in this exclusive fashion. Several experiments have considered the possibility that the subsets were nested in the observer's perceptual scheme. Good patterns would belong to the smallest subset, which itself would be included in the next larger subset of less good patterns, and so on up to the largest subset of poor patterns. Handel and Garner (1966) used all 126 possible patterns of five dots in a 3×3 matrix, presented one at a time, and asked subjects to draw a pattern suggested by the stimulus but different from it. Responses from sets of the same size made up 69% of all responses (39% were from the same equivalence set as the stimulus pattern), 22% were from smaller sets, and only 9% were from sets larger than that of the stimulus pattern. Thus, there is strong evidence that subjects perceive the smaller subsets as being nested within the larger subsets, and a tendency to provide better patterns as associates leads to a strong asymmetry in the data. Bear (1973) used the same 126 patterns with a single dot removed from each, and asked subjects to place a fifth dot in a position implied by or suggested by the other four dots. The placement of the dots was much more predictable for good patterns (small subsets) than for poor patterns. In a subsequent paper, Bear (1974) suggested that the placement of a fifth dot to produce a pattern that is the same, equivalent, or better—but rarely poorer—than the original pattern indicates that the subjects may be using a schema-plus-correction encoding. The implication is that a relatively good pattern may be encoded as a schema, and poorer patterns that differ from it by the placement of one dot are "recorrected" to the better pattern (made closer to the schema) in this kind of generation task. Bear's studies also lend support to the idea of a population of patterns being encoded in nested fashion.

7. Temporal Patterns

The relationships between pattern redundancy (size of equivalence set) and performance are not limited to visual stimuli. For example, Royer and Garner (1966, 1970) and Preusser, Garner, and Gottwald (1970) used temporal patterns consisting of the presentation of 2 alternative auditory elements at a very rapid rate, the length of each pattern being 8 elements (Royer & Garner, 1966, 1970) or 7–10 elements (Preusser *et al.*, 1970).

The subject was required to reproduce the pattern as soon as it was perceived. Several matters of perceptual organization were investigated in these studies, but of most relevance here is the finding that the number of specific patterns that are perceived differs from one pattern to another, and the difficulty of pattern perception varies directly with the number of perceived alternative organizations.

C. Learning of Structure

In keeping with the general orientation of this chapter, I shall focus upon the learning of correlational and dimensional information or structure. This context is one in which process-oriented psychologists will feel easier about my use of the word *encoding,* as this certainly can refer to the process of getting information into some retention place within the observer. My own difficulty in discerning the boundary between perception and learning (or even between iconic memory and short-term storage) has led me to a less precise use of the term *encoding,* including most any kind of process that puts stimulus information into the organism for any nonzero length of time.

1. Free-Recall Learning

Whitman and Garner (1962) used geometric figures with four dimensions (form, location of a gap, number of vertical lines, and location of a dot) and three values on each dimension, describing a total stimulus set of 81 figures. Different subsets of nine patterns were given in a free recall task, the subject drawing the figures as responses. The nine figures were selected to represent different forms of redundancy (total redundancy for the subset remained constant with size of the subset). Simple correlations between pairs of dimensions led to much faster learning than more complex forms of redundancy (interactions among the dimensions), but this was a function of the *subset,* not the individual patterns. There were few differences in ease of learning among patterns within a subset and large differences between subsets, even though some of the same patterns appeared in different subsets with different forms of redundancy. The difficulty of learning of these identical patterns was completely determined by the redundancy characteristics of the subset! The individual pattern was relatively unimportant, whereas the structure of the subset (determined from the inferred total population of 81 figures), effectively changed learning difficulty. A more forceful demonstration of the difficulties that complex forms of redundancy can provide for subjects comes from a study (Garner & Whitman, 1965) in which subjects learned nonsense words from a stimulus population of 16 different words (two possible letters for each of four letter positions within the word). A subset of four stimuli with redundancy in the form of two simple correlations between letter positions was learned faster than a subset of 8 stimuli with redundancy in the form of one simple correlation between letter positions. This was to be expected as the former had higher redundancy than the latter. However, the subset of 8 stimuli with simple contingent redundancy was as hard to learn as the total set of 16 stimuli, even though the former had redundancy and the latter did not. Furthermore, a set of 8 stimuli in which the constraint was complex in nature (all interaction redundancy) was much harder to learn than the *entire set* of 16 stimuli. Garner and Whitman (1965) interpreted the results as indicating that the subject learned the dimensional structure of the total set, *plus* the correlational structure of the subset. Thus, very small subsets with simple redundancy are learned faster than the whole set, primarily due to the small size of the subset. As subsets increase in size, the size advantage is overcome by the increased encoding necessary to learn dimensional structure plus subset redundancy rules, and the more complex the form of the redundancy, the harder the subset is to learn. Garner's (1962) argument that it should be more difficult to learn a redundant subset than to learn a total set of patterns of the same size was demonstrated by Nelson, Garland, and Crank (1970) with redundant subsets and nonredundant subsets of 8 stimuli each. Thus, redundancy, although effective as a determinant of perceptual structure, is only facilitative under some conditions and not under others. Or perhaps, the form of the redundancy must be considered before predicting ease of perceptual processing. This suggests a metalanguage approach in considering good patterns to be more redundant than poor patterns, but also in considering that redundancy can be described as good or poor depending upon its form and upon the situation in which it is to be used.

2. Concept Learning

A task of concept learning essentially requires that a subject learn which stimuli belong together in one class, and which in another class or classes. Thus, the subject must recognize appropriate similarities, appropriate discriminations, or both among stimuli. One way of approaching this is in terms of defining the subsets of stimuli belonging to a specified response. Stated in this way, it is reasonable to consider the effect of the amount and form of redundancy within each stimulus subset. Whitman and Garner (1963) used the same type of stimuli as described in their 1962 paper, eliminating levels of dimensions to obtain a population, or total set, of 16 patterns. The 16 patterns were divided into subsets of 8 for classification in one of two ways—the subsets had either a simple form of constraint (simple correlation) or a complex form of constraint (interaction). If the two subsets were presented in a sequence that mixed them together, subjects could learn to sort them into two classes at about the same rate, whether the stimulus subset structure was simple or complex. However, if the subsets were presented in segregated fashion (first all members of one, then all members of the other), learning to classify the 16 stimuli was facilitated for the simple structure, but was no different from intermixed presentation for the complex structure. Concept learning apparently can be influenced by the form of redundancy, but only under conditions that facilitate the subject's perception of the structure (i.e., with all of one kind of exemplar, then all of another kind). The observer can make the best use of information about subset structure when she is best able to process the subset entirely without interference. Clearly, the influence of stimulus structure is more difficult to isolate in concept-learning tasks than in free recall or other kinds of performance previously discussed.

3. Discrimination Learning

Just as with concept learning, and unlike free recall, discrimination learning requires that subjects be able to separately encode individual patterns, and thus simple forms of redundancy that assist in clustering stimuli together into discriminable *sets* will be helpful only when the emphasis is upon discrimination between sets (or between stimuli from different sets) rather than upon discrimination between stimuli from within sets. This has been demonstrated in a previously discussed context in which sorting stimuli of two kinds was easier with increasing subset disparity (Clement & Varnadoe, 1967; Clement & Weiman, 1970). The direct comparison of the learning *of* patterns with learning *to discriminate* patterns by Whitman (1966) used consonant–vowel–consonant syllables in lists of the same length and same total redundancy, but with differences in the form of constraint. Under free-recall instructions, simple structure was learned much more easily than complex structure. However, when serial recall was used (requiring discrimination of which stimulus belonged to each serial position), the complex structure was actually easier than the simple structure. Garner (1974) emphasized the importance of the kind of cues that stimulus structure presents in such instances. Complex structure provides more discriminable cues between any pair of stimuli, whereas simple structure provides less discriminable cues between pairs of stimuli with simple dimensional correlation.

4. What Is Being Enclosed and Where?

I have alluded before to my casual use of the word *encoding*. Although extensive discussion of this term and its position in psychological hypotheses is beyond this chapter's scope, a brief mention of a few papers should give some idea of the directions such discussion might take. Johnson (1973) compared models in which it was assumed that information is registered in memory and then encoded versus models that assumed that information is already in an encoded state when it is registered in memory. He found support for the latter. This certainly is consonant with the consideration of structure as existing in stimuli to be perceived, rather than as imposed by the perceiver. Restle (1973) spoke even more directly to this orientation, suggesting that encoding processes (implying observer encoding) occur only with learning of nonsense materials and not with meaningful materials. With the latter, subjects learn the pattern already existing in the stimulus materials. In fact, to reduce misconceptions in the future, Restle suggested moving away from considerations of encoding toward emphasis on learning of organization. A less directly relevant study by Briggs (1974) focused upon the problem of auditory and visual encoding in memory, and resulted in the conclusions that auditory and visual confusions in the typical modality-encoding kind of experiment reflect *recoding* rather than *encoding* of stimuli. This brief selection of papers is not intended to prove that encoding should be defined in one way or another. Rather, it is intended to add support to the orientation that the human is an active observer in perceiving structure, but that this activity is involved with identification or learning of structure rather than with creating structure where none exists.

D. Limitations Imposed by Structure upon Performance

1. Integral and Separable Dimensions

Prior discussion has included mention of instances in which redundant structure can assist some kind of performance, and instances in which the same structure can hinder performance. How so? A productive approach to this question has been taken by Garner (1974), who discriminated between different functional classes of dimensions in perceptual tasks. *Integral* dimensions (a term taken from Lockhead, 1966) are two or more physical dimensions that are processed in a unitary fashion by human observers (e.g., the brightness and saturation of a single stimulus); *separable* dimensions are two or more physical dimensions that are processed in concert but (more or less) independently by human observers (e.g., the size and shape of a single stimulus). When scaled for similarity, stimuli differing on integral dimensions typically combine to yield a Euclidean metric (i.e., the scaled distance between two stimuli differing in two dimensions is the square root of the sum of the squares of distance in each dimension); separable dimensions combine to yield a city-block metric (i.e., the scaled distance between two stimuli differing in two dimensions is the sum of the distances between the stimuli in each dimension). A brief discussion of these metrics is in Garner (1974), with a more complete discussion in Torgerson (1958). Perceptual classification tasks yield interesting interactions between types of dimensions and the effects of stimulus structure. Classification of stimuli with integral dimensions into different categories yields discrimination based upon similarities; classification of stimuli with separable dimensions yields discrimination based upon dimensional structure (e.g., Handel & Imai, 1972). In fact, Garner (1974, p. 119) suggested that the concept of dimensions is not meaningful with integral dimensions, as the objectively separate dimensions are really perceived as a single dimension by the observer.

2. State and Process Limitations

In addition to the integral–separable distinction made by Garner (1974) in classifying stimulus dimensions, a dichotomous treatment of the sources of performance errors described by the same author is necessary to an adequate approach to the effects of redundancy on performance. Garner has classed limitations upon performance into *state* and *process* limitations. *State* limitations are those that are a function of the observer's state (e.g., alertness, individual sensitivity, adaptation level); *process* limitations are those that are a function of the stimuli themselves, given a normally functioning observer. Thus, a too-brief presentation of stimuli that can be recognized perfectly at longer durations can lead to a state limitation; a 10-min presentation of a linear pattern against an extremely "noisy" background of linear segments, with less than perfect identification, would be due to a process limitation.

3. Improvement of Performance with Redundancy

What happens when redundancy is added to stimuli in a perceptual task? A casual consideration of the relation between ease of encoding (and other kinds of performance) and redundancy, might lead to the conclusion that redundancy always helps unless performance is perfect to begin with. This idea is entirely wrong. The kind of redundancy, the dimensional structure of the stimuli, and the type of performance limitation are all important in determining whether redundancy helps or not. Garner and Felfoldy (1970) found that redundancy of two variable integral dimensions (brightness and saturation) improved performance in a task requiring sorting on the basis of one dimension, compared to the same task in which the stimuli varied in only one dimension (either saturation or brightness). Furthermore, if the two dimensions varied orthogonally (and thus the second dimension was an irrelevant variable), performance worsened. This makes sense when one considers that the two dimensions are processed as a single dimension. Variability in both physical dimensions in a correlated fashion increases the discriminability of the stimuli; variability in an orthogonal fashion in both physical dimensions obscures perception of the differences in the required dimension. Garner and Felfoldy repeated the experiment, but used a pair of color chips as each stimulus with the two dimensions varying on separate chips. Thus, the same dimensions were used as before, but physical separation of the variability made these separable dimensions (the constant brightness of the chips varying in saturation was integral with saturation on those chips; the constant saturation of the chips with varying brightness was integral with brightness on those chips; but this meant each chip alone was similar to those of the control condition of the first experiment, and only pairs of chips considered together were equivalent to the earlier experimental conditions).

When sorting based upon one dimension was required, the presence of variability in the second dimension made no difference, whether the variability was correlated (redundant) or uncorrelated (orthogonal) between the two dimensions. This again makes sense, since separable dimensions can be separated, and subjects appropriately pay attention only to the dimension of importance. Thus, dimensional redundancy helps with integral dimensions (and irrelevant dimensional variation hinders), whereas dimensional redundancy or irrelevant dimensional variation neither facilitates nor hinders performance in this type of task with separable dimensions. It should be noted that subjects can be affected by redundancy in tasks involving separable dimensions, but the task must be such that they have to process both dimensions to arrive at correct responses (e.g., Gottwald & Garner, 1972). When performance is less than perfect, and dimensional redundancy improves it, we typically have an instance of integral dimensions. As mentioned before, redundancy with integral dimensions actually increases the discriminability of the integrated dimension. Thus, this improvement in performance overcomes a *process* limitation. Process limitations, discriminability *among* stimuli, and facilitation with redundancy of integral dimensions all belong together conceptually.

State limitations are a different matter. As they derive from discriminable stimuli that for some reason have not been completely (or ideally) processed by the observer, redundancy in the form of repetitions of the stimuli should help by increasing the chances for the subject to process sufficient information to perform the task. This might involve sequential presentations of the same stimulus or, as in Garner and Flowers (1969), might involve simultaneous multiple presentations of the same stimulus. When questions of the nature of performance limitations arise, comparison of the effects of increasing dimensional discriminability (or adding redundant dimensions) with the effects of adding element redundancy can clarify whether the limitation is a process or state one. Flowers and Garner (1971) found that element redundancy facilitated performance only for state-limited, not process-limited, situations. As Garner (1974) has pointed out, repetition, or element redundancy, can be considered as redundancy of multiple variables. It is not unreasonable to consider repeated elements as involving separable dimensions of a sort. If this notion were arbitrarily adopted, we could consider that state limitations, discriminability *of* stimuli, and facilitation with redundancy of separable dimensions belong together conceptually.

E. Context, Instructions, and What the Perceiver Does

I have emphasized before that structure is available in the stimulus for processing, and that the perceiver selects from this existing structure, thus stressing both the importance of the environment in defining structure and the importance of the observer in implementing the possible processing of this structure. I wish now to emphasize that *observer* and *experimenter* should not be confused in this context. Studies in which the experimenter *thinks* that a particular thing is being processed can yield surprising results if no steps are taken to assess what structure actually is being used by the subjects in the experiment. An *effective* universe of stimulus alternatives is important in defining such terms as *redundancy,* and this effective universe is what exists for the subject rather than for the experimenter. One might consider the distinction between experimenter-defined and subject-defined stimulus alternatives as analogous to the distinction between nominal and functional stimuli (e.g., Underwood, 1963). Of course, the latter member of each pair is the important one.

The possible difference between what the experimenter intends to be the stimulus and what the subject actually processes as the stimulus can be extremely difficult to eliminate. Varnadoe and I (Clement & Varnadoe, 1967) performed experiments in which we presented subjects with a deck of cards, each card containing one of two stimuli (five-dot patterns), with the task being to sort the deck on the basis of the patterns. We were interested in the relative discriminability of patterns containing different amounts of redundancy, as discussed earlier in this chapter. In one of the experiments, each of the two patterns always was in the same orientation on the cards in the deck (fixed orientation); in the second experiment, the patterns were randomly oriented in each of the four directions possible with square cards (random orientation). The results were similar but not identical in the two experiments, and we were confident that subjects had processed the whole five-dot pattern only in the experiment with random pattern orientation. When patterns were presented in fixed orientations, subjects *could* have focused upon a single location that had a dot in one pattern but no dot in the other pattern and achieved perfect discrimination. The similarity of the results in the two experiments did not suggest that many subjects did this; but if even some subjects had used a single-dot strategy rather than a whole-pattern strategy, the supposed independent variables of subset identity and subset size (redundancy) were not functioning for those subjects.

Weiman and I (Clement & Weiman, 1970) decided to see how easy it was for subjects to use a single-dot strategy by giving increasingly biased instructions to different subjects, attempting to cause them to use element-processing rather than whole-pattern processing strategies. Each condition required subjects to sort decks of 50 cards into different bins according to which one of two patterns was on each card; sorting time was the dependent variable, and different subjects were used in each condition. In all conditions, the cards were in a fixed orientation (each of the two patterns was oriented in the same way on all cards). In all conditions, as in the earlier study (Clement & Varnadoe, 1967), a sample card for each pattern (criterion card) was taped on a vertical surface behind the bin into which cards of that kind were to be sorted.

The first condition was a replication of the fixed-orientation experiment run earlier, with subjects instructed to "look at the entire pattern, not just a few of the dots, for many times the choice is much easier to make if you look at the whole pattern [Clement & Weiman, 1970, p. 334]." The differences in sorting times were the same function of redundancy of pattern subset as previously obtained.

The second condition instructed subjects to "try to find a single dot which is different in the two patterns . . . when you sort each card, look for the key dot and do not waste time looking at the rest of the pattern [Clement & Weiman, 1970, p. 334]." We thought this would reduce the sorting-time differences among tasks involving different pattern redundancy to chance levels. It did not. The differences were reduced only a little, and many subjects reported difficulty in ignoring the whole pattern, although all claimed to have learned to do so by the end of the first of eight sorting tasks they completed. Obviously, they had not learned to do so, and were not aware that they still were influenced by whole-pattern (and subset) characteristics. We decided to help the subjects follow instructions a little more.

The third condition was like the second, except that a mask was placed over the criterion patterns, with one corner cut out of the mask so that the subjects could see (as a criterion) only a corner of the pattern that contained a dot for one pattern and no dot for the other. We thought that requiring subjects to look at a single-dot criterion when deciding where to place each card (on which the whole pattern was visible) would help them completely eliminate whole-pattern processing. Again, it did not. Differences due to the redundancy of patterns were attenuated from those found with the second condition, but were still significantly different. Frustrated, we decided to design conditions that would make it extremely unlikely or impossible for the subjects to accomplish the sorting task by use of whole-pattern processing.

The fourth and fifth conditions were similar to the third condition in instructions and in having a mask over the criterion cards that exposed only one corner dot location. The fourth condition used decks of 48 cards, with the task being to sort each deck into two groups of 24. However, eight different patterns were used instead of just two. Each of the eight patterns came from a different equivalence set of the same size; all were presented in fixed orientation, and four of the patterns had a dot in the critical location, while four did not. Thus the subject would have to learn which four patterns (from different sets) went into one bin and which four patterns went into the other bin in order to sort on the basis of some kind of whole-pattern processing. This seemed so difficult that we thought the subjects might follow our single-element instructions instead. The fifth condition also used decks of 48 cards, with different patterns in each deck. However, in this condition all patterns in a deck were from the same equivalence set (either size four or size eight) and, in fact, were rotations of the same pattern. The rotations were selected so that 24 had a dot in the critical location and 24 did not. In this condition, processing of the whole pattern, which was the same for all cards in a deck, was absolutely useless for accomplishing the desired discrimination. At last, in the fourth and fifth conditions there were no differences in sorting time due to pattern redundancy. The power of these patterns to evoke whole-pattern processing on the part of subjects was extraordinary, and extreme experimental restrictions were required to eliminate this tendency. The point is not that patterns always are processed in one way, but rather that subjects under the influence of particular stimulus characteristics, context, and instructions perform in a manner jointly determined by all of these, instead of just in a manner consonant with instructions alone.

The influence of unintended variables on performance is certainly well established (e.g., Rosenthal & Rosnow, 1969). However, the convenience of ignoring uncontrolled factors and the tendency to accept experimenter perceptions as identical to subject perceptions are difficult to overcome. Many current lines of research in perceptual structure and perceptual processing recognize this by considering multiple sources of influence upon the performance of subjects. For example, Reed (1974) studied the ability of subjects to determine whether or not the second of two sequentially presented patterns is a part of the first pattern. His results suggested the presence of whole-pattern processing that interfered with recognition of pattern segments. His interpretation of the results was that subjects stored a structural description that was a combination of visual and verbal codes, and that these codes were of prime importance in subsequent part-recognition, rather than the also-remembered visual image. His consideration of several possibilities for subsequent comparisons, rather than just a single possibility (e.g., a language description of the pattern) was in keeping with sensitivity to possibilities of subject variation in perceptual processing. Klatzky and Stoy (1974) required subjects to indicate whether or not two pictures of common objects had the same name. They used several conditions in two experiments that involved positive instances of stimulus identity, mirror image, or name match only. The experiments varied as to whether only one kind of positive instance or several kinds could occur in a sequence of trials. Their results indicated that the subjects used visual codes in comparing nonidentical pictures *and* that these codes varied with experimental context and task demands.

Context and similar variables are not limited to experimentally concurrent conditions either, as learning studies, in particular, rely upon sequential effects for demonstration of behavior change. Frith (1974) found that it takes subjects longer to find a normal *N* embedded in a context of reversed *N*s than it takes to find a reversed *N* embedded in a context of normal *N*s. His interpretation was that subjects use a schema derived from prior experience for the normal form of a letter and that this schema is a flexible, rather than stable, construct. As observers have learned to recognize mirror images as to letter identity, both normal and reversed letters may be processed as the same letter, and this tendency is pronounced in an experimental context with many reversed letters.

What I am trying to indicate with these diverse examples is that humans are complex creatures perceptually, as well as in other regards. We investigate perceptual structure and selection in very constrained settings in order to obtain some basic consistencies in processing. We must be careful in generalizing these consistencies in an uncritical fashion to more realistic and interactive settings.

III. ONTOGENY OF PERCEPTUAL SELECTION

The developmental sequence or sequences of processing of perceptual structure are never studied and seldom observed in the most direct way possible: by following the development within the same human or humans from birth through adolescence. We each may have anecdotal material from observations of our own children, and on rare occasions scientists may deliberately investigate certain aspects of behavior of their own children under well-controlled conditions, but no one controls the presentation of stimuli and contexts to a child even in one aspect of a single perceptual modality during the entire development of that child into an adult. The obvious reason is the practical one of duration of the study. The effort involved would enormous, perhaps requiring the full-time service of several experimenters for 15 years. To justify such immense effort, advance guarantees would have to be made that all procedures to be used would be perfect in terms of reliability and that the infant to be studied would be a modal-type of perceptual processor—both are impossible. A less obvious, but to my way of thinking more important, reason for not doing such an "ideal" experiment is the lack of generality such a study would have. Of course, the subject might not be typical in some fashion, but this minor problem could be solved by having two subjects or by other means. Of critical importance would be the lack of generality of *developmental situations*. If precise control were to be instituted over stimuli available to the subject, the reality of the developmental context for nonexperimental children would be lost. In fact, such stimulus control might necessarily lead to a nonnormal human information processor (not necessarily worse than normal; just different).

Thus, I suggest that the best way to study perceptual development is the way it is done now: cross-sectional studies, in which humans of different ages at a given point in time are compared on some kind of task. The task may not be very realistic (i.e., complex), but the experiential context prior to the task for all observers will be an actual one, interacting with normal human behavior. No matter what the limitations are upon the generalization of age differences in performance on the task to other tasks, there will be few limitations upon the generalization of age differences in performance on the same task to other observers of the same ages, and with normal and uncontrolled prior experiences. One may wonder at this point why this section is headed *ontogeny* rather than *age differences*. It is simply because I share with other investigators the fond speculation that age differences do in fact reflect different stages in the same general developmental sequence. If this gratuitous assumption is unacceptable, you may write in your preferred title. Even if the assumption is correct, we must remember the one truism of human behavior: All individuals differ from each other. Therefore, anything we learn about development of perceptual processing in the nomothetic studies I shall be describing should be considered the basic matrix from which individuals will diverge in individualistic ways.

A. Prospecting versus Croesian Riches

The heading of this section reflects my own encoding upon first reading about the perceptual learning controversy between Gibson and Gibson (1955a, 1955b) and Postman (1955). The view of Postman has been termed *enrichment* and that of the Gibsons has been termed *differentiation*. That is, perceptually, the organism may start with John Locke's blank slate and by learning stimulus and response connections *enrich* perception with increasing experience until reaching some accomplished state of perceptual competency. On the other hand, the organism may have all the necessary perceptual material available at first blush and, by extended experience, learn to *differentiate* important from unimportant aspects until reaching the desired state of competence. The enrichment approach is one of association of stimuli and responses in a rather basic learning fashion. In fact, Postman (1955) called this approach *associationism* rather than *enrichment*. The differentiation approach is one of learning to give specific responses to increasingly specific stimuli, and the Gibsons (1955a) referred to their approach as involving the *specificity theory*. To return to my heading, an enrichment procedure is akin to a prospector gradually finding more and more perceptual valuables, whereas the differentiation approach grants the observer initial ownership of all possible perceptual wealth, with the task being to sort the embarrassing largess into meaningful and coherent chunks. People do learn in all situations, including those of interest to the study of perception. The question is whether it is of more importance to focus upon the learning process or upon what is to be learned. As I am oriented toward considering structure as existing in the stimulus surround and as being perceived rather than as being created, it seems to me that the learning orientation is less important to perceptual structure than the selection orientation. Were the Gibsons–Postman argument going on today, I would be placed squarely in the Gibsons' camp because of my emphasis upon stimulus characteristics, and this should be obvious from the examples I have chosen as most germane to this chapter. Human observers *can* make up their own procedures for organizing their perceptual environment regardless of the composition of that environment, but they probably only do so under very unusual circumstances (most of which would be created in the experimental laboratory). Normally, humans select and make use of the logic and relationships that exist in the external world, and their perceptual structure faithfully mimics stimulus structure.

B. Comparisons across Age Groups

How do younger children differ from older children and adults in their selection of perceptual structure? The reader is referred to the excellent book by Gibson (1969) and the review article by Bond (1972) for coverage of broader content and broader theoretical considerations than is provided by the limited material in this section. Both these authors cover data collected from infants as young as a few weeks old to adolescents and adults. These data have been collected through ingenious methods by investigators such as Fantz (1956) and Bower (1966), who have dealt with very young infants not susceptible to the usual experimental laboratory blandishments.

My focus is upon studies that relate to the effects of pattern redundancy (and its obverse, complexity) in different age groups. A large body of data indicates that preference for more complex forms over simpler forms increases with increasing age (e.g., Hershenson, Munsinger, & Kessen, 1965; Munsinger & Weir, 1967; Thomas, 1965), thus suggesting that older children are better able to process more complex patterns. The data do not suggest whether this is due to ability to make better use of structure, such as redundancy, that exists in the complex figures, or to some other factor or factors (e.g., see Berlyne, 1957). That human observers are better able to recognize certain kinds of equivalencies at increasing ages has been shown by data such as those of Gibson and her associates (Gibson, Gibson, Pick, & Osser, 1962), which indicated improvement in recognizing identity under reversal, rotation, perspective, and other transformations as functions of age. The functions they found were different for the different kinds of transformations, indicating that certain organizational strategies become manageable at different ages. The ability of subjects at different ages to process the kinds of equivalency in the previously mentioned dot patterns was the subject of a study I did with Sistrunk (Clement & Sistrunk, 1971). We had subjects of four different age groups (ages 9–10, 13–14, 17–18, and 20–21) rate five-dot patterns for pattern goodness, and obtained highly correlated ratings across all age groups, with clear discrimination of levels of redundancy even in the youngest group. There were age differences in the proportion of variance in ratings attributable to different patterns within the same equivalence set. Subjects from ages 13–21 years had less than 10% of the variability in ratings derived in this fashion, while 9–10-year-old subjects had almost 19% of the variability due to differences within equivalence sets. The implication was that all age groups used about the same basis for rating, but the youngest group had not yet attained adult levels of equivalency encoding. Gibson *et al.* (1962) had found children to reach adult levels of reflection and rotation encoding by age 7, so our results were rather consistent with that earlier study and indicated that the treatment of reflection and rotation transformations as providing effective redundancy was viable for different age groups. In addition, our study controlled for degree of redundancy independent of total amount of undertainty, and these factors were often confounded in prior published studies. This particular approach is of obvious importance to the learning that goes on in such tasks as reading (e.g., Gibson, 1970). Of course, there is very little information available as to *how* humans come to process structure in increasingly competent fashion, and studies oriented in this direction will be of great value.

C. Comparisons across Cultures

Without knowledge of the mechanisms by which humans learn to process perceptual structure, it is difficult to predict what kinds of differences might be expected in different cultures. Rather than speculate about these differences, I shall give some evidence of developmental similarities and differences in comparisons of three cultural backgrounds. We used the ubiquitous dot patterns and rating task described previously (i.e., Clement & Sistrunk, 1971) in two studies that compared the same four age ranges of subjects from the United States with subjects from Brazil (Clement, Sistrunk, & Guenther, 1970) and with subjects from Japan (Iwawaki & Clement, 1972). Instructions were modified in asking subjects to rate patterns as to *how well-formed* they were, as the term *goodness* has primarily ethical connotations in Portuguese. Of course, the U.S. groups received the same instructions in English as the Brazilian groups did in Portuguese, and the Japanese groups did in Japanese. The instructions were equivalent in both form and semantic content. The results with the Japanese subjects were quite similar to those for the U.S. subjects, with high correlations (greater than $r = .80$) for each of the age groups between ratings of pattern goodness and pattern redundancy, and with the three older groups having less than 10% of the variance in ratings associated with differences among patterns from within the same equivalence set. The youngest age group in both the U.S. and Japanese samples had 15–19% of the variance attributable to intraequivalence-set differences. The Brazilian subjects were similar for the two oldest age groups, but the 13–14-year-old group had 12% of the ratings variability associated with intraequivalence-set differences, and the 9–10-year-old group had 43% of the ratings variance associated with this factor. In fact, the correlation for 9–10-year-old Brazilian subjects between pattern ratings and pattern goodness was only $r = .40$, the sole such correlation for all three cultures that fell below a value of $r = .80$. The happy result of these studies was support for the idea that patterns such as these dot patterns, and tasks such as the rating of goodness, are relatively free of cultural bias. However, the results with the Brazilian subjects of the youngest age group remain a mystery. Differences existed that were interpreted as a slower rate of development of adult strategies for perceptual encoding of structure in Brazilians, as compared to Japanese and U.S. subjects. The developed strategies were not different, as indicated by the results with older subjects, but simply came about in a different temporal fashion. Perhaps further cross-cultural studies using rather culturally independent stimuli could elucidate the culture-specific differences.

D. Behavioral and Physiological Hints

The processing of perceptual structure has been treated in this chapter with little commentary on what neural events underlie it. The omission is deliberate, reflecting the emphasis upon stimulus structure and recognizing contributions in other chapters of the *Handbook of Perception* that treat the topic in some detail (e.g., Sutherland, 1973). The human observer will be limited in perceptual processing by neural limitations and will be assisted by "prewired" kinds of systems that might encode stimuli immediately at hierarchical levels above the simple-element stage. Development of the selection of perceptual structure must of necessity be facilitated and constrained by these internal structures. This brief section is intended to suggest some neurophysiological and behavioral data that should be considered in any studies related to the organism *qua* organism and its interaction with external events.

Hebb's (1949) speculation on the neural events underlying perceptual learning, and subsequent studies of single cells and cell clusters that respond to stimulus characteristics (as contrasted to punctate stimulus elements), have suggested ways in which the brain may be "wired" to process such things as line segments, angle orientation, and so forth. Hubel and Wiesel (e.g., 1962) represent a major line of research with nonhuman subjects in investigating the neural organization underlying perception. Their work has been extended to the human cortex by Marg, Adams, and Rutkin (1968). As with the human subjects in the visual cliff experiments (e.g., Gibson & Walk, 1960), the existence of certain perceptual processing at a very early age does not eliminate the possibility of this processing being learned rather than innate. Whether learned at an early age or preexistent, such evidence of early perceptual organization, together with implications of neural organization at any age, is important in determining the kinds of flexibility or plasticity that human observers have in the processing of perceptual structure. Milner (1974) has taken a direction suggested by the existence of feature-detection units in the visual cortex to develop a model for visual shape recognition that posits angle and length-ratio feature detectors in the human brain. He argues that many examples of stimulus equivalence may be explained in this fashion, and it is possible that some of the effects of simple redundancy discussed in this chapter are due to similar detector units (or perhaps higher-order detector units). Although he does not suggest an underlying neural specification, Fox (1975) has argued that diagnostics of a structural or relational nature are the bases for performance in recognition and visual matching tasks. His ideas might suggest, in turn, the development of a neural organization designed to test for certain diagnostic properties, such as symmetry.

One class of experiments that focuses upon both behavioral data and assumed neural functioning has been that concerned with the fading of fixated visual images. Pritchard, Heron, and Hebb (1960) found that as a fixated image faded, it tended to do so (according to reports by the subjects) in a holistic or in a meaningful partial fashion. That is, a line figure faded in such a manner as to leave visible segments that formed meaningful organizations. Subsequent work has often assumed the existence of this perceptual phenomenon and has been oriented toward investigation of factors that might influence the effect, such as eye movements (Coren & Porac, 1974), or toward mimicking the effect by other procedures, such as the use of tachistoscopic presentation (Johnson & Uhlarik, 1974). The dangers inherent in assuming neural or physiological processes on the basis of inference from behavior have been made salient by the recent work of Schuck (1973) with stabilized images. He used conditions involving both simulation and actual stabilization of visual images and concluded that there was a significant bias toward the reporting of meaningful fragments when verbal reports were used throughout each trial. Perhaps response bias was of more importance in the earlier studies than the effects of neural organization.

The reasonable conclusions, in terms of research on the physiological basis of perceptual development, are that converging operations must be used with behavioral data to ensure that a process is present. Then, and only then, is it sensible to look for the cortical events which underlie the behavioral events.

IV. GENERAL DISCUSSION AND CONCLUSIONS

What do we know and where should we go in studying perceptual structure and selection in human observers? The data and conclusions cited in various parts of this chapter are a good start, but much more needs to be accomplished. It still seems reasonable to me, at this stage of our knowledge, to concentrate on stimulus structure as what is being processed. Structure exists in stimuli and stimulus populations, and the observer learns to extract and select this structure, only imposing arbitrary structure on rare occasions. The selection of structure is highly influenced by many factors, including situational demands and individual preferences, and it would be reasonable to concentrate more effort on determining the specific influences of these factors on the types of tasks I have discussed.

The work which has been done by Garner and his associates is important for its emphasis upon entire stimulus populations and subsets of equivalent stimuli, which is in contrast to more traditional work on the single stimulus. One can deal with stimulus and perceptual structure in a meaningful way only by considering what *could* have occurred as stimuli as well as what actually *did* occur. The concept of redundancy, both distributional and correlational, and its relation to equivalence sets of stimuli, provide a rational and objective basis for the prediction of performance characteristics. Redundancy has been shown to be related to ease of encoding, judgments of pattern goodness, labeling of stimuli, paired-associate learning, and discriminability of stimuli. Most of the work has been with visual perception and with very simple stimulus patterns, less has been done with auditory perception and more complex patterns, and very little with other perceptual modalities. Obviously we need to investigate the extent to which principles already learned generalize to other modalities and to more complex stimuli and to carry out these investigations in more realistic and interactive settings. I expect that increasing numbers of explanatory concepts will be needed as the experimental situation becomes more naturalistic. The consideration of organization of structure in complex hierarchies almost certainly will be necessary, as prior work has shown reason to believe that even redundancy of simple patterns orders them into nested and partially nested subsets rather than into mutually exclusive subsets.

The way in which structure is learned as a function of task (e.g., free-recall learning, concept learning, discriminative learning) is closely related to whether the variable dimensions of stimulus populations are integral or separable in nature. Limitations upon performance in such tasks determine the ways in which redundancy may be used, with state limitations requiring redundancy of separable dimensions for improvement of the discrimination of stimuli (their recognition or identification), and with process limitations requiring redundancy of integral dimensions for improvement of discriminability among stimuli. Do these same relationships hold with complex stimuli? To what extent may limitations be either those of state or of process, and to what extent may dimensions be processed as integral or separable, dependent upon task and other situational demands? Only additional research can provide the answers.

How do people learn to process perceptual structure, and how do changes in perceptual selection come about? We do know that humans learn what structure to extract, but we do not have much idea about the developmental process nor the relative contributions of maturation and learning. Cross-cultural comparisons may prove helpful in this regard. As to the ultimate level of analysis, neural structure and functioning, we really will not know what to look for until we have sufficient convergence of studies of perceptual behavior under different conditions. Prior to that point, we may learn a great deal more about neural organization in isolation, but will probably learn little about the correlation between neural organization and perceptual organization.

In summary, we need a mass of additional information about the perception of structure, with extended ranges of context, instructions, and prior learning in increasingly complex and interactive settings. Our current status is just the beginning.

References

Attneave, F. Some informational aspects of visual perception. *Psychological Review*, 1954, **61,** 183–193.

Attneave, F. *Applications of information theory to psychology*. New York: Holt, 1959.

Bear, G. Figural goodness and the predictability of figural elements. *Perception & Psychophysics*, 1973, **13,** 32–40.

Bear, G. Implicit alternatives to a stimulus, difficulty of encoding, and schema-plus-correction representation. *Memory & Cognition*, 1974, **2,** 360–366.

Berlyne, D. E. Conflict and information-theory variables as determinants of human perceptual curiosity. *Journal of Experimental Psychology*, 1957, **53,** 399–404.

Bond, E. K. Perception of form by the human infant. *Psychological Bulletin*, 1972, **77,** 225–245.

Bower, T. G. R. Slant and shape constancy in infants. *Science*, 1966, **151,** 832–834.

Briggs, R. Auditory and visual confusions: Evidence against simple modality encoding hypotheses. *Memory & Cognition*, 1974, **2,** 607–612.

Clement, D. E. Uncertainty and latency of verbal naming responses as correlates of pattern goodness. *Journal of Verbal Learning and Verbal Behavior*, 1964, **3,** 150–157.

Clement, D. E. Paired-associate learning as a correlate of pattern goodness. *Journal of Verbal Learning and Verbal Behavior*, 1967, **6,** 112–116.

Clement, D. E., & Carson, D. H. Multivariate uncertainty analysis of symmetric prediction. *American Psychologist*, 1961, **16,** 465. (Abstract)

Clement, D. E., Guenther, Z. C., & Sistrunk, F. Incertidumbre y percepcion de patrones: Reseña y perspectiva. (Uncertainty and pattern perception: A review and prospectus.) *Revista Latinoamericana de Psicologia*, 1972, **4,** 177–188.

Clement, D. E., & Sistrunk, F. Judgments of pattern goodness and pattern preference as functions of age and pattern uncertainty. *Developmental Psychology*, 1971, **5,** 389–394.

Clement, D. E., Sistrunk, F., & Guenther, Z. C. Pattern perception among Brazilians as a function of pattern uncertainty and age. *Journal of Cross-Cultural Psychology*, 1970, **1,** 305–313.

Clement, D. E., & Varnadoe, K. W. Pattern uncertainty and the discrimination of visual patterns. *Perception & Psychophysics*, 1967, **2,** 427–431.

Clement, D. E., & Weiman, C. F. R. Instructions, strategies, and pattern uncertainty in a visual discrimination task. *Perception & Psychophysics*, 1970, **7,** 333–336.

Coren, S., & Porac, C. The fading of stabilized images: Eye movements and information processing. *Perception & Psychophysics*, 1974, **16,** 529–534.

Dodwell, P. C. Pattern and object perception. In E. C. Carterette & M. P. Friedman (Eds.), *Handbook of perception* (Vol. 5). New York: Academic Press, 1975.

Fantz, R. L. A method for studying early visual development. *Perceptual and Motor Skills*, 1956, **6,** 13–15.

Flowers, J. H., & Garner, W. R. The effect of stimulus element redundancy on speed of discrimination as a function of state and process limitations. *Perception & Psychophysics*, 1971, **9,** 158–160.

Fox, J. The use of structural diagnostics in recognition. *Journal of Experimental Psychology: Human Perception and Performance*, 1975, **104,** 57–67.

Frith, U. A curious effect with reversed letters explained by a theory of schema. *Perception & Psychophysics*, 1974, **16,** 113–116.

Garner, W. R. *Uncertainty and structure as psychological concepts.* New York: Wiley, 1962.

Garner, W. R. The stimulus in information processing. *American Psychologist*, 1970, **25,** 350–358.

Garner, W. R. *The processing of information and structure*. Hillsdale, New Jersey: Erlbaum, 1974.

Garner, W. R., & Clement, D. E. Goodness of pattern and pattern uncertainty. *Journal of Verbal Learning and Verbal Behavior*, 1963, **2,** 446–452.

Garner, W. R., & Felfoldy, G. L. Integrality of stimulus dimensions in various types of information processing. *Cognitive Psychology*, 1970, **1,** 225–241.

Garner, W. R., & Flowers, J. H. The effect of redundant stimulus elements on visual discriminations as a function of element heterogeneity, equal discriminability, and position uncertainty. *Perception & Psychophysics*, 1969, **6,** 216–220.

Garner, W. R., Hake, H. W., & Eriksen, C. W. Operationism and the concept of perception. *Psychological Review*, 1956, **63,** 149–159.

Garner, W. R., & McGill, W. J. Relation between information and variance analyses. *Psychometrika*, 1956, **21,** 219–228.

Garner, W. R., & Sutliff, D. The effect of goodness on encoding time in visual pattern discrimination. *Perception & Psychophysics*, 1974, **16,** 426–430.

Garner, W. R., & Whitman, J. R. Form and amount of internal structure as factors in free-recall learning of nonsense words. *Journal of Verbal Learning and Verbal Behavior*, 1965, **4,** 257–266.

Gibson, E. J. *Principles of perceptual learning and development*. New York: Appleton, 1969.

Gibson, E. J. The ontogeny of reading. *American Psychologist*, 1970, **25,** 136–143.

Gibson, E. J., Gibson, J. J., Pick, A. D., & Osser, H. A developmental study of the discrimination of letter-like forms. *Journal of Comparative and Physiological Psychology*, 1962, **55,** 897–906.

Gibson, E. J., & Walk, R. D. The "visual cliff." *Scientific American*, 1960, **202**(4), 64–71.

Gibson, J. J., & Gibson, E. J. Perceptual learning: Differentiation or enrichment? *Psychological Review*, 1955, **62,** 32–41. (a)

Gibson, J. J., & Gibson, E. J. What is learned in perceptual learning? A reply to Professor Postman. *Psychological Review*, 1955, **62,** 447–450. (b)

Glanzer, M., & Clark, W. H. Accuracy of perceptual recall: An analysis of organization. *Journal of Verbal Learning and Verbal Behavior*, 1963, **1,** 289–299.

Glanzer, M., & Clark, W. H. The verbal-loop hypothesis: Conventional figures. *American Journal of Psychology*, 1964, **77,** 621–626.

Glanzer, M., Taub, T., & Murphy, R. An evaluation of three theories of figural organization. *American Journal of Psychology*, 1968, **81,** 53–66.

Glushko, R. J. Pattern goodness and redundancy revisited: Multidimensional scaling and hierarchical clustering analyses. *Perception & Psychophysics*, 1975, **17,** 158–162.

Gottwald, R. L., & Garner, W. R. Effects of focusing strategy on speeded classification with grouping, filtering, and condensation tasks. *Perception & Psychophysics*, 1972, **11,** 179–182.

Haber, R. N. Information processing. In E. C. Carterette & M. P. Friedman (Eds.), *Handbook of perception* (Vol. 1). New York: Academic Press, 1974.

Handel, S., & Garner, W. R. The structure of visual pattern associates and pattern goodness. *Perception & Psychophysics*, 1966, **1,** 33–38.

Handel, S., & Imai, S. The free classification of analyzable and unanalyzable stimuli. *Perception & Psychophysics*, 1972, **12,** 108–116.

Hebb, D. O. *The organization of behavior*. New York: Wiley, 1949.

Hershenson, M., Munsinger, H., & Kessen, W. Preferences for shapes of intermediate variability in the newborn human. *Science*, 1965, **147,** 630–631.

Hochberg, J. Organization and the Gestalt tradition. In E. C. Carterette & M. P. Friedman (Eds.), *Handbook of perception* (Vol. 1). New York: Academic Press, 1974.

Hochberg, J., & McAlister, E. A quantitative approach to figural "goodness." *Journal of Experimental Psychology*, 1953, **46,** 361–364.

Hock, H. S. The effects of stimulus structure and familiarity on same-different comparisons. *Perception & Psychophysics*, 1973, **14,** 413–420.

Hubel, D. H., & Wiesel, T. N. Receptive fields, binocular interaction, and functional architecture in the cat's visual cortex. *Journal of Physiology* (*London*), 1962, **160,** 106–154.

Iwawaki, S., & Clement, D. E. Pattern perception among Japanese as a function of pattern uncertainty and age. *Psychologia*, 1972, **15,** 207–212.

Johnson, N. F. Higher-order encoding: Process or state? *Memory & Cognition,* 1973, **1,** 491–494.

Johnson, R. M., & Uhlarik, J. J. Fragmentation and identifiability of repeatedly presented brief visual stimuli. *Perception & Psychophysics*, 1974, **15,** 533–538.

Klatzky, R. L., & Stoy, A. M. Using visual codes for comparisons of pictures. *Memory & Cognition*, 1974, **2,** 727–736.

Levy, R. M., & Kaufman, H. M. Sets and subsets in the identification of multidimensional stimuli. *Psychological Review*, 1973, **80,** 139–148.

Lockhead, G. R. Effects of dimensional redundancy on visual discrimination. *Journal of Experimental Psychology*, 1966, **72,** 95–104.

Marg, E., Adams, J. E., & Rutkin, B. Receptive fields of cells in the human visual cortex. *Experientia*, 1968, **24,** 345–350.

Milner, P. M. A model for visual shape recognition. *Psychological Review*, 1974, **81,** 521–535.

Munsinger, H., & Weir, M. W. Infants' and young children's preference for complexity. *Journal of Experimental Child Psychology*, 1967, **5,** 69–73.

Nelson, D. L., Garland, R. M., & Crank, D. Free recall as a function of meaningfulness, formal similarity, form and amount of internal structure, and locus of contingency. *Journal of Verbal Learning and Verbal Behavior*, 1970, **9,** 417–424.

Postman, L. Association theory and perceptual learning. *Psychological Review*, 1955, **62,** 438–446.

Preusser, D., Garner, W. R., & Gottwald, R. L. Perceptual organization of two-element temporal patterns as a function of their component one-element patterns. *American Journal of Psychology*, 1970, **83,** 151–170.

Pritchard, R. M., Heron, W., & Hebb, D. O. Visual perception approached by the method of stabilized images. *Canadian Journal of Psychology*, 1960, **14,** 67–77.

Reed, S. K. Structural descriptions and the limitations of visual images. *Memory & Cognition*, 1974, **2,** 329–336.

Restle, F. Coding of nonsense vs. the detection of patterns. *Memory & Cognition*, 1973, **1,** 499–502.

Rosenthal, R., & Rosnow, R. L. (Eds.), *Artifact in behavioral research*. New York: Academic Press, 1969.

Royer, F. L., & Garner, W. R. Response uncertainty and perceptual difficulty of auditory temporal patterns. *Perception & Psychophysics*, 1966, **1,** 41–47.

Royer, F. L., & Garner, W. R. Perceptual organization of nine-element auditory temporal patterns. *Perception & Psychophysics*, 1970, **7,** 115–120.

Rubin, E. *Synsoplevede figurer: Studien i psykologisk analyse*. (*Visual perception of figures: Studies in psychological analysis*.) Copenhagen: Gyldendal, 1915.

Schuck, J. R. Factors affecting reports of fragmenting visual images. *Perception & Psychophysics*, 1973, **13,** 382–390.

Sutherland, N. S. Object recognition. In E. C. Carterette & M. P. Friedman (Eds.), *Handbook of perception* (Vol. 3). New York: Academic Press, 1973.

Thomas, H. Visual-fixation responses of infants to stimuli of varying complexity. *Child Development*, 1965, **36,** 629–638.

Torgerson, W. S. *Theory and methods of scaling*. New York: Wiley, 1958.

Underwood, B. J. Stimulus selection in verbal learning. In C. N. Cofer & B. S. Musgrave (Eds.), *Verbal learning and behavior: Problems and processes*. New York: McGraw-Hill, 1963.

Wertheimer, M. The problem of perceptual structure. In E. C. Carterette & M. P. Friedman (Eds.), *Handbook of perception* (Vol. 1). New York: Academic Press, 1974.

Whitman, J. R. Form of internal and external structure as factors in free recall and ordered recall of nonsense and meaningful words. *Journal of Verbal Learning and Verbal Behavior*, 1966, **5,** 68–74.

Whitman, J. R., & Garner, W. R. Free-recall learning of visual figures as a function of form of internal structure. *Journal of Experimental Psychology*, 1962, **64,** 558–564.

Whitman, J. R., & Garner, W. R. Concept learning as a function of form of internal structure. *Journal of Verbal Learning and Verbal Behavior*, 1963, **2,** 195–202.

Zusne, L. *Visual perception of form*. New York: Academic Press, 1970.

Chapter 3

SORTING, CATEGORIZATION, AND VISUAL SEARCH

PATRICK RABBITT

I. INTRODUCTION

It is probable that no two signals ever received by a human sense organ are ever precisely identical.

This truism has nontrivial consequences for our understanding of

ISBN 0-12-161909-5

human information processing. It implies that signals from the same source will have considerable variation and that this variation may indeed be easily discriminable. Since most of this variation will be redundant to decisions that we have to make in order to interpret the perceptual world at any moment in time, it follows that we cannot afford to respond to unique sensory events, distinguishing them from all the other similar events we have ever encountered in our life histories. We must rather learn to assign events to categories or equivalence classes to which particular responses may be appropriate at particular times. So, for example, the responses *cow*, *dog* or *the third signal light from the left on the experimental console* will each be a categorical response to a class of very diverse sensory inputs.

It is important to recognize that such categorizations may, or may not, be described in terms of some commonality of *perceptual* features. We may indeed attempt to define the perceptual characteristics that identify dogs or cows as such, but dogs, cows, and signal lights can all be collapsed together into such valid categories as *anxiety-arousing stimuli*, which are more pertinently defined in terms of our responses to them than in terms of any common intrinsic physical attributes.

This example also makes the point that categorizations are shifting and arbitrary things. I may, if I choose, classify the objects on my desk at one moment as letters, sheets of paper, paperclips and ballpoints. The next moment I may classify them as ''junk'' that has to be swept away before I can make room to write.

Nevertheless, recent evidence shows that we cannot *necessarily* take for granted that all categorizations of sensory input are voluntarily capable of modification on a moment's notice.

A valuable series of experiments by Cutting, Rosner, and Foard (1976) and Cutting and Rosner (1976) have developed work on the classification of vowel-like sounds begun by Liberman, Harris, Hoffman, and Griffith (1957). These authors show that although subjects can be trained to make more or less continuous distinctions along continua of acoustic variations between complex sounds, they do not spontaneously do so. They have learned to make step-function distinctions *between* particular parameter ranges, and to ignore continuous distinctions *within* these ranges. In experiential terms, without special practice, they cannot ''hear'' such distinctions until they are taught to do so. The categorization system they use may be arbitrary in the sense that it may be determined by their exposure to spoken language, but it is hardly voluntary, they can modify it only with difficulty and special training.

These considerations may seem too general to concern practical scientists, but robust errors have occurred because they have been neglected. In June, 1865, F. C. Donders reported a series of experiments that were intended to obtain separate measures for the temporal durations of processes involved in discriminations between signals and choices between responses. As is well known, Donders measured an "a" reaction time when a subject made a single response to a single signal, a "b" reaction time when the subject made one of five different responses to five different signals, and a "c" reaction time when the subject might hear any one of the same five signals but had to respond only to one and ignore the rest.

Donders argued that in case "b" the subject had to discriminate among five different signals and to choose among five different responses. In case "c" he had to discriminate among the same five signals but had only to make one response, as in case "a." Thus RT(c) − RT(a) gave a measure for the extra time needed to select among five responses, and RT(b) − RT(c) a measure of the extra time required to discriminate among five different signals. The previous discussion has shown that this is a misleading inference. In case "b" subjects had to distinguish every signal from every other signal in order to respond correctly. In case "c" they had to distinguish one signal from four others, but did not have to distinguish among the four signals to which they made no response. In brief, Donders assumed that in both cases "b" and "c" subjects distinguished between five classes of one signal each, whereas it is possible to argue that in case "c" they discriminated between two classes, one of one signal and the other of four signals.

Similar failures to discriminate between possible assumptions are still very much part of the literature on visual search, and a review must begin by making the alternatives explicit.

A visual search task may be defined as a categorization task in which a subject has to distinguish between at last two classes of signals—*target* signals, which must be located and reported, and *background* signals, which must be ignored.

A first (class A) extreme assumption is that the subject identifies every signal on the display as a particular member of the set of all possible signals (target plus background signals). In effect this would mean that he discriminates between N target plus N background classes of signals. Note that this is the assumption Donders made when subtracting "c" from "b" reaction time.

A second (class B) assumption is that the subject will identify signals on a display only in terms of two classes, categorizing them as members of the target or of the background class. We may assume that the subject discriminates between, but not within, these classes.

If distinctions between members of the target class of signals are important to the subject, we have a choice of two assumptions as to what may follow.

The subject may make a search (under our second assumption) to locate *any* member of a target set and, having done so, then make a second classification in the manner outlined in our first assumption to determine which member of the target set has been found. Models implying such two-stage processing, in which target localization precedes identification, are very common in the literature and will be reviewed in detail in the following text.

A third (class C) alternative assumption to a two-stage process might be that if a subject has to search a display for N different target signals and to distinguish each of them from the rest, the signals on the display will be categorized into N classes of one signal each and one class comprising all the possible background signals (between which no distinctions are necessary, cf. Rabbitt, 1967).

A fourth (class D) and final extreme assumption is that target and background signals are not categorized or identified in the same way. We may suppose that the individual can set up a perceptual test, comparing perceptual input against a memory representation of one or more signals for which he may have to search. If input matches representation he can identify a target. However, if input does not match representation no identification takes place, and the process of scanning and comparison continues until it does.

On such assumptions background signals are never identified, but merely classified by exclusion as not being members of the target set. A discussion by Prinz (1977) of control processes in visual search shows some of the consequences of this assumption for experimental predictions. It is also an assumption implicit in theoretical discussions by Neisser (1967) and by later authors who have put his work into various theoretical contexts (Nickerson, 1966; Sternberg, 1969). For historical reasons a review of the literature should begin with Neisser's important contributions, and some consequences of the experimental technique he used.

II. NEISSER'S EXPERIMENTS REEVALUATED

Neisser (1963) and Neisser, Novik, and Lazar (1963) instructed their subjects to search for 1–10 different letters of the alphabet, designated as targets, in columns of 20 groups of 5 letters drawn from the remainder of the alphabet. Only one target was present on each display, and might occur in any group. All other background letters had to be ignored. A display was presented and a timer simultaneously started. The subject scanned the display from top to bottom and closed a switch as soon as he located any target. Times taken to locate targets gave a measure of scanning rate per item on a display. Early in practice scanning time varied with the size of the target set, but late in practice subjects took no longer to find 10 targets than to find 1.

Neisser interpreted these results in a way conformable to a "Pandemonium" processing model (e.g., Selfridge & Neisser, 1960) in which systems of feature analyzers ("demons") checked perceptual inputs in parallel to determine the presence or absence of critical states. Neisser inferred that his subjects must have memory representations of features necessary to identify any of up to 10 targets, and that these representations could all be simultaneously compared against current input.

At about the same time a number of studies of human performance at stimulus categorization (Rabbitt, 1962; Pollack, 1963; Sternberg, 1966) and visual search (Rabbitt, 1962, 1964) gave discrepant results, showing increases in classification or search time when the number of stimulus classes discriminated or the number of targets sought were increased. It became important to compare studies for idiosyncrasies of procedure. The methodological points discussed in the following text seemed to be especially important.

A. Logical Structure of Categorization Task

In Neisser's (1963) experiments, subjects made the same response to all target symbols. Rabbitt's (1962, 1964) subjects were also highly practiced, but made different responses to each of two to eight target symbols. Rabbitt, but not Neisser, found that search time increased with target set size.

These experiments thus framed in operational terms the distinction made previously between class-B or class-D assumptions (Neisser's experiments) and class-C assumptions (Rabbitt's experiments; see also Rabbitt, 1971, pp. 259–260).

This distinction had been found important in another context. Rabbitt (1959) and Pollack (1963) had measured RTs to classes of signals, independently varying both the number of signals in each response class and the number of response classes among which subjects had to choose. They both found that when subjects discriminated between only two classes of signals, large increases in the number of signals in each class had no effect on RT. However, when subjects had to discriminate among six or more response classes, the same increases in signal entropy produced sharp increases in RT. Neisser's (1963, 1967) studies can be interpreted as examples of tasks in which subjects distinguished between only two classes of signals (class-B assumption, target and background classes). Rabbitt's (1962, 1964) experiments can be interpreted as tasks in which subjects had to distinguish between *N* target classes of one signal each and one additional background class of many signals (assumption C).

Thus the effects of variations in the number of targets sought may differ with the experimental tasks employed. More importantly, we have the suggestion that the techniques of classification that subjects use (i.e., B or D on the one hand or C on the other) may be flexible and may change with experience of particular tasks for which they may be more or less optimal.

B. Effects of Stimulus Frequency

It was interesting that in Neisser's tasks target signals were very rare and background signals very frequent. Early attempts to replicate Neisser's experiments (e.g., Kaplan & Carvellas, 1965; Kaplan, Carvellas, & Metlay, 1966) found slight but significant increases in scanning time with increases in the number of targets sought. In these studies, and in a replication by Shurtliff and Marsetta (1968) carried out to eliminate the confounding of factors of target frequency of occurrence and target set size, targets occurred much more frequently on displays than in Neisser's (1963, 1967) experiments.

Rabbitt (1962, 1964), who also found significant effects of target set size, also used displays on which the ratios of targets to nontargets were relatively high. To check this point of difference Rabbitt (1966) systematically varied ratios of targets to nontargets on each display and found that when targets were rare scanning times were relatively fast, failures to detect targets were increasingly common and the effects of target set size on RT were much reduced. This suggested that subjects could employ more or less stringent criteria for target detection when scanning displays and that target probability determined these criteria, as in conventional psychophysical decision tasks (Green & Swets, 1966).

C. Speed–Accuracy Trade-Off

This last suggestion was elegantly extended by Wattenbarger (1969). In early experiments by Neisser (1963) and Neisser *et al.* (1963) error rates appear to have been very high (see Neisser, 1967, and Neisser & Beller, 1965), often exceeding 20%. It is probable that most of these errors were failures to detect targets on displays, though some seem to have been false identifications (Neisser & Beller, 1965).

Wattenbarger (1969) replicated Neisser's experiments, instructing his subjects either for high speed or high accuracy. He found the expected inverse trade-off between RT and accuracy (Schouten & Bekker, 1967; Pew, 1969). Most interestingly, when his subjects had low error rates they showed a marked increase in scanning time with target set size. At error rates comparable to those tolerated in Neisser's experiments, target set size had little or no effect.

These data are consistent with either of two explanations: We may interpret speed–accuracy trade-off data, as did Schouten and Bekker (1967), by suggesting that subjects take more or less perceptual evidence from a display and accordingly make more or less rapid and accurate judgments. A reformulation in terms of the mathematics of decision processes would be a useful advance on such assumptions (Swets, 1964; Taylor, 1967). An alternative kind of explanation would be simply to suggest that when high error rates are tolerated, or scanning is heavily paced, subjects simply omit to test for some members of a large set of target symbols. Unfortunately these explanations are not mutually exclusive, but checks on error patterns should reveal how far the second applies in any particular task. Such work would be very useful indeed.

D. Effects of Practice

In discussing the effects of extended practice on visual search, we must consider data and arguments based on two other kinds of signal-categorization tasks. These have been treated as logically similar to search tasks, but points of experimental detail make distinctions necessary.

In visual search tasks, subjects discriminate between at least two classes of items, the target set and the background set. They respond, however, only to items in the target set. Typically, they are not required to report, or to respond to, background set items. Visual search tasks, therefore, closely resemble Donders's 1865 "c" reaction tasks, in which one signal must be identified but all others may be ignored.

A second class of categorization tasks requires the subject to make one overt response to all signals in one learned set, and another overt response to all signals in another learned set (Rabbitt, 1959; Pollack, 1963). Such tasks most closely resemble Donders's (1865) "b" reaction tasks, in which one response must be made to one signal and other responses to others.

A third category of tasks, now in very wide use, was first developed by Sternberg (1963; 1966). The subject knows that all signals will come from a particular vocabulary (often decimal digits or letters of the alphabet). On each trial the subject is given a subset of items from this vocabulary to remember. The size of this subset is typically varied from trial to trial. After presentation of the subset he is then given a probe item from the same vocabulary. He then classifies the probe, making one response if it is a member of the subset given to him (positive set) and another response if it is not (negative set).

During the 1960s, differences in the effects of variations in the size of signal sets on classification time in these three paradigms encouraged much speculation and research. Some early studies of visual search had specifically made the point that early in practice scanning time varied proportionately to the size of the target set, and indifference to target set size was only attained late in practice (Neisser, 1963; Neisser, Novik, & Lazar, 1963; Neisser, 1967). Replication studies again suggested that early in practice subjects adopted a serial strategy, comparing each item on a display to memory representations of all target set items in turn, but that this strategy altered with practice to a parallel mode of comparison in which memory representations of all target items were simultaneously compared against each symbol on a display. In other words, search proceeded serially over the display, but in parallel for the memory set (Kaplan & Carvellas, 1965; Kaplan, Carvellas, & Metlay, 1967; Shurtliff & Marsetta, 1968).

In conformity with these assumptions when subjects were trained to make one overt response to all signals in one set and another response to all signals in another, RTs varied with the number of signals in each set early, but not late, in practice. A change from a serial to a parallel memory set-comparison strategy seemed a possible hypothesis (Rabbitt, 1959, 1962; Pollack, 1963).

In contrast, some visual search tasks failed to show very marked reduction in effects of target set size with extended practice (Rabbitt, 1962, 1964). And, in particular, Sternberg's experiments with his classification task continued to suggest that even with extended practice subjects retained a serial, rather than a parallel, memory set-comparison strategy (Sternberg, 1963, 1966, 1967, 1969, 1975).

Sternberg (1967) directly compared these two types of tasks by combining them in the same experiment. Both the number of items for which the subject had to search (memory set) and the number of items on each display (display set) were varied from trial to trial. Reaction time increased linearly both with the size of the memory set and with the size of the display set. It seemed that sequential rather than parallel comparisons were being made at about the same rate in both cases (i.e., at a rate of 30–45 msec per item). Sternberg suggested that whereas comparisons against members of the memory set were exhausative (i.e., each display item scanned was always compared against all items of the memory set in turn), comparisons against items on the display were self-terminating (i.e., they stopped when a target had been located).

Sternberg's (1967) subjects were not highly practiced, and in any case they experienced different memory sets on each trial. Neisser's (1963) subjects and subjects Neisser, Novik, and Lazar (1963) searched for nested memory sets on each trial (i.e., smaller memory sets were always subsets of larger memory sets). Nickerson (1966) had used Sternberg's procedure, giving his subjects 22 days of practice in one of his experiments. However, once again, memory sets were not nested. This point was finally tested by Burrows and Murdock (1969), who used both constant and varied set procedures in the same paradigm employed by Sternberg and Nickerson. After 14 days of practice with 192 trials per day they found that error rates were extremely low and that a scanning rate of approximately 30.6 msec per memory set item reduced to one of about 8.4 msec per item. There was, however, no difference in the effects of practice between the fixed or varied set procedures. Burrows and Murdock (1969) concluded that the memory scanning function described by Sternberg was robust at even high levels of practice.

Although it is true that the slope of the memory scanning function was the same in fixed or varied set procedures, the interesting point about the Burrows and Murdock (1969) study seems to be the considerable reduction in the slope of the memory-set search function in both conditions. Indeed their results suggest that with further practice no significant change in RT with memory load would have been observed and an assumption of parallel processing would have been possible. Since memory sets never exceeded three items, it is also possible that the experiment was insensitive with respect to differences between fixed and varied sets, which might have been very obvious had memory sets of as many as six items been used (Sternberg, 1966, 1967, 1969).

In the late 1960s and early 1970s further data accummulated suggesting that, with enough practice, changes in the sizes of fixed or nested memory set sizes cease to have any effect on scanning time for displays. Chase (1969) pointed out that the speed with which subjects search through mixed lists of targets and nontargets increases by about 25% after as few as 40 trials with the same memory set. He concluded that "the practice effect is important because it underlies the transition from serial to parallel processing in search tasks." Graboi (1971) also directly tested the effects of practice with nested memory sets, using words (randomly chosen five-letter surnames) rather than letters. After seven daily sessions of 1½ hr practice, Graboi was able to show that although practice with varied memory sets did not abolish the set-size effect, specific practice with particular nested memory sets did.

When interpreting such experiments a caveat is that a very great deal of practice may be necessary before the effects of memory-set size disappear. Rabbitt, Cumming, and Vyas (1977a) found that variations in set size between two and eight items (letters of the alphabet) produced small but significant variations in search time after 15 days of intensive practice, though these later entirely disappeared. Error rates were as low or lower than in Nickerson's (1966), Sternberg's (1966, 1967, 1969), or Burrows and Murdock's (1969) experiments, so that fast performance cannot be explained in terms of speed-error trade-off.

It now seems that even with Sternberg's original paradigm (1966, 1969, 1975), in which a memory set is probed on each trial by a single test item, if sufficient practice is given in a fixed-set experiment the effects of set size on RT are completely abolished (Kristoffersen, 1976).

This is a comfortingly consistent conclusion, since it means that in *all* paradigms in which subjects are required to distinguish between two classes of signals after sufficient practice large variations in the number of signals in each class cease to have any effect on choice RT. (See, for visual search, Neisser, 1963; Rabbitt, Cumming & Vyas, 1977; for the Sternberg memory-set probe paradigm, Kristoffersen, 1976, and for two-choice signal categorization tasks, Rabbitt, 1959, 1962; and Pollack, 1963).

As with all practice effects the interesting point is not that subjects learn to perform a task faster and more accurately than when they first attempt it, but rather that they apparently learn to perform in a different way. The conclusion that with practice subjects shift from a serial to a parallel processing mode may be accurate, but it is insufficient. We need to know precisely what it is that subjects learn in order to allow such a change to take place. The effects of practice on different aspects of visual search tasks will be discussed.

III. LEARNING AND REMEMBERING THE TARGET SET

It is not a trivial point that in order to carry out visual search tasks subjects have to be told what the target set is and must continue to remember to search for *all* the items within it. Early in practice, Rabbitt (1962) found that particular items in large target sets were apparently forgotten (they were systematically missed on displays on which they occurred). It is possible that the high error rates obtained by Neisser (1963) were partly due to neglect of some target-set members.

When targets are arbitrary sumbols which cannot be designated by name it is obviously necessary for subjects to be pretrained in order to be able to identify them. The nature, and the duration, of this pretraining will affect the efficiency of their search. A more interesting situation occurs in the usual style of experiment, in which sets of familiar, nameable symbols (letters, digits, words, etc.) are designated to subjects as target-set members.

Fitts and Switzer (1962) found a cognitive effect in conventional choice reaction tasks in which digits were used as signals. If a choice was necessary between two symbols, it was made more rapidly if symbols formed a natural subset (e.g., a set of adjacent digits, such as 1 and 2) rather than an arbitrary subset (e.g., 2 and 7).

Egeth, Marcus, and Bevan (1972) tested this point in visual search tasks where subjects searched for sets of adjacent digits (e.g., 1, 2, and 3—"natural" sets) or for sets of nonadjacent digits (e.g., 1, 4, and 7—"unnatural" sets). For natural sets, memory scanning time was apparently as low as 2 msec per character at levels of practice at which it was 24 msec per character for unnatural sets.

Egeth, Marcus, and Bevan's (1972) paper raises a further important point about the relationship of categorization strategy to experimental methodology. In their Experiment I subjects made a response only if a target item was presented (i.e., Donders "c" reaction time, or our class D classification procedure, see Section I). When, in another of their experiments subjects were required to make a positive response to acknowledge presence of a target item, and a different, negative response to indicate absence of any target, memory search time, even for natural sets, increased to 35 msec per character. In the unnatural set condition there was also a sharp increase from 24 to 45 msec per character scanned.

The definition of what constitutes a *category* of items in visual search, or in any other task, is an elusive business that forms the main topic of this paper.

The simplest description of a perceptual category of signals would be in terms of some unique physical attribute shared by all its members. It is clear that this is not a sufficient description, since the categorical effects discussed here seem to have little to do with the respective physical properties of target and background sets. They appear to be established by memory associations between some, but not other, subsets of the *names* of items (i.e., in a digit series), irrespective of their physical attributes.

There have been similar demonstrations of category effects in tasks using the Sternberg memory search paradigm. Here the presentation of lists of words which are categorizable in terms of one or another system of semantic associations has been shown to increase the speed, and perhaps affect the order, of search through immediate and long-term memory (De Rosa & Tkacz, 1975; Morrin, De Rosa & Stultz, 1967).

In these cases, explanations seem to be of four distinguishable but nonexclusive kinds.

First, the presentation of a categorically associated memory set of items may allow faster rehearsal of a list in immediate memory and so contribute to faster comparisons.

Second, the presentation of lists of familiar items, especially if they are categorically related to each other, may result in faster comparisons of each item in memory with a presented probe.

Third, when memory-set items are all drawn from one obvious semantic category (e.g., letters) and a negative probe is drawn from another (e.g., a digit) it may take less time for a subject to decide that the probe is not a list member by recognizing that it is from a discrepant class than it takes to compare it with all positive set items (whether serially or in parallel).

Finally, if the memory set consists of items from two sets, learned separately, or otherwise mutually identifiable, subjects may recognize that a probe belongs to one such class but not to the other, and then carry out comparisons only with items in the appropriate subset of reference.

A further discussion of categorization processes in search and classification experiments follows. For the present, the only point required is that if subjects are practiced with a constant, arbitrary, target set, this set can become definable as a category of associated items. We see that irrespective of variations in physical features the use of such categories may offer advantages in speed of information processing. It is relevant to consider evidence on the various ways in which practice leads to different kinds of categorization.

IV. CATEGORIZATION IN TERMS OF CRITICAL PERCEPTUAL DISTINCTIONS

The benefits of categorization in visual search are most easy to understand when the categories in question may be defined in terms of the presence or absence of a single common attribute.

Green and Anderson (1956) studied a task in which subjects searched for symbols that might be of many different shapes and might appear in one of several colors. If subjects were instructed to search in terms of symbol shape alone, search time was proportional to the number, and to the variance, of *all* symbols (shapes) on a large display. If they were instructed to find a critical shape of a specified color, search time was proportional to the number and variance of other symbols *of that particular color only*. Green and Anderson (1955) reported similar results when symbols varied in size and shape. Neisser (1969) and Willows and McKinnon (1973) showed that if alternate lines of text were printed in black and red subjects could read lines printed in one color without interference from, or apparently even awareness of, the content of lines in the other. In all these cases it seems that symbols were first scanned in terms of one critical attribute (e.g., color or size). Once any symbol possessing that attribute was located it would be further processed to determine its shape. Symbols *not* printed in the critical color or size were apparently not processed for shape.

A number of studies have shown information gain in the processing of tachistoscopic displays when subjects were cued to scan a subset of items defined by a particular physical characteristic (e.g., location—Sperling, 1963; color—Brown, 1960; von Wright, 1968; size or brightness—von Wright, 1970). Such cues are most useful if they are presented immediately before the onset of a display, and their effectiveness is reduced if they are delayed much beyond 500 msec after the display has disappeared (a useful study of the relative latencies at which different cues retain their advantages has been published by Dick, 1969). In all cases the advantage seems to imply that subjects can efficiently process subsets of display items identified in terms of a common characteristic, but are less efficient if this is not possible.

In all the experiments described above it seems that subjects *first* make a class D classification (our taxonomy of Section I) in terms of which items that meet a test on a particular attribute are further processed, whereas a subset of other items, defined by exclusion on this test, do not further consume processing time or capacity.

Garner (1974) has shown that not all pairs of stimulus attributes are equally useful for classifications of this kind. He distinguishes between pairs of attributes, like shape and color, that may be efficiently processed separately, and other pairs, like hue and brightness, that are apparently jointly, or integrally, processed (see Felfoldy and Garner, 1971). Thus categorizations made in terms of one stimulus attribute may be used to lower the information load of identifications made in terms of another attribute only when these attributes, by Garner's definition, are separable rather than integral.

It is conceptually easy to see how symbols of diverse shapes can be treated as separable classes if they are defined in terms of some independent critical attribute such as color. It is less easy to see how complex stimuli (e.g., shapes), for which separate continua of classification are not easily specified, can be ranked into categories. It is, however, apparent from Cutting and Rosner (1976) and Cutting, Rosner, and Foard (1976), in their experiments on complex sounds, that subjects achieve and use such classifications.

An experiment by Rabbitt, Clancy, and Vyas (1978) illustrates at least one of the problems facing subjects who have to make such classifications of signals varying along a single dimension of physical difference. If a set of pure tones varying in steps of 100 Hz are assigned to two classes, high and low, with respect to some arbitrary frequency (e.g., 1500 Hz), subjects can make correct assignments very quickly and accurately. Large increases in the number of different tones to be classified do not then affect RT. In contrast, when all pairs of tones adjacent in frequency are assigned to different categories, classification becomes very slow and inaccurate. This experiment suggests that some rules, or principles, of classification may be, for functional reasons, much easier for subjects to use than others. With complex stimuli (such as shapes), which may logically be rank-ordered along many dimensions of the difference, it must be borne in mind that the existence (or even the relative discriminability) of particular dimensions of difference may not be the most important factor in determining whether they can be used in classification schemes.

When considering classifications of letters, words, and digits, which are the most frequently used symbols in visual search experiments, there are other caveats to be considered. It is certainly true that some pairs, or sets, of letters are visually more confusable than others (Townsend, 1971). It is dubious, however, whether such confusions can be related to the presence or absence of particular *features* or *feature sets*. Rather, it seems that discriminations between such symbols as Nevada cattle brands or letters of the alphabet require the specification of rules determining the relationships between discriminative characteristics (Naus and Shillman, 1976).

Bearing this in mind, Neisser (1963) and Rabbitt (1967) have shown that visual search is especially fast and accurate when the subsets of target and background letters of the alphabet may be defined by the presence or absence of one of two critical attributes (e.g., straight lines or curved lines). Rabbitt (1967) found that subjects detecting straight-line letters among curved-line letters, or vice versa, improved with practice, but showed perfect transfer when shifted to discriminations between other subsets of letters that were similarly defined. This was not the case when target and background letters were assigned by random selection. Five different groups of subjects were given different amounts of practice in searching for a particular arbitrary target set among a particular background set. They were then transferred to a new search task in which the background set changed but the target set did not. The amount of negative transfer increased both with the duration of initial practice and with the size of the target set for which search was made. It seemed that subjects learn cue systems critical for distinctions between a *particular* target set and a *particular* background set. Improvement with practice is at least partly attributable to this specific cue learning. The larger the target set, the more complex the necessary optimal cue system seems to be, the longer it takes to learn, and the more a subject is inconvenienced when deprived of it on transfer.

The empirical definition of categories therefore seems to be in terms of the cue systems, discriminative generative rules, or whatever, which subjects learn and use in order to make distinctions between particular sets of stimuli. A further question is whether such specific discriminative systems form part of the transient, or relatively permanent, equipment which the individual may use to make sense of his visual world. Rabbitt, Cumming, and Vyas (1979b) practiced four groups of subjects on a particular target–background discrimination and then retested them 0, 2, 4, or 6 weeks later on either the same search task or on one in which they looked for the same targets among new background items. Learning that was specific to the practiced discrimination apparently lasted for at least 4 weeks. Subjects apparently developed, and retained in long-term memory, analytic programs specific enough to facilitate discriminations between particular subsets of symbols and not others.

The existence of such programs, which are often of a very general and abstract kind, has been demonstrated by earlier experiments. Corcoran and Rouse (1970) showed that subjects can better recognize words written in manuscript, or typed, when they know in advance which of these two formats to expect. They also point out that no advantage is obtained from advance information as to whether words are to be presented in upper- or in lower-case symbols, so that the facilitating programs that are deployed must be of a very general kind. A similar point may be made from a number of studies that show a *word superiority effect* (WSE) in character recognition. The WSE is enhanced, and sometimes only obtained at all, if subjects know in advance that words are to be presented.

In sum, people clearly can and do use particular physical differences between signals, or even complex and abstract stimulus-encoding rules in order to efficiently categorize target and background items in visual search. But even within the literature supporting this point of view there are suggestions that not *only* physical characteristics determine efficient categorization. Rabbitt, Cumming, and Vyas (1977) practiced three subjects for 30 days on discriminations between particular nested target sets and a particular background set of letters. After this amount of practice, transfer to an unfamiliar *background* set did *not* affect performance. Very similar results have been recently reported by Kristoffersen (1976), who practiced her subjects for over 20 days with particular positive and negative sets in the Sternberg (1966, 1969, 1975) paradigm. She also found no negative transfer when sets were altered. These data at least raise the question as to whether some factors other than specific perceptual discrimination learning are involved in categorizations of overlearnt stimulus sets. It is therefore not surprising that when we consider experiments with categories of symbols, such as letters and digits, overlearnt outside the laboratory, the same difficulties of interpretation regularly arise.

V. DISCRIMINATIONS BETWEEN CLASSES OF LETTERS AND DIGITS: CATEGORIZATION INDEPENDENT OF NAMING

There are some visual cues, probably those involving relationships between sets of features rather than single critical attributes, which may be common to all Arabic numerals but not to Latin letters and vice versa. Arabic digits were designed for the brush, whereas the Latin alphabet was for lapidary or for inscription with a stylus. There are some other obvious, common distinguishing features. For example, the (asymmetrical) digits 2, 3, 5, 7, and 9 (along with 4 in some fonts), all have concavities to the left, whereas no (asymmetrical) capital letters except *J, S,* and *Z* do so.

It is important to distinguish between two kinds of argument based on such possibilities of categorical identification by particular physical features. The first is that the subsets of cues that may be used to distinguish all digits from all letters may *be more or less numerous* than the subsets of cues used to distinguish any given letter or digit from all other symbols in both sets (i.e., to name it). The second possibility is that the cue systems used to categorize letters and digits are simply *different* from those used to name them. The literature supports the second statement. The first involves a complicated series of assumptions and finds no solid empirical evidence.

Particular ensembles of symbols may be deliberately constructed to make categorization easier than naming (e.g., symbols with curved lines versus symbols made up of straight lines). But it is important to recognize that for any arbitrary natural set of symbols categorization is likely to require *more* and not *fewer* cues than naming. For any given set of N such symbols there will be $N^{C}(N - X)$ possible ways in which categorizations into two subsets of X and $N - X$ symbols may be made. Thus, the number of cues which may be necessary to assign a symbol correctly to one set or the other will, in general, be proportional to $N^C(N - X)$. The number of cues necessary to identify a symbol as a particular member of the total ensemble will be proportional to the number N, which will be smaller for all $N > 4$ and all $X < 1$. This problem is common in applications of information theory to choices for which we cannot exactly specify the total ensemble of *possible* states between which a receiver *actually* makes a choice of alternatives. Unless we can be sure that there are good operational reasons for other assumptions we must be guided by theoretical upper and lower limits of ensemble size.

With this point in mind, we may consider three kinds of experiments made to discover whether subjects can categorize letters and digits more easily than they can name them.

The first possibility is to tachistoscopically expose symbols on the assumption that loss of contrast, superimposition of noise, or a postexposure masking field will partly degrade them, obliterating some cues but not others at random. If a man can name a symbol correctly he can, from memory, correctly assign it to its category. Thus, such experiments can never test whether naming thresholds are lower than categorization thresholds. The question, therefore, always is whether symbols may offer sufficient cues for categorical discriminations when cues adequate for naming have been lost. Nickerson (1973) found that degraded characters could be named as easily as they could be identified when both types of recognition were imperfect. However, Butler (1975) found that tachistoscopic masking functions obtained when characters were cued by both a bar marker and a category name are significantly different from those obtained with a bar marker only.

Tachistoscopic recognition experiments do not test whether *fewer* cues are necessary to distinguish between categories than between individual items. Fitts, Weinstein, Rappaport, Anderson, and Leonard (1956) and Anderson and Leonard (1958) showed that when displays differed only in terms of a few critical features they were poorly recognized at brief exposures because the critical discriminative features were often obscured. In contrast, when displays could be distinguished in terms of many different cues, tachistoscopic recognition thresholds were low, since if some cues were lost other alternatives might be available. Thus if categorical decisions between letters and digits were possible at lower recognition thresholds, we might assume that more *alternative* cues were available for such choices than for naming. The fact that *more* cues may be necessary to categorize than to name items in fact makes it less likely that more *alternative* categorical cues are available.

A different kind of experiment has been described by Connor (1971), who replicated experiments by Donderi and Case (1969) and Donderi and Zelnicker (1970). These authors had found that when a single target was embedded on a display among a set of mutually identical background items, direction times did not vary with the total N of items on the display. Connor (1971) showed that this was possible for digit targets exposed among sets of (physically different) letters. This seems to show that subjects can detect discrepancies between target and background *classes* of signals, and that the classification of letters versus digits can be used in this way to facilitate localization of particular items. What seems to be germane here is that *different* cue systems are used for categorization and naming, rather than that one such cue system is more elaborate than another.

A second kind of experiment tests whether latencies for naming letters and digits are different from latencies for categorizing them. A possible argument would be that whichever discrimination requires *fewer* cues should take less perceptual processing time and so be faster. This begs questions of perceptual quantification of input, and the upper limits to which parallel processing of features may be possible (Rumelhart, 1970). Tests by Nickerson (1973) and by Dick (1971) showed that alphanumeric symbols were named faster than they were categorized. However, since Fraisse (1969) has shown that the symbol *O* is named faster when called *oh* than when called *zero* and more slowly still when called *circle*, naming latencies may reflect familiarity of response associations more than the relative durations of perceptual processing.

A final class of experiments examines RT as it relates to detecting the *difference* between two or more symbols. Posner (1971) has presented data from same–different classification tasks, including a finding that letter–digit pairs are classified as being nominally different faster than are either letter–letter or digit–digit pairs. This is consistent with Connor's (1971) finding in a tachistoscopic detection task (i.e., categorization may be used for detecting *differences*, though not for *recognition*).

In all visual search experiments, reported times for detection of targets among backgrounds are shorter when they require distinctions between, rather than within, digit and letter categories. This was reported by Rabbitt (1962) and confirmed by Brand (1971), who further showed that search between categories was equally efficient whether a target was designated by class (e.g., any letter) or by name (e.g., the letter *K*). Ingling (1971) replicated these results and suggested that categorization should be regarded as an especially efficient means of information processing, but that it should not be assumed that categorization effects were based on specific cue structures in discriminations between symbols.

It seems better to first consider what can be done without discarding the simple-minded assumption that perceptual discriminations must be based on perceptual cues.

Jonides and Gleitman (1972) showed that subjects detected the same symbol, *O*, faster among letters when they were instructed to search for *zero* and faster among digits when instructed to search for *oh*.

Gleitman and Jonides (1976) report that with 150-msec exposure durations subjects in a four-target, alternative forced-choice task located digits among letters with mean RTs of 421 msec and letters among letters with mean RTs of 564 msec. On subsequent recognition tests, memory for background items was at chance level for displays with digit targets but was above chance when letter targets were used.

Jonides and Gleitman (1976) extended this task to include catch trials in which subjects were told to locate specified digit targets among letters, on occasions nontarget digits were included in the display. They still found an improvement in target *identification* in between-category as compared to within-category search. They explained this by the attractive comment that "If one had to decide whether a cowboy in a herd of cows were John Wayne or Henry Fonda it wouldn't matter how many cows were in the herd [p. 291]."

In other words, their subjects distinguished between cues necessary for localization and those necessary for naming of particular symbols.

When RT was plotted against the *N* of targets present in a display, the addition of catch trials affected the intercept, but not the slope, of the function obtained. This may be interpreted as the addition of a constant to RT for the time taken to identify a character already located on the display. Since catch trials increase the set of symbols among which discriminations have to be made, this is a reasonable assumption.

A further experiment compared RTs for target localization and target naming in between- and within-category search. RT for target *localization* was reduced more than RT for target naming when search was between categories. This is puzzling, since if the *only* benefit conferred by between-category search is faster target *localization*, the RT reduction for target identification in between-category search should simply reflect this advantage. That is, it should be the same as the reduction for localization.

In a final experiment, a predisplay target localization cue was shown to be effective in searching within, but not between, categories. This would suggest that the benefits to target localization conferred by categorical cueing are as great as those conferred by more direct methods.

Gleitman and Jonides (1976) discuss categorization as *partial processing* and suggest that it differs in no way from within-category search except that it is faster, and so allows target items to be rapidly *located* by extraction of attribution common to all members of a class. Such a process also entails the loss of possible information, since fewer distinctions between background items are registered during within-category search (and so remembered afterwards).

Many other experiments have shown that functions relating target-detection RT to the number of symbols on a display have shallower slopes when search is within rather than between categories (e.g., Egeth, Atkinson, Gilmore, & Marcus, 1973; Egeth, Jonides, & Wall, 1972). Sperling, Budianski, Spivak, and Johnson (1971) ingeniously paced visual search so that sequences of displays followed each other at very fast rates. Here the limiting rates for detections of digits among letters were as fast as one letter scanned every 8–13 msec. Search for letter targets was very much slower. Early in practice a specified digit was better detected than an unspecified digit, but this difference disappeared as subjects became accustomed to the task (see Brand, 1971).

For all these experiments Gleitman and Jonides's (1976) and Jonides and Gleitman's (1976) hypothesis that targets may be located by categorical cues and subsequently identified, provides an adequate account. We may call this the two-stage location–identification hypothesis, to stress the assumption that these transactions represent different, and successive, acts of information processing. The assumption of serial, rather than parallel, processing will be later challenged, but the assumption that the perceptual cues used in these processes may be *different* appears to be necessary. To paraphrase Jonides and Gleitman's (1976) example, we must agree that we use *different* cues to distinguish John Wayne from Henry Fonda than we use to distinguish either from a herd of cows, or the cows within the herd from each other. It does not necessarily follow that we use more or fewer cues for one kind of discrimination than for the other.

This distinction becomes less pedantic when we consider recent experiments on visual search for printed words defined in terms of the semantic class to which they belong. It seems very unlikely that all words in a particular class of semantic association have any particular subset of physical features in common.

It is useful to remember that categorization of targets and nontargets involves processes of memory search that, at least in some experiments with the Sternberg paradigm, seem to be independent of processes concerned with extraction of features from displayed symbols (Sternberg, 1967, 1969). Memory search time affects the slope of the RT–target set size function, whereas display discriminability apparently affects the intercept only.

In such tasks, when target lists can be categorized, whether as letters or digits or in other ways (Lively & Sanford, 1972; Naus, Glucksberg, & Ornstein, 1972; Underwood, 1976, pp. 154–159) the slope of the memory scanning function is reduced. The mutual discriminability of classes of *names* of target and background items in memory may thus facilitate categorical decisions independently of the perceptual discriminability of their symbolic representations.

VI. SEMANTIC CLASS AND VISUAL SEARCH

Recent experiments by Karlin and Bower (1976) have shown that when subjects search a display for a single, defined target word, search time does not differ whether target and background words belong to the same or to different semantic categories. However, if subjects search for any of a set of three or six defined targets they can do so faster if target and background items belong to different semantic categories. Henderson (1976) reports that target words specified by class (e.g., color name) are more rapidly located if background items all belong to another, common class than if they are drawn from the lexicon at random. Fletcher and Rabbitt (1976) reported identical results for a two-choice categorization task in which each of a list of successively presented words was classified in turn as belonging, or not belonging, to a specified target class. In this case, differences in frequencies of occurrence of nontarget words affected RT in the random, but not in the constant, nontarget-class condition.

Karlin and Bower's (1976) results suggest that where a *single* target is defined subjects can detect it (perhaps) on the basis of physical cues without necessarily processing it, or the background items, for meaning. This is what we might expect from Neisser and Beller's (1965) finding that subjects search lists of random words faster for single, defined target items (e.g., *monkey*) than for words defined by meaning class alone (e.g., names of animals or days of the week). It is also consistent with Schulman's (1971) demonstration that subjects remember background words better after searches for targets defined by meaning class than after searches for named targets. In the first case, subjects must presumably have to identify the meanings of background words before classifying them as nontargets. In the second case, they need consider no more than the initial letters of background words.

However, Karlin and Bower's (1976) experiments suggest that when a named target class is as large as three or six items the range of physical cues necessary to discriminate target from background words may be so large that subjects can conduct search more easily by scanning words on the display to identify their meaning. The usual two problems for a subject conducting search may contribute to this difference: First, with several target words, the range of critical cues necessary to discriminate all targets from nontargets may be quite large, so that search will be relatively slow; second, subjects have the problem of learning and remembering which are the critical cues. Early in practice they may find it easier to search for words by exercising highly practiced reading skills because, among other reasons, it is easier to remember the target names than to learn and remember a critical feature list.

A further experiment by Fletcher (1978) apparently conflicts with Karlin and Bower's (1976) results. He found that search for a single word (e.g., *monkey*) is slower if background items are drawn from a similar semantic class (e.g., animal names). Background items are classified more slowly as nontargets, presumably because their meaning is recognized.

In my view, these results must be related to a wider context of demonstrations that subjects can extract several different kinds of information from words, apparently using independent cue systems. Work on dyslexic patients by Marshall and Newcombe (1967) has shown that brain-damaged patients may be able to extract meaning from words that they cannot pronounce or identify. They may therefore produce as errors phonemically dissimilar words that are related in meaning to the words they try to read. Recent work on normal subjects by Allport (1977) and by Coltheart (1976) similarly suggests that both facilitation and interference in word recognition may arise, independently, from orthographic or phonemic similarities, on the one hand, and semantic relationships, on the other.

It is not necessary to suggest that the meaning of a word can be recognized as fast as one or more of its component letters may be identified. Indeed, Neisser and Beller's (1965) result suggests that the reverse is generally the case. We need only stress that the extraction of these different kinds of information may require processing of different sets of cues, and that these cues may be processed independently in parallel. In this case, Fletcher (1978) subjects may indeed scan words for critical features, but also simultaneously process them for meaning. The latter process may be slower, so that the meaning of one word may be recognized only after the subject begins to scan the next word for critical features. It may be this delayed recognition that a word similar in meaning to the target word has just been scanned that interferes with categorization. On this assumption, Fletcher's (1978) results would not be obtained if words were presented, one at a time, for classification at long R–S intervals.

This particular point is best dealt with in detail in reviewing experiments on rapid, paced, serial visual search by Lawrence (1971), Fischler

(1975), and Frankish (1977). We may first consider other evidence that subjects scanning letter strings or words to find individual target letters also, apparently involuntarily, extract other kinds of information that may speed or slow their search.

VII. THE USE OF ACOUSTIC PROPERTIES OF LETTER NAMES TO SEPARATE TARGET AND BACKGROUND CATEGORIES IN VISUAL SEARCH

People can obviously be set to search for arbitrary symbols which have no names, and in such cases search time will vary only with the visual discriminability of target and background items. Letters of the alphabet have names, and most individuals are highly practiced at extracting and using these names when reading. The question therefore arises whether distinctions between target and background letters are made solely in terms of visual cues, or whether subjects also name the symbols that they scan so that distinctions between background and target names provide either main or ancillary cues in terms of which these classes of items can be discriminated from each other.

Conrad (1964, 1967) showed that when letters of the alphabet were visually presented for immediate recall, subjects nevertheless made errors that indicated confusions among letters with similar sounding names (acoustic confusions). This raised the possibility that the so-called internal representations of target items in categorization tasks, like the memory representations of letters in Conrad's experiments, might reflect their acoustic properties as well as their visual distinguishing features. It was, therefore, possible that discriminations between sets of letters might take longer if their names were acoustically confusable as well as visually similar.

Work in the early 1970s has shown that this is indeed so if the letters to be compared are printed in different cases, so that they cannot be judged to be different from visual cues alone, and subjects must extract and compare names before they can make confident decisions (Dainoff & Haber, 1970, Posner, 1971). Early experiments in visual search did not show such effects, but make some useful methodological points.

Chase and Posner (1965) found no effects of acoustic confusability of target and background items in visual search, or of acoustic confusability between positive and negative sets in memory search. In their experiments, target items were exposed for long periods of time before displays were presented for scanning, and it is possible that such preexposure to the specific visual characteristics of particular symbols encouraged search in terms of visual cues alone.

Kaplan, Yonas, and Shurtliffe (1966) found no effects of either acoustic or visual confusability between target and background sets of letters in visual search. But their subjects searched for a constant set of only two target letters, *e* and *k*. It is likely that they thus rapidly became skilled in the use of critical visual features alone (see Rabbitt, 1967), and did not employ acoustic representations of targets.

Gibson and Yonas (1966) found that speed of visual search among letters was not affected by a concurrent task of auditory distraction involving letters with acoustically similar names. They took this as evidence that subjects did not search by naming the letters that they scanned. It is probable that the distracting information was very easy to ignore because it was presented on a different modality from that used for the main task (Broadbent, 1971).

In contrast to these experiments, a study by Krueger (1970) replicated Neisser's (1963) experiment, using large displays on which an average of 50 items had to be scanned before a target could be located. Subjects searched for large sets of target items, and were slower if target and background letters had acoustically similar names. It seems that subjects are more likely to store and use acoustic representations of target letters when they are relatively unpracticed, and when the target set is sufficiently large as to be difficult to remember. In such conditions background items which, when named, *sound like* target items, result in longer decisions and perhaps in false detections. The effects of such confusions are more apparent when displays are large, so that many items have to be scanned, and many possible successive confusions and delays contribute to overall observed increases in scanning time. These effects may be expected to disappear as subjects become very familiar with the target set through extended practice and have learned to optimize their visual scanning in terms of an optimal subset of critical cues (Rabbitt, 1967; Rabbitt, *et al.*, 1977).

In such tasks we may assume that subjects search by considering one letter at a time. That is to say, the unit of perceptual processing may be a single letter. For most subjects the task of scanning random lists of letters is rather unfamiliar. Letters are most usually encountered as components of words, and a very large literature attests to the fact that words are recognized as perceptual units in their own right, rather than being perceptually constructed from their individual component letters. When subjects scan lists of words, or continuous text, to find individual target letters, we would therefore expect that practice at word recognition might interfere with detection of single-letter targets.

Corcoran (1966, 1967) and Corcoran and Weening (1968) showed that when subjects scan English text, particular target letters (e.g., *p* or *e*) are detected more efficiently in words in which they are voiced than in words in which they are silent (e.g., *psychology, pneumonia*). They concluded that subjects processed text by encoding it into phonetic representations, using the shapes of entire words rather than the shape of individual letters to do so. It is possible that subjects can also scan text to detect individual letters (otherwise, of course, these would *never* be detected *unless* they were voiced, whereas voicing confers, at most, a slight advantage). The phonetic representations of entire words and the visual representations of particular letters may thus provide two separate, but ancillary, kinds of information used to detect targets in visual search.

Subsequent experiments by James and Smith (1970) on comparisons between letter strings showed that comparisons between pairs of words, or between pronounceable nonwords, were faster than comparisons between random letter sequences. While there were systematic list position effects with all these kinds of material, these appeared to interact with factors influencing voicing. For example, RTs for the detection of differences in vowels were shorter than RTs for the detection of differences in consonants, and RTs in the detection of differences for both vowels and consonants were faster if these symbols occurred to the right of a consonant, rather than to the right of a vowel (with a correspondingly higher probability of being voiced).

All these experiments suggest that the detectability of individual letters in the text may be a function of the probability that the words in which they are embedded will be processed as perceptual units. Healy (1976) has taken up this point, replicating Corcoran's (1966, 1967), and Corcoran and Weening's (1968) finding that the probability of detection of the target letter *e* in text is lower when it occurs in the word *the* than in other contexts. Healy has shown that there are clear efforts of target pronounceability, but that these effects also depend on the fact that the word *the*, being both short and very common, is likely to be processed as a single graphemic unit, rather than synthesized from its individual letters.

At this point several additional questions occur. First, we need to know whether or not there is a word superiority effect (W.S.E.) in visual search and, if there is, under what conditions words may conceal or reveal their component letters. We also need to know whether the detection of letters in words is facilitated or inhibited only because words are recognized as perceptual units and then transduced into phonemic or acoustic representations, or whether the presence of redundancy in words or in text may allow more efficient detection of the visual cues by means of which individual letters may be distinguished from each other.

VIII. THE BASES OF WORD SUPERIORITY EFFECTS IN VISUAL SEARCH FOR INDIVIDUAL LETTERS

Gibson, Tenney, Barron, and Zaslow (1972) failed to find any difference in the speed of visual search for individual target letters whether subjects scanned displays of words or nonwords. Most other studies show faster search through word lists, and suggest several distinct explanations for this effect. Since these explanations are not mutually exclusive, it seems better to consider that there are a series of distinct sources of word superiority effects that may operate together or individually depending on the tasks subjects are asked to perform.

Four kinds of explanation seem to have some empirical support. First, as we have seen, there is the suggestion that words may be recognized by transduction of groups of letters into pronounceable units (the *vocalic center group* theory of Spoehr & Smith, 1973, 1975). By this hypothesis, faster search in word lists is due to the fact that words, unlike nonwords, can be processed in pronounceable units. By such an explanation, internal *acoustic* representations, or words, or parts of words, are then compared against acoustic representations of target letters. Thus, silent letters would be detected less easily than voiced letters (Corcoran & Weening, 1968). Second, the redundancy implicit in words, especially words in context, may be supposed to aid the *visual* recognition of component letters in words. On this hypothesis such factors as orthographic regularity or contextual predictability facilitate the recognition of *visual* cues in terms of which letters differ from each other. Third, the fact that words are visually or phonemically redundant may allow subjects to process large groups of letters very rapidly, because they extract cues necessary to identify only some of these letters and guess or infer the rest. According to such a theory, subjects can rapidly *infer* the presence of incompletely processed target letters from context and so make rapid detection responses at some risk of inaccuracy (i.e., redundancy allows a speed–accuracy trade-off).

A final hypothesis is also a statement about the form of redundancy that is useful in word recognition, rather than a specific statement about the way in which this redundancy facilitates perceptual comparisons. This hypothesis is based on the fact that a letter string may have some, or all, of the lower-order properties of a word (e.g., it may be orthographically regular, it may have bigram and higher-order frequencies of letter clusters similar to English, it may be pronounceable, it may even have the same pronounciation as a real word) but nevertheless not be a proper, or real, word in English. Subjects will therefore, in appropriate contexts, attempt to fit partially sampled or degraded perceptual input against their learnt lexicon of real words. To this extent a target letter in a real word will have a higher probability of recognition or correct inference than a target letter in an artificial word. To this extent also we may suppose that textual context at the level of phrases, or even sentences, may facilitate the correct recognition (or inference) of target letters.

There is some evidence for all these hypotheses, and no clear evidence why any of them should be rejected.

Apart from the experiments of Spoehr and Smith (1973, 1975), Krueger (1970) and Krueger and Weiss (1976) found that targets were detected faster in pronounceable than in nonpronounceable nonwords. There is also a large body of evidence on the use of phonemic cues in word recognition, excellently reviewed elsewhere (Krueger, 1975; Barron, 1975).

The assumption that contextual cues facilitate the recognition of purely *visual* (i.e., unnameable) differences in component letters of words seems necessary to explain Snyder's (1970) demonstration that fragmented letters are more easily recognized in tachistoscopic exposures of words than nonwords. It also directly follows from Krueger and Weiss's (1976) finding that proofreaders targets (e.g., mutilated letters) are more easily detected in visual search of words than nonwords. McClelland (1976) mixed upper- and lower-case letters in strings and found that this slowed recognition of individual component letters in both real and artificial words but not in random letter groups. Baron (1973) found that the speed of discriminations between semantically and syntactically valid and invalid phrases were made as quickly when they employed identical sound patterns (e.g., *tie the not* and *tie the knot*) as when they were irregular in both sound pattern and orthography. Henderson (1974) showed that unpronounceable and orthographically irregular sequences of three or four letters (e.g., FBI, CBE, IBM) were more easily recognized than other unfamiliar sequences of the same length. This suggests that contextual effects need not operate *only* through the extraction of information about the sound patterns that words represent. The work of Marshall and Newcombe (1967) and, later, of Shallice (1976), Allport (1977), and Coltheart (1976) also suggests that under certain circumstances, or for subjects with certain kinds of brain damage, the visual pattern of a word may act as an idiogram by means of which a class of words of similar meaning but phonetically different structure may be accessed by the reader. Henderson and Henderson (1975) have elegantly shown that the order in which letters in a string are processed is different for words and non-words.

This last demonstration, and work by Krueger and Weiss (1976), also shows that subjects may operate detection of component letters in familiar strings on the basis of incomplete or possibly inferential cues, thus adopting a speed–accuracy trade-off. That is, subjects will adopt a more stringent criterion for letter detection when processing unfamiliar than when processing familiar letter strings, and will seem faster in the latter case.

The last assumption, that the familiarity of large strings of letters may speed detection, or inference, of component letters (which may include targets in search) seems necessary to explain Krueger and Weiss's (1976) finding that targets are located faster in real words than in orthographically regular, pronounceable nonwords. This is also endorsed by Chambers and Forster's (1975) demonstrations that common words are more rapidly processed than rare words, and by Broadbent and Gregory's (1968, 1971) demonstrations that orthographic effects in word recognition may have different results depending on word frequency. The most direct demonstration in visual search is Krueger and Weiss's (1976) finding that individual letter targets are more rapidly located in common than in rare words.

There is thus excellent supportive evidence for repeated findings that individual target letters can be found faster in words than in nonwords (Krueger, 1970; Krueger and Weiss, 1976; Novik and Katz, 1971). There is also, as yet, no evidence that this advantage can be attributed to only one of the possible sources of facilitation described above.

A more useful series of questions is whether these different kinds of perceptual processing can be undertaken together in parallel, whether the use of one kind of processing prevents the simultaneous use of others, or whether some kinds of processing may be employed to locate the presence of members of a target class of symbols, while other kinds of processing are subsequently undertaken to identify target symbols once they are detected.

Allied to these questions is the possibility that, whether they are undertaken serially or in parallel, some of these processing strategies may take more time than others, so that the latency of target detection may vary with the particular strategy the subject chooses, or is obliged, to use.

These questions require evidence about the times taken to detect individual targets during continuous search of successive items on a display. Recent developments in the technique of rapid serial visual presentation (RSVP) of lists of items offers the best evidence now available (Sperling *et al.*, 1971, & Lawrence, 1971).

IX. RAPID SERIAL VISUAL PRESENTATION OF DISPLAYS FOR VISUAL SEARCH

Visual search tasks of the kind used by Neisser (1963) have the limitation that subjects must make a succession of rapid eye movements to scan a display. Since we can only estimate inspection time per item from total scanning time, estimates of the time taken to process individual symbols must include time taken to initiate, make, and control eye movements. Such techniques also do not allow us to directly determine whether subjects need different times to dismiss background symbols and to recognize targets. If this could be measured, it might provide evidence that would allow us to decide whether scanning and recognition were different types of operations.

Ericksen and Spencer (1969) first used the technique of rapidly displaying successive symbols in such a way that the subject scanned successive displays with a stationary gaze. They found that very wide variations in presentation rate had no effect on efficiency of target detection, so that scanning times per item on a display were much less than the 100 msec per item estimated from Neisser's data.

Sternberg and Scarborough (1969) made a similar experiment, in which they independently varied the size of the set of items for which subjects searched (memory set size, or target set size) and the rate at which successive items were presented. They found that the slope of the function for detection RT against memory set size was about 40 msec per item at all presentation rates; that is, about the same as in Sternberg's (1966, 1967) experiments in which single probe items were presented as displays for judgment. Target detection was possible even when items were presented at rates faster than the subject could exhaustively scan the entire target list (on Sternberg's [1966, 1967] assumptions). There were two ways in which subjects might have done this. They might have continued the input of new items while scanning the preceding items. If this were the case, during the course of a long series of presentations they would have built up a backlog of items awaiting scanning. Errors would thus have become more frequent as sequences of displays became longer. This was not the case. From this evidence it seems likely that a target is judged as being increasingly probable as a run continues and that target *detection* is a different, faster process than target *identification*. The latter process requires serial, exhaustive scanning of a target set in memory.

Fischler (1975) presented sequences of five-letter words at rates of 6, 9, or 12 words per second. Target set sizes varied from one to three words. Fischler asked his subjects to *detect* a target, but to *report* the first word they could *after* it. Unlike Sternberg and Scarborough (1969), Fischler found that the effects of memory set size interacted with the rate of presentation (scanning appeared to be faster and errors more numerous at fast rates). As sequences became longer, RTs for target detections grew significantly faster, again suggesting expectancy effects.

The distance, in number of words, of the reported word from the target word gave the time taken to initiate recognition of one item following detection of another. Subjects sometimes did not wait until the target word appeared, since reported words often preceded targets in a list. This suggested that latencies for target *detection* responses and latencies for posttarget *recognition* responses could be studied independently if different kinds of cues were necessary for the two responses. Fischler (1975, Exp. 2) set subjects to search for a designated target word and report a subsequent word (WW), to locate a designated initial letter of a word and report a subsequent word (LW), and to detect an initial letter and report a subsequent initial letter (LL). The LL and LW conditions gave very similar results, while the WW condition gave faster and more accurate performance than either. This suggested that performance was determined by the latency of the response for target detection, rather than by the time taken to recognize a subsequent item (i.e., letter or word). Somewhat surprisingly, detection responses made to entire words seemed to be more efficient than those made to individual letters.

Previous experiments had shown that distinctions between detection and recognition responses might be necessary. Sasaki (1970) had found that latencies in the target detection of words increased with the size of the target set (as had Sternberg & Scarborough, 1969). However, if target words were all drawn from the same semantically defined set (e.g., items of furniture), variations in the number of target words actually used had little effect on detection RT.

Sperling *et al.* (1971) found that subjects could detect digit targets at very high rates of presentation of successive displays and that detection of targets specified by class was as efficient, after practice, as detection of targets specified as individual items.

The most compelling evidence for dissociation of detection and recognition processes in this paradigm comes from experiments by Lawrence (1971), who set his subjects to detect target words printed in capitals, set among a list of other words printed in the lower case. Nearly all of the subjects' errors were intralist confusions. They usually reported a lowercase word that had occurred *after* the target word in the display sequence. A possible hypothesis is that the detection of characteristics of capitalized letters triggered a recognition response, but that the latency of this response was so long that it could only become effective after one or more further displays had appeared. By this time the target word had been masked, and the subject had to do as best as he could with the information then available to him.

A series of experiments by Frankish (1977) suggest that a different explanation is probably correct. He first set subjects to detect pairs of word targets designated by semantic class (animal names or parts of the body), which occurred one after the other during a sequence of displays. Subjects could sometimes, but not always, detect both of a pair of animal names. The categorization cue was not useful with parts of the body, which did not appear to be a common classification employed by most subjects. This indicated, with Fischler's (1975) data, that the time taken to detect the first of two items was the critical variable of the experiment. Since the task was possible with one categorical cue and not the other, time for detection of the targets, rather than competition between successive responses, was evidently the limiting factor in this experiment.

Frankish next set subjects to search for capitalized words among lower-case background items. At a constant rate of presentation the probability of detection of a single capitalized word was about 60%. When two words followed each other in a list, the probability of detection of both was about the same, at approximately 34%. (The overall probability of detection was therefore only slightly better than for a single word.) If two successive targets were separated by a single "filler" background item, the probability of detection of the first rose to 54% and the probability of detection of the second fell to 17%. In this case, the overall probability of detection of any target, at 71%, was not significantly better than the same probability when targets occurred in adjacent positions.

Frankish has suggested that detection of one target acts as a signal to the subject to stop sampling the display and to further process the information available to him. The efficiency of target recognition would thus depend on how fast subjects could "close the window." If they are slow at this, they admit further items from the display and lose efficiency. In the case of a single target, subjects stop their sampling as fast as possible and then face the task of recognizing the target from among others that had entered before the sampling stopped. When two targets are immediately adjacent, the subject might (rarely) get both, but failing this may get either, equally often, from a sample which includes the second. When targets are separated by a filler item the sampling is usually stopped before the second target appears, so that the first is more usually reported.

As a further test Frankish (1977) introduced pseudotargets made up from Cyrillic or Greek letters, or from characters used in phonetic script. These had some of the characteristics of capitalized targets, but were illegible to his subjects. A first hypothesis was that if a pseudotarget *preceded* a real target, recognition of the real target would be facilitated.

This occurred, but it was not clear that target detection was better than in the case when only a single real target was presented. A second hypothesis was that if a pseudotarget immediately *followed* a real target it should not (unlike a second real target) reduce the probability of target detection. This hypothesis was confirmed.

All these experiments on rapid serial visual presentation (RSVP) strongly suggest that visual search should be regarded as a control process, in which the initiation of some analytic operations may be triggered by the outcomes of others. In particular, a process of target *detection* may initiate a second process of target *identification*. Target detection acts as a signal for the subject to cease sampling new information from a display and to process any information momentarily available to him. Without such a signal the subject will continue sampling the display, and information briefly held in buffer storage will be overwritten by new input. It seems logical to suppose that target detection is more rapid than target identification. This does not mean that the latency of target detection responses is independent of the complexity of symbolic units to be scanned. Fischler (1970) reported that latencies for detection of target words significantly increased with the number of syllables in each word that had to be scanned. It does, however, seem possible that target detection can be carried out using categorical identification of signals as belonging, or not belonging, to *classes* of items (e.g., digits rather than letters—Sperling *et al.*, 1971; Rabbitt, 1962; Brand, 1971; Ingling, 1971; Jonides & Gleitman, 1972, 1976; Gleitman & Jonides, 1976—or even words of one semantic class rather than another—Sasaki, 1970; Lawrence, 1971; Karlin & Bower, 1976; Rabbitt & Fletcher, 1976; Fletcher & Rabbitt, 1977; Frankish, 1977). Target *recognition*, in contrast, at least sometimes seems to require serial self-terminating memory scanning and so varies in speed with the number of items in the target set for which search is conducted (Sternberg & Scarborough, 1969; Fischler, 1976).

It is evident that if such control processes are useful when the rate of exposure of displays is not in the subject's control, they will be all the more necessary when subjects must pace the rate at which they scan displays, directing eye movements and controlling durations of successive fixations.

X. CONTROL PROCESSES IN SELF-PACED VISUAL SEARCH AND SCANNING OF TEXT

When scanning a large, static display, as in RSVP, there is a demand for a "stop rule," which allows subjects to halt their continuous scan at a particular moment and at a particular critical display location. It may be possible for subjects to determine by an initial, quick scan that a particular item is certainly *some* member of a target set, and the stop rule may then have to be exercised so that they can decide which particular member of the target set they have located.

Control processes are necessary not only to guide the locations at which information is sampled, but to determine the *kind* of information that is sought at these locations. An ingenious study by Neisser and Becklen (1976) has shown that when two separate scenes, both involving continuous movements and each presented by a different television-monitor screen, are optically super-imposed, subjects can continuously follow and commit to memory information about one of these episodes with little apparent interference from the other. Information extracted at one moment can therefore be used to control the *kind* of information subsequently sampled, as well as to control the order of interrogation of spatial locations at which it is sampled and the moments at which successive samples are taken.

Two kinds of arguments about such control processes must therefore be separated. The first argument is that subjects can process the visual world in a preliminary way, using the information they extract to control both the duration and the localization of succeeding samples. A different argument, which may, but does not necessarily, involve the assumption of preliminary sampling, is that the kind of information extracted on one sample may control or affect the kind of information extracted on the next.

Evidence for two-stage models of information extraction in visual search will be considered first.

A. Preattentive Processing, the Detection of Novelty, and the Categorization of Background Items in Visual Search

Neisser (1963, 1967) first put forward a two-stage theory of visual search hypothesizing very rapid, preattentive processes that located targets that could be subsequently subjected to more elaborate and analytical attentive processing. Preattentive processing was defined in terms of two distinct characteristics. First, while attentive processing might be a slower process, possibly involving seriatim comparisons against representations of potential targets held in memory, preattentive processing allowed simultaneous parallel tests to be made for a very wide range of possibly informative targets. This might explain the discrepant effects of target set size in the Neisser (1963) and Sternberg (1966, 1967, 1969, 1975) paradigms. In Neisser's tasks, a target might be *located* by parallel, preattentive processing, and once located might be *identified* as a specific member of the target set by serial memory search. This *identification* process might take longer for large than for small target sets, but such variations in time would be minor in relation to the recorded times for scanning entire large displays. As we have seen, Sternberg's (1966) and Nickerson's (1966) and Burrows and Murdoch's (1969) experiments found evidence for serial scanning of displays as well as serial scanning of target sets, offering no support for this conceptualization. Nevertheless, it may now be argued that under conditions of very extended practice (Neisser, 1963; Rabbitt *et al.*, 1979b), or familiarity with particular restricted "categorizable" target sets such as digits and letters (Jonides & Gleitman, 1976), this may be what subjects do.

A second characteristic attributed to preattentive processing was very rapid detection of *novel* stimuli in the visual field. Note that a subject cannot detect a novel background item unless he has built up some internal representation of the class of familiar background items.

This second point was taken up in an experiment by Neisser and Lazar (1964). Subjects were given 22 days' practice at searching among background letters for the following targets: any digit, any unfamiliar symbol (i.e., not a letter), the digit 3, and a particular symbol not a letter. Searches for any digit were faster than searches for any unfamiliar symbol, and searches for specified digits or symbols were faster still. There was therefore no evidence that novelty, per se, critically determined the speed of visual search, though there was clear evidence that subjects could very efficiently locate unspecified targets in terms of their differences from *any* of a familiar set of background items.

Prinz (1977) has reviewed evidence from earlier experiments by Prinz and Ataian (1973) and Prinz, Tweer, and Feige (1974), arguing that since subjects process displays in terms of large numbers of successive fixations, they must be able to decide that they have *not* located a target in a particular fixation in order to initiate the next saccade and sample. He and his associates have shown that subjects trained on visual search for particular target letters among a particular background set halt their search when occasional new background letters occur on displays. Prinz argues that detection of discrepancies between these unfamiliar background items and the practiced subset of letters invoke the stop-rule, so that a sample in which they are detected is prolonged until the subject has reassured himself that these are not the targets he seeks.

Note that this is an entirely different description of a stop-rule from that used by Frankish (1977). Frankish's (1977) data suggest that the stop rule is invoked when any display that has characteristics similar to a target item (e.g., a pseudotarget) is located. Frankish's (1977) conceptualization thus relates to our taxonomy of *type* categorization above. That is, items that do not possess target characteristics are passed, by exclusion, as members of the nontarget set. Prinz (1977) argues that the stop rule is invoked by the detection that an item is *not* a member of the background set. Thus he might speak for our type categorization on the principle that items are categorized as being members of the background set, thus invoking the decision to continue a scan, or are defined by exclusion as "non-background-set members," thus invoking the stop rule and further processing. If this were the case, the speed of visual scanning would vary with the size of the vocabulary of background items, rather than of target items. The reverse is known to be the case (Rabbitt, 1962).

Alternatively, Prinz's (1977) argument can be taken as a statement that subjects process each item to determine whether it belongs to one of two classes, target or background (the process of type categorization, in our taxonomy). In this case, search time would be equally dependent on the sizes of target and background vocabularies—which is also known not to be the case (Rabbitt, 1962).

Prinz (1977) has not yet shown that subjects do not use Frankish's (1977) stop rule in continuous scanning. Subjects may take samples of regular size and duration, initiating new samples after the lapse of a determinate time period, unless features appropriate to a target set member have been located. In such a case, subjects could also exercise a speed–accuracy trade off, reducing sampling duration, and so accuracy, to increase scanning speed, as suggested by Wattenbarger (1969). The detection of new background items on displays can be explained by Rabbitt's (1967) demonstration that subjects practiced at visual search learn to detect and use cues critical for discriminations between a particular target set and a particular background set of items. It is not clear from Prinz's (1977) experiments that subjects do not detect new background items merely because they may share cues that have been established as critical for the target set alone. Both Neisser (1963) and Rabbitt (1967) found that when cues for distinctions between target and background items were very obvious (straight-line as against curved letters) changes of entire background sets were undetected so long as these critical differentiating characteristics were preserved.

There are, however, many demonstrations that efficiency of search may be controlled by the commonality of features *within* a background set of items. Gordon (1968, 1971) showed that visual search was faster if background items were homogeneous, rather than disparate, so that the effects of increases in the vocabulary of background items from two to four words in a visual search task are not directly predictable from the effects of these items when they occur as the *only* background items employed.

Tachistoscopic recognition tasks by Donderi and Zelnicker (1969) and Donderi and Case (1970) found that detection probability and recognition times were invariant with the number of items on a display (up to 12) provided that all nontarget items were identical. Connor (1971) reported that RT for target detection may be invariant if all background items are drawn from the same class (e.g., letters or digits) though they are not physically identical. Similarly, Frith (1974) and Reicher, Snyder, and Richards (1976) have reported that search for an upright target letter (e.g., a capital *T*) embedded among vertically reversed background letters is slower than search for a reversed target among upright background letters. Reicher *et al.* (1976) commented that their results could be interpreted in terms of a mechanism sensitive to "unusual or informative" display characteristics [p. 530], as Neisser's (1963, 1967) description of preattentive processes might suggest.

It seems to me that the evidence is still incomplete, and that a critical condition would be visual search for upright targets among mixed upright and reversed background letters. If subjects are indeed delayed by inspection of novel background items, then search should be faster when few rather than all background items are novel. If search is slower in the mixed condition, this would be another demonstration that commonality among background items aids search (Gordon, 1968, 1971; Donderi & Zelnicker, 1969; Donderi & Case, 1970).

A point to be considered is that subjects can scan displays faster if they repeat identical or similar acts of information processing. Thus the kind of information taken in during a particular sample of a display may be controlled by the kind of information processed in the immediately previous sample.

B. Control of Information Processing by Sequences of Events

A large literature shows that particular acts of perceptual analysis are facilitated by immediate repetition. Rabbitt *et al.* (1977a) tested this in visual search by presenting displays of five to nine letters on a computer-controlled CRT. Each display might or might not contain one of a pair of designated targets. When successive displays both contained targets these might or might not appear in the same display location. Individual targets might or might not be repeated.

At R–S intervals of 150 msec (too short for a saccadic eye movement) target detection RTs were very fast when identical targets recurred in identical display positions. Variations in the vocabulary of background letters did not affect detection RT. An obvious assumption is that background letters were never scanned because subjects detected a physical identity match between successive targets. When successive targets were different, variations in the vocabulary of background items significantly affected detection RT. If R–S intervals were as long as 1500 msec this privileged processing for identical targets in the same location was lost.

If identical targets recurred in different spatial locations on successive displays there was evidence that background items were scanned, since RT increased significantly with the disparity of target location on successive displays. Nevertheless, detection RTs were faster than if different targets occurred on successive display locations, with equivalent degrees of locational displacement.

Rabbitt *et al.* (1977) have suggested that target detection in visual search can be achieved by two processes. When R–S intervals are short, subjects can retain a memory representation of the physical characteristics of the last target detected, achieving a physical identity match (Posner & Mitchell, 1967) with a sequent target inspected in the same location. If the target moves, so that the first symbols scanned are background items, a physical identity match is no longer possible; and a more analytic, and slower, process is required to detect a target. This process also may be facilitated by target repetition.

Rabbitt *et al.* (1977c) tested the effects of variations in the nature and locations of background items on successive displays at short R–S intervals (150 msec). Search was fast if successive displays contained the same background items in similar display locations, even when the target shifted position. Search was also faster when successive displays contained the same background items in new locations than when they contained different sets of background items.

The experiments discussed in the immediately preceding paragraphs illustrate two points that have recurred throughout these discussions. First, target detection appears to be possible by at least two different kinds of processing. One of these (matching) seems to be faster than the other (analytic comparison). It remains to be determined whether these processes are successive in time so that analytic processing only occurs contingent on failure of initial matching, or whether they occur simultaneously (although they are functionally different and may use different display attributes). Second, the nature of information processing is modulated by moment-to-moment events, some under the control of subjects and others not (Posner and Boies, 1971). The question of the way in which attention to one display is modulated or primed by exposure to a previous display is a lively topic of research in experiments on binary classification (Posner & Klein, 1973; Posner & Snyder, 1975) and experiments on word recognition (Meyer *et al.*, 1975; Meyer & Schvaneveldt,

1971, 1975; Barron, 1973, 1975). Its role as a control process in visual search demands further investigation.

From the question as to whether, and how, perceptual processing of one display may affect processing of the next, we may turn to consider whether subjects may simultaneously make identifications of some symbols on a display while taking in different kinds of information from other parts of the same display to guide their further search. This question is most directly encountered in experimental studies of reading.

C. Control Processes in Reading and Visual Search

Early investigations such as that by Morton (1964) revealed adaptive control principles in operation, since subjects flexibly varied the number of samples which they took from a line of text with the information load of material read. Distant approximations to English required many more samples than first-order approximations, or regular text. Contextual redundancy thus controlled the amount of evidence that subjects extracted from a display in order to identify the words on it.

A series of studies by McConkie and Rayner (1973) and Rayner and McConkie (1973) introduced a very ingenious technique. Continuous text was exposed on a computer CRT. Information about the subject's momentary fixation point on this display was recorded and used by a controlling computer to arrange that a "window" of regular text should be available at the point of fixation, while text outside this window was systematically degraded so as to offer no useful cues. The size of the window could then be altered, and the kinds of peripheral cues used in reading could be determined by using different kinds of degradation with different window sizes.

Rayner and his associates found that good readers pick up information of *any* kind from not more than a total area of 17–19 letter spaces. Within this range, word length affects reading more than does word shape or information about the nature of specific letters within words. The question was therefore partially answered: Cues extracted from the periphery of the usable visual sample were *different* in nature from cues extracted and used to recognize letters at the fixation point itself.

In a later study, Rayner (1975) found no evidence that subjects could recognize that a letter string was not a word when these subjects were fixating only four to six character spaces to the left of it. They could nevertheless use information about string length and about the initial and terminal letters in an identified string. They thus appeared to use physical features of a letter string as a preliminary to the extraction of semantic information about it. This suggests that the extraction of attributes of letter strings proceeds in an orderly fashion. The extraction of *some* physical attributes seems to antecede the extraction of semantic information, and this information is useful to control the extent of subsequent eye movement used to obtain the appropriate sample.

Fisher (1975) followed Smith (1969) and Smith, Lott, and Cromwell (1969) in using systematically degraded text to directly compare the use of peripheral cues in reading and in visual search. Text was printed with or without spaces between words (the spaces being filled with dummy characters or left vacant). Text was also printed either in the same case (upper or lower) or in mixed case (upper and lower in random alternation). Reading with normal (unfilled) spacing was faster and more accurate than reading with filled or absent spacing. Capitalized text was read more slowly than normal text, and text with alternating cases more slowly still. These effects did not interact, again perhaps suggesting that information about word length is extracted and employed independently of information about word shape or letter shape.

Comparing speed of visual search and reading on displays of these various kinds, Fischer found that the units in which segments of text were processed were larger for visual search. This suggests, in turn, that during visual search subjects do not pause to extract all possible information about the semantic properties of individual words, as they are obliged to do when reading. Within this restriction both visual search and reading were very rapid when word-boundary cues were available, and much slower when they were not. The interesting point was that without word-boundary cues both reading and visual search appeared to be conducted on a letter-by-letter rather than on a word-by-word basis.

The larger question is how contextual information, including information about the meaning of words which have been read, interacts with the extraction of information from words which are about to be read. In other words, can we speak of the mechanisms by which semantic context operates (as it very evidently does)? Rayner and Osgood (1972) have also raised this question by showing that tachistoscopic recognition of semantically ambiguous words, such as *conduct*, is facilitated by priming with short phrases giving context for one or the other possible meaning. Such priming facilitates both recognition and the determination of which of two form classes is used as a response.

In summary, the evidence leaves no doubt that when subjects identify words on crowded displays, whether their task is defined as visual search or reading, they use peripheral cues to determine the extent of each successive saccade and so the size of the sample of text that they process during a single fixation. There is also evidence from Morton (1964) that control is also exercised on the basis of an advance assessment of the mean redundancy of the material to be sampled. It seems likely that people can simultaneously extract the information which they need to identify words they currently fixate and different *kinds* of information that guides their choice of the point of their next fixation. The point of emphasis is that the cues used to extract information necessary to control fixations are different from the cues used to recognize individual symbols or words. Nevertheless, both kinds of cues are apparently simultaneously processed to make different kinds of decisions in the same task.

XI. CONCLUSIONS

Early discussions of signal categorization and of visual search centered around discussions as to whether perceptual processing could best be regarded as a serial or a parallel process. In the event, both characteristics were claimed for perceptual processing in different tasks and at different stages of practice in each task (e.g., Neisser, 1963, 1967; Sternberg, 1966, 1967, 1969; Kaplan *et al.*, 1966; Nickerson, 1966; Burrows & Murdock, 1969; Egeth, 1967). There have been very illuminating discussions of the extent to which data from particular experiments can be expected to provide evidence for the operation of systems of one of these classes, or of the many subclasses (e.g., self-terminating systems, exhaustive systems, stop-rule systems, distributed processing systems, parallel interactive systems, parallel independent systems, distributed termination systems, and other systems [Egeth, 1967; Hawkins, 1969; and particularly Townsend, 1971]).

The present discussion has not considered these distinctions. It will be argued that while such distinctions are important, and while the distinction between serial and parallel systems is still helpful, certain problems of definition must be resolved before they are applied to particular sets of data.

A. The Problem of Perceptual Analysis

The statement that a particular system processes input in parallel is meaningless in the abstract. It is essentially a statement that if we, as experimenters, classify perceptual information into particular discrete units or quanta of evidence, then this system will take no longer and be no less accurate as the number of such discrete units is increased from one to N. We usually do not know whether this is true for *all* values of N. We usually assume that for any given system it is not, and that at some limiting value of *N* processing time will increase or accuracy will be reduced.

The first problem is to define these units, but it is also important to recognize that, when we have done so, we have only made a statement relevant to *this* particular classification of input. For example, an implicit assumption in Neisser's (1963, 1967) and Sternberg's (1966, 1969) experiments is that units can be defined as nameable symbols such as letters of the alphabet or digits. The statement that a system processes letters in parallel or in series does not either allow any conclusions about processing of subunits (features, cues, etc.) comprising each letter, or any necessary conclusion about processing of higher-order units that may be composed of letters (e.g., words). In either type of system, such different units may be processed in series or in parallel, for all this statement implies.

A comparative statement between a serial and a parallel system implies only that with respect to the system of classification *in which such units are defined as quanta*, the processing efficiency of one system is affected by the *N* of such quanta, whereas the efficiency of the other (up to some unknown limit) is not. To this extent the question of whether either kind of system employs categorization is trivial. In making a distinction between types of systems we have already assumed a particular categorization of input, defined by our assumptions as to the nature of the perceptual quanta we suppose to be common to both.

In this respect it is convenient to speak of systems as being either serial or parallel only with respect to a particular level of processing, or kind of perceptual categorization. This level of processing is defined in empirical terms by a demonstration that for particular discriminations between inputs the arbitrary units have *functional significance*.

It is not useful to say of a particular system that, early in practice, it operates in a serial mode and later in practice in a parallel mode. Both these statements are necessarily relevant to particular implied arbitrary units. (See Kaplan & Carvallas, 1965). The most that can be said is that at one stage in practice a system appears to be sensitive to variations in the *N* of a *particular* perceptual unit. This is only a statement that the unit appears to be functionally significant to the operation of the system at that point in time (the system, like the experimenter, appears to classify perceptual input in terms of specific units at this point in practice). If, at some other point in practice, variations in the unit have no effect, we do not know (from this evidence alone) whether the system has learned to process many units simultaneously rather than one at a time, or whether it is now operating in terms of some classification system other than the particular unit system that it once used.

As a simple example, if a man is set to detect the occurrence of any of the letters *o, c, q,* or *g* among other letters, he may first use internal representations of these symbols as the perceptual units in which he classifies probes or letters on a display as target or nontarget items. If he discovers that only straight-line letters will occur as nontargets he may then classify letters in terms of their being straight-line or curved. His performance in the first case may well suggest serial processing, and in the latter case, parallel processing. However, it is the functional unit of perceptual classification that has changed, rather than the characteristics or limitations of the perceptual processing system he employs.

Thus, distinctions between serial and parallel processing must entail distinctions between the types of classification a subject is required to make, or can make, when categorizing input. The preliminary taxonomy for classification systems devised in the introduction to this chapter allows such questions to be framed. It is important to recognize that this taxonomy is relevant to distinctions between serial and parallel processing systems in that it defines the unit of comparison in terms of which such distinctions must be discussed.

B. Single and Multiple Categorizations, and Variations in Unit of Analysis

A large literature attests to the necessary distinctions between processes by which relevant items can be located on displays and other processes by means of which items that have been located may be identified.

For present purposes, the simplest example is from the study by Green and Anderson (1956), who showed that subjects can apparently locate relevant target symbols by color and, having done so, inspect only these located symbols for other attributes, such as shape. This is a demonstration that two different functional units of analysis are employed, perhaps successively, to process a display. Thus we may speak of these successive categorizations of the display as being either serial or parallel with respect to each other. Each distinct categorization may itself be either serial or parallel with respect to whatever functional units it idiosyncratically uses (e.g., color in one case and discriminative shape features in the other). There is evidence that subjects learn to optimize in their selection of shape cues when discriminating between symbols such as letters (Neisser, 1963; Rabbitt, 1967; Rabbitt, Cumming, & Vyas, 1979b). Thus there may be independent effects of practice on either categorical transaction.

C. Serial and Parallel Categorical Transactions

The necessity for distinctions between serial and parallel processes is further emphasized when we consider the possibility of processing in terms of more than one categorization scheme, and so in terms of more than one distinct kind of unit of perceptual analysis. We may then ask whether subjects can simultaneously categorize input in terms of more than one such scheme, or whether they are restricted to making only one such categorization at a time. We may also ask whether they can simplify categorization tasks in the latter case by making an initial categorization in terms of one type of unit of analysis (e.g., color) so as to reduce the complexity of subsequent decisions on other bases (e.g., decisions on shape).

Garner (1972) has done much to clear up questions of this kind by demonstrating that particular pairs of stimulus attributes may be independently processed, while other pairs must be jointly, or integrally, processed in categorization tasks.

A large body of literature has shown that word recognition does not necessarily employ the same subsets of cues as are used to distinguish the component letters of words from each other. Both of these kinds of cue patterns will, or course, be precisely correlated in Garner's (1972) terms, since we *may* always recognize a word by recognizing each individual letter within it, and we *may* choose to recognize component letters by phonetic or other derivations from the names of recognized words. We may then ask, to the extent that different cues are processed to recognize letters and words, whether such cue systems are *integral* or *independent*, in Garner's terms; that is, whether they need not or must be processed together.

If they need not be processed together, we may then ask whether nevertheless they *can be* processed together, so that a subject can simultaneously extract information about the nature of a word and information about the individual letters within it. If subjects can do this then they will achieve, simultaneously, more than one classification of a pattern of symbols on a display. To the extent that these classification systems are related, classifications of words on a display may provide information that either helps or hinders subjects in their search for individual letters. We have seen that this appears to happen in practice (e.g., Krueger, 1970, 1975; Corcoran, 1966).

A general moral to this review, therefore, is that subjects may have an arbitrarily high number of different categorization systems that they *can* employ in a task, although the experimenter may wish to study only one.

In visual search, especially, a subject can use one categorization system to locate relevant parts of the display and another in order to inspect them more closely. This is especially relevant in reading, where a control process based on use of one kind of cue may guide other processes, based on other cues, that allow word recognition and the extraction of meaning.

It also seems possible that the subject can simultaneously use more than one system to classify symbols which he has localized. Insofar as these classification systems provide correlated information, identification of particular items can be achieved by more than one simultaneously. Insofar as they provide orthogonal kinds of information, the output of one classification system may interfere with the output of others. Improvement with practice may therefore involve not merely optimization of use of any one particular classification, based on particular perceptual evidence. It may involve either the selection of one optimal classification system with the consequent disuse of others or the employment of one or several joint-classification systems, where this is most convenient for the distinctions to be obtained. More interestingly, it may involve the development of some categorization systems for the specific purpose of control, guiding search, or reading of continuous text. The fact that further investigation of these mechanisms promises to be extremely exciting and interesting increases our debt to the investigators whose work has been reviewed above and who have so patiently and lucidly developed methodologies and terminologies that now allow us to begin to analyze these problems.

References

Allport, D. A. On knowing the meaning of words we are unable to report. In S. Dornic (Ed.), *Attention and Performance VI*. Erlbaum Associates, Potomac, Maryland. 1977. Pp. 505–533.

Anderson, N. S., and Leonard, J. A. The recognition, naming and reconstruction of visual figures as a function of contour redundancy. *Journal of Experimental Psychology*, 1958, **56,** 262–270.

Baron, R. Perceptual dependence: evidence for an internal threshold. *Perception & Psychophysics*, 1973, **13,** 527–33.

Barron, J. Successive stages in word recognition. In: S. Dornic & P. M. A. Rabbitt (Eds.), *Attention and performance V*. New York: Academic Press. 1975.

Brand, J. Classification without identification in visual search. *Quarterly Journal of Experimental Psychology*, 1971, **23,** 178–186.

Broadbent, D. E. *Decision and stress*. Academic Press: London and New York. 1971.

Broadbent, D. E., & Gregory, M. Visual perception of words differing in letter digram frequency. *Journal of Verbal Learning and Verbal Behaviour*, 1968, **7,** 569–571.

Broadbent, D. E., & Gregory, M. Effects on tachistoscopic perception from independent variation of word probability and letter probability. *Acta Psychologica*, 1971, **35,** 1–14.

Brown, J. Evidence for a selective process during perception of tachistoscopically presented stimuli. *Journal of Experimental Psychology*, 1960, **59**, 176–81.

Burrows, D., & Murdock, B. B., Jr. Effects of extended practice on high speed scanning. *Journal of Experimental Psychology*, 1969, **82**, 231–237.

Butler, B. Selective attention and target search with brief visual displays. *Quarterly Journal of Experimental Psychology*, 1975, **27**, 467–477.

Chambers, S. M., and Forsters, K. L. Evidence for lexical access in a simultaneous matching task. *Memory & Cognition*, 1975, **3**, 549–559.

Chase, W. H. Parameters of visual and memory search. Unpublished Ph.D. dissertation. University of Wisconsin. 1969.

Chase, W. G., & Posner, M. I. The effect of auditory and visual confusability on visual and memory search tasks. Paper presented at the Annual Meeting of the Midwestern Psychological Association, Chicago, April 1965.

Coltheart, M., Davelaar, E., Jonarren, J. T., & Besnev, D. Access to the internal lexicon. In S. Dornic (Ed.), *Attention and Performance VI.* Potomac, Maryland: Erlbaum, 1976. Pp. 535–556.

Connor, J. M. Factors affecting parallel processing of visual displays. Unpublished doctoral dissertation, Univ. of Wisconsin. 1971.

Conrad, R. Acoustic confusions immediate memory. *British Journal of Psychology*, 1964, **55**, 75–83.

Conrad, R. Interference or decay over short retention intervals? *Journal of Verbal Learning and Verbal Behavior*, 1967, **6**, 49–54.

Corcoran, D. W. J. An acoustic factor in letter cancellation. *Nature*. London: Macmillan, 1966, **210**, 658.

Corcoran, D. W. J. Acoustic factors in proof reading. *Nature*, 1967, **214**, 851–852.

Corcoran, D. W. J., & Rouse, R. O. An aspect of perceptual organisation involved in reading typed and hand-written words. *Quarterly Journal of Experimental Psychology*, 1970, **22**, 526–530.

Corcoran, D. W. J., & Weening, D. L. Acoustic factors in visual search. *Quarterly Journal of Experimental Psychology*, 1968, **20**, 83–85.

Cutting, J. E., & Rosner, B. S. Categories and boundaries in speech and music. *Perception & Psychophysics*, 1976, **16**, 564–570.

Cutting, J. E., Rosner, B. S., & Foard, C. F. Perceptual categories for music-like sounds: implications for theories of speech perception. *Quarterly Journal of Experimental Psychology*, 1976, **28**, 361–379.

Dainoff, J. M., & Haber, R. N. Quoted as personal communication by Posner, M. I. (1970), p. 282. Similar communication to present author.

De Rosa, D. V., & Morrison, R. E. Recognition reaction time for digits in consecutive and non consecutive memorised sets. *Journal of Experimental Psychology*, 1970, **83**, 472–479.

Dick, A. O. Relations between the sensory register and short term storage in tachistoscopic recognition. *Journal of Experimental Psychology*, 1969, **82**, 279–284.

Dick, A. O. Processing time for naming and categorization of letters and numbers. *Perception & Psychophysics*, 1971, **9**, 350–352.

Donderi, D., & Case, B. Parallel visual processing: constant same-different latency with two to fourteen shapes. *Perception & Psychophysics*, 1970, **8**, 373–375.

Donderi, D., & Zelnicker, D. Parallel processing in visual same-different categorizations. *Perception and Psychophysics*, 1969, **5**, 197–200.

Donders, F. I. Proceedings of the Royal Dutch Academy of Sciences, Department of Natural Sciences. 24 June 1865. (trans. W. Koster) *Acta Psychologica*, 1865, **30**, 409–411.

Egeth, H. E. Parallel versus serial processes in multidimensional stimulus discrimination. *Perception & Psychophysics*, 1966, **1,** 245–252.

Egeth, H. E., Atkinson, J., Gilmore, G., & Marcus, N. Factors affecting processing rate in visual search. *Perception & Psychophysics*, 1973, **13,** 394–402.

Egeth, H. E., Jonides, T., & Wall, S. Parallel processing of multi-element displays. *Cognitive Psychology*, 1972, **3,** 674–698.

Egeth, H. E., Marcus, N., & Bevan, W. Target-set and response-set interaction: implications for the study of human information processing. *Science*, 1972, **176,** 1447–1448.

Ericksen, C. W., & Spencer, T. Rate of information processing in visual perception: some results and methodological considerations. *Journal of Experimental Psychology Monographs*, 1969, **79** (No. 2) part 2.

Felfoldy, G. L., & Garner, W. R. The effects on speeded classification of implicit and explicit instructions regarding stimulus dimensions. *Perception & Psychophysics*, 1971, **9,** 289–292.

Fischler, I. Detection and identification of words and letters in simulated visual search of word lists. *Memory & Cognition*, 1975, **3,** 175–182.

Fisher, D. F. Reading and visual search. *Memory & Cognition*, 1975, **3,** 188–196.

Fitts, P., & Switzer, G. Cognitive aspects of information processing I. The familiarity of S-R sets and sub-sets. *Journal of Experimental Psychology*, 1962, **63,** 321–329.

Fitts, P. M., Weinstein, M., Rappaport, M., Anderson, N., & Leonard, J. A. Stimulus correlates of pattern recognition. *Journal of Experimental Psychology*, 1965, **51,** 1–11.

Fletcher, C. E., & Rabbitt, P. M. A. Categorisation and visual search. Paper presented at meeting of Experimental Psychology Society, Durham, July 1976.

Fletcher & Rabbitt (1977)

Fletcher, C. E. Manuscript submitted to *Quarterly Journal of Experimental Psychology*, 1978.

Fraisse, P. Why is naming longer than reading? In W. Koster (Ed.), *Attention and performance II.* North-Holland Publ., Amsterdam, 1969.

Frankish, C. Paper submitted for publication. February 1977.

Frith, U. A curious effect with reversed letters explained by a theory of schema. *Perception and Psychophysics*, 1974, **16,** 113–116.

Garner, W. R. The processing of information and structure. Erlbaum, Potomac, Maryland. 1974.

Gibson, E. J., Tenney, Y. T., Barron, R. W., & Zaslow, M. The effect of orthographic structure on letter search. *Perception & Psychophysics,* 1972, **11,** 183–186.

Gibson, E. J., & Yonas, A. A developmental study of search behavior. *Perception & Psychophysics,* 1966, **1,** 169–171.

Gleitman, H., & Jonides, J. The cost of categorisation in visual search: incomplete processing of targets and field items. *Perception & Psychophysics,* 1976, **20,** 281–288.

Gordon, I. Interactions between items in visual search. *Journal of Experimental Psychology,* 1968, **76,** 348–355.

Gordon, I. E. Irrelevant item variety and visual search. *Journal of Experimental Psychology,* 1971, **88,** 295–296.

Green, B. F., & Anderson, L. K. Size coding in a visual search task. *M.I.T. Internal Research Reports,* **38,** 1955.

Green, B. F., & Anderson, L. K. Colour coding in a visual search task. *Journal of Experimental Psychology,* 1956, **51,** 19–24.

Green, D. M., & Swets, J. A. Signal detection theory and psychophysics. New York: Wiley, 1966.

Hawkins, H. L. Parallel processing in complex visual discrimination. *Perception & Psychophysics,* 1969, **5,** 56–64.

Healy, A. Detection errors on the word *the:* evidence for reading units larger than letters. *Journal of Experimental Psychology: Human Perception and Performance,* 1976, **2,** 235–242.

Henderson, L. A word superiority effect without orthographic assistance. *Quarterly Journal of Experimental Psychology,* 1974, **26,** 301–311.

Henderson, L. Semantic effects in visual reward through word lists for physically defined targets. Paper presented at the British Psychological Society Annual Meeting, York, 1976.

Henderson, L., & Henderson, S. Visual comparison of words and random letter strings: effects of number and position of letters different. *Memory & Cognition,* 1975, **3,** 97–101.

Ingling, N. W. Categorisation: A mechanism for rapid information processing. *Journal of Experimental Psychology,* 1971, **94,** 239–243.

James, C. T., & Smith, D. E. Sequential dependencies in letter search. *Journal of Experimental Psychology,* 1970, **85,** 56–60.

Jonides, J., & Gleitman, H. A conceptual category effect in visual search: O as letter or as digit. *Perception & Psychophysics,* 1972, **12,** 457–460.

Jonides, J., & Gleitman, H. The benefit of categorization in visual search: Target location without identification. *Perception & Psychophysics,* 1976, **20,** 289–298.

Kaplan, I. T., Carvellas, T., & Metlay, W. Visual search and immediate memory. *Journal of Experimental Psychology,* 1966, **71,** 488–493.

Kaplan, I. T., & Carvellas, T. Scanning for multiple targets. *Perceptual and Motor Skills,* 1965, **21,** 239–243.

Kaplan, G. A., Yonas, A., & Shurcliffe, A. Visual and acoustic confusability in a visual search task. *Perception & Psychophysics,* 1966, **1,** 172–174.

Karlin, M. B., & G. H. Bower. Semantic category effects in visual word search. *Perception & Psychophysics,* 1976, **19,** 417–424.

Kristoffersen, M. Categorisation times in choice RT experiments. Paper to Friday seminar. Department of Psychology, University of Oxford, 1976.

Krueger, L. E. The effect of acoustic confusability on visual search. *American Journal of Psychology,* 1970, **83,** 389–400.

Krueger, L. E. Familiarity effects in visual information processing. *Psychological Bulletin,* 1975, **82,** 949–974.

Krueger, L. E., & Weiss, M. E. Letter search through words and nonwords: the effect of fixed, absent or mutilated targets. *Memory & Cognition,* 1976, **4,** 200–206.

Lawrence, D. H. Two studies of visual search for word targets with controlled rates of presentation. *Perception & Psychophysics,* 1971, **10,** 85–89.

Liberman, A. M., Harris, K. S., Hoffman, H. S., & Griffith, B. C. The discrimination of speech sounds within and across phoneme boundaries. *Journal of Experimental Psychology,* 1957, **54,** 358–368.

Lively, B. L., & Sanford, B. J. The use of category information in a memory-search task. *Journal of Experimental Psychology,* 1972, **93,** 379–385.

McClelland, J. L. Preliminary letter identification in the perception of words and non-words. *Journal of Experimental Psychology: Human Perception and Performance,* 1976, **2,** 80–91.

McConkie, G. W., & Rayner, K. Identifying the span of the effective stimulus in reading. Final report OEG 2–71–0531 submitted to the Office of Education, July 1974.

Marshall, J. C., & Newcombe, F. Immediate recall of sentences' by patients with unilateral cerebral lesions. *Neuropsychologia,* 1967, **5,** 329–334.

Meyer, D. E., & Schvaneveldt, R. W. Facilitation in recognizing pairs of words: Evidence of a dependence between retrieval operations. *Journal of Experimental Psychology,* 1971, **90,** 227–234.

Meyer, D. E., Schvaneveldt, R. W., & Ruddy, M. G. Loci of contextual effects on word recognition. In P.M.A. Rabbitt & S. Dornic (Eds.), *Attention and performance V.* Academic Press, London, 1975.

Morris, R. E., De Rosa, D. V., & Stultz, V. Recognition memory and reaction time. *Acta Psychologica,* 1967, **27,** 298–305.

Morton, J. The effects of context upon speed of reading, eye movements and eye-voice span. *Quarterly Journal of Experimental Psychology,* 1964, **16,** 340–354.

Naus, M. J., Glucksberg, S., & Ornstein, P. A. Taxonomic word categories and memory search. *Cognitive Psychology,* 1972, **3,** 643–654.

Naus, M. J., & Shillman, R. J. "Why a Y is not a V: A new look at the distinctive features of letters. *Journal of Experimental Psychology: Human Perception and Performance,* 1976 **2,** 396–400.

Neisser, U. Decision time without reaction time: Experiments in visual scanning. *American Journal of Psychology*, 1963, **76,** 376–385.

Neisser, U. *Cognitive psychology.* Appleton, New York, 1967.

Neisser, U. Selective reading: A method for the study of visual attention. Paper presented to the 19th International Congress of Psychology, London, 1969.

Neisser, U., & Becklen, R. Attending to visually specified events. *Cognitive Psychology*, 1975, **7,** 480–494.

Neisser, U., & Beller, H. Searching through word lists. *British Journal of Psychology*, 1965, **56,** 349–358.

Neisser, U., & Lazar, R. Searching for novel targets. *Perceptual and Motor Skills,* 1964, **19,** 427–432.

Neisser, U., Novik, R., & Lazar, R. Searching for ten targets simultaneously. *Perceptual and Motor Skills,* 1963, **17,** 955–961.

Nickerson, R. S. Response times with a memory dependent decision task. *Journal of Experimental Psychology,* 1966, **72,** 761–769.

Nickerson, R. S. Can characters be classified directly as digits vs. letters or must they be identified first? *Memory & Cognition,* 1973, **1,** 477–484.

Novik, N., & Katz, L. High speed visual scanning of words and nonwords. *Journal of Experimental Psychology,* 1971, **91,** 350–353.

Pew, R. The speed-accuracy operating characteristic. *Acta Psychologica,* 1969, **30,** 16–26.

Pollack, I. Speed of classification of words into superordinate categories. *Journal of Verbal Learning and Verbal Behavior,* 1963, **2,** 159–165.

Posner, M. I. On the relationship between letter names and super-ordinate categories. *Quarterly Journal of Experimental Psychology,* 1970, **22,** 279–287.

Posner, M. I., & Boies, S. Components of Attention. *Psychological Review,* 1971, **78,** 391–408.

Posner, M. I., & Klein, R. M. On the functions of consciousness. In S. Kornblum (Ed.), *Attention and Performance IV* London and New York: Academic Press, 1973.

Posner, M. I., & Mitchell, R. A chronometric analysis of classification. *Psychological Review,* 1967, **74,** 394–409.

Posner, M. I., & Snyder, C. R. R. Facilitation and inhibition in the processing of signals. In P. M. A. Rabbitt & S. Dornic (Eds.), *Attention and Performance VI.* London: Academic Press, 1975.

Prinz, W. Memory control of visual search. In S. Dornic (Ed.), *Attention and Performance VI.* Potomac, Maryland: Erlbaum Associates, 1977.

Prinz, W., & Ataian, D. Two components and two stages in search performance: A case study in visual search. *Acta Psychologica,* 1973, **37,** 255–277.

Prinz, W., Tweer, R., & Feige, R. Context control of search behaviour: Evidence from a "hurdling" technique. *Acta Psychologica,* 1974, **38,** 73–80.

Rabbitt and Fletcher. Paper at E.P.S. meeting, Dusham, July 1976.

Rabbitt, P. M. A. Effects of independent variations in stimulus and response probability. *Nature,* 1959, **183,** 1212.

Rabbitt, P. M. A. Perceptual discrimination and the choice of responses. Unpublished Ph.D. thesis, Univ. of Cambridge, 1962.

Rabbitt, P. M. A. Ignoring irrelevant information. *British Journal of Psychology,* 1964, **55,** 403–414.

Rabbitt, P. M. A. In Vinozradova O. S. & Noroselova, V. V. (XVIII International Congress of Psychology, Moscow), 1966.

Rabbitt, P. M. A. Learning to ignore irrelevant information. *American Journal of Psychology,* 1967, **80,** 1–13.

Rabbitt, P. M. A. Times for analysing stimuli and selecting responses. *British Medical Bulletin,* 1971, **27,** 259–265.

Rabbitt, P. M., Clancy, M., & Vyas, S. M. Manuscript submitted to the Quarterly Journal of Experimental Psychology, 1978.

Rabbitt, P. M. A., Cumming, G., & Vyas, S. M. An analysis of visual growth, entropy and sequential effects. In S. Dornic (Ed.), *Attention and Performance VI.* Potomac, Maryland: Erlbaum Associates, 1977.

Rabbitt, P. M., Cumming, G. & Vyas, S. M. Improvement, learning, and retention of skill at visual search. To appear in *Quarterly Journal of Experimental Psychology*, 1979. (a)

Rabbitt, P. M. A., Cumming, G., & Vyas, S. M. Modulation of selective attention by sequential effects in visual search tasks. To appear in *Quarterly Journal of Experimental Psychology*, 1979. (b)

Rayner, K. The perceptual span and peripheral cues in reading. *Cognitive Psychology*, 1975, **7,** 65–81.

Rayner, K., & McConkie, G. W. A computer technique for identifying the perceptual span in reading. Paper presented at the Eastern Psychological Association Meeting, Washington, D.C. 1973.

Rayner & Osgood (1972)

Reicher, G. M., Snyder, C. R. R., & Richards, J. T. Familiarity of background letters in visual scanning. *Journal of Experimental Psychology: Human Perception and Performance*, 1976, **2,** 522–530.

Roydes, R. L., & Osgood, C. E. Effects of geometrical form-class set upon perception of grammatically ambiguous words. *Journal of Psycholinguistic Research*, 1972, **1,** 165–174.

Rumelhart, D. E. A multicomponent theory of the perception of briefly presented visual displays. *Journal of Mathematical Psychology*, 1970, **7,** 191–218.

Sasaki, E. H. The influence of the number and type of targets upon rapid scanning of word lists. Unpublished Ph.D. thesis. Stanford Univ., 1970.

Schouten, J. F., and Bekker, J. A. M. Reaction time and accuracy. *Acta Psychologica*, 1967, **27,** 143–153.

Schulman, A. A. Recognition memory for targets from a scanned word list. *British Journal of Psychology,* 1971, **62,** 335–346.

Selfridge, O. T., and Neisser, U. Pattern recognition by machine. *Scientific American*, 1960, **203** (August), 1960, 60–68.

Shallice, T. Paper presented at Department of Psychology Annual Seminars, Univ. of Oxford, May 1976.

Shurtliff, D. A., & Marsetta, M. Y. Visual search in a letter cancelling task re-examined. *Journal of Experimental Psychology*, 1968, **77,** 19–23.

Smith, E. E. Effects of familiarity on stimulus recognition and categorisation. *Journal of Experimental Psychology*, 1967, **74,** 324–332.

Smith, F. Familiarity of configuration versus discrimination of features in the visual identification of words. *Psychonomic Science*, 1969, **14,** 261–262.

Smith, F., Lott, D., and Cromwell, B. The effect of type-size and case alternation on word identification. *American Journal of Psychology*, 1969, **82,** 248–253.

Snyder, C. R. R. Familiarity and processes of visual search. Unpublished M.A. thesis, Univ. of Oregon, 1970.

Snyder, C. R. R. Selection, inspection and naming in visual search. *Journal of Experimental Psychology*, 1972, **92,** 428–441.

Sperling, G. A model for visual memory tasks. *Human Factors*, 1963, **5,** 19–31.

Sperling, G., Budianski, J., Spivak, J. G., & Johnson, M. C. Extremely rapid visual search: the maximum rate of scanning letters for the presence of a numeral. *Science*, 1971, **174,** 307–310.

Spoehr, K. T., and Smith, E. E. The rule of syllables in perceptual processing. *Cognitive Psychology*, 1973, **5,** 71–89.

Spoehr, K. T., and Smith, E. E. The rule of orthographic and phonotactic rules in perceiving letter patterns. *Journal of Experimental Psychology: Human Perception and Performance*, 1975, **1,** 21–34.

Sternberg, S. Estimating the distribution of additive reaction-time components. Paper presented at the meeting of the Psychometric Society, Niagara Falls, Ontario, October 1964.

Sternberg, S. High speed scanning in human memory. *Science*, 1966, **153,** 652–654.

Sternberg, S. Two operations in character recognition: some evidence from reaction time measurements. *Perception and Psychophysics*, 1967, **2,** 45–53.

Sternberg, S. The discovery of processing stages: extensions of Donders' method. *Acta Psychologica*, 1969, **30,** 276–315.

Sternberg, S. Memory scanning: new findings and current controversies. *Quarterly Journal of Experimental Psychology*, 1975, **27,** 1–32.

Sternberg, S., & Scarborough, D. L. Parallel testing of stimuli in visual search. Paper presented at the International Symposium on Visual Information Processing and Control of Motor Activity, Bulgaria, July 1969.

Swets, J. *Signal detection and recognition by human observers*. New York: Wiley, 1969.

Taylor, M. M. Detectability theory and the interpretation of vigilance data. In A. F. Sanders (Ed.), *Attention and performance I*. Amsterdam: North Holland, 1967.

Townsend, J. T. A note on the identifiability of parallel and serial processes. *Perception & Psychophysics*, 1971, **10,** 161–163.

Townsend, J. T. Theoretical analysis of an alphabetic confusion matrix. *Perception & Psychophysics*, 1971, **9,** 40–50.

Underwood, G. *Attention and memory*. Oxford: Pergamon, 1976.

von Wright, J. M. Selection in visual immediate memory. *Quarterly Journal of Experimental Psychology*, 1968, **20,** 62–68.

von Wright, J. M. On selection in visual immediate memory. *Acta Psychologica*, 1970, **33,** 280–292.

Wattenbarger, B. L. Speed and accuracy set in visual search performance. Unpublished doctoral dissertation, Univ. of Michigan, 1969.

Willows, D. M., & McKinnon, G. E. Selective reading: Attention to the unattended lines. *Canadian Journal of Psychology*, 1973, **27,** 292–304.

Part II

Pattern Processing

Chapter 4

SCHEMES AND THEORIES OF PATTERN RECOGNITION

STEPHEN K. REED

I. INTRODUCTION

This chapter reviews some of the current empirical and theoretical approaches to the problem of understanding human pattern recognition. It emphasizes recent studies and does not pretend to be exhaustive or historically motivated. Rather, I have chosen a limited number of studies to illustrate what psychologists are currently doing and thinking about as they try to learn more about a complex process. In a recent book (Reed, 1973), I attempted to place pattern recognition within a broad cognitive framework that included the representations of patterns, sensory storage, temporal aspects of recognition, the modality and structure of memory codes, perceptual classification, and response aspects of recognition. Although the acquisition of knowledge in all of these areas will facilitate our understanding of pattern recognition, the scope of this chapter is more

HANDBOOK OF PERCEPTION, VOL. IX

ISBN 0-12-161909-5

limited and focuses primarily on the central problem of representation. The different sections discuss theories of templates, features, structure, analysis by synthesis, topology, and prototypes.

Before looking at these different theories in more detail, let us briefly preview how they might apply to the pattern shown in Fig. 1. The pattern, which can be seen as either a stingray or a full-blown sail, was chosen by Clowes (1969) to illustrate the advantages of structural theories. The first problem we face in trying to formulate a theory of pattern recognition is that of finding ways to describe a pattern. The template, feature, structural, and topological methods are alternative ways of describing patterns.

The *template theory* proposes that patterns are not really "described" at all, but are holistic entities that can be compared to other patterns by measuring the degree of overlap between pattern and template. Such a theory is very limited and inflexible. It is not clear, for example, how we could distinguish the two alternative interpretations of Fig. 1 if degree of overlap is the only operation we can use. Although template theories have limited application, there are a few cases where they might apply, and we will consider these cases in Section II.

The *feature theory* proposes that patterns are analyzed into their various features, which are then used as the basis for identifying the pattern. The features of Fig. 1 would most likely be the four line segments; *a, b, c,* and *d*. Patterns that are composed of line segments like *a, b, c,* and *d* would likely be either a sail or a stingray, but which one? In order to answer this question, we have to be more explicit about how the features are related.

Structural theories emphasize the relationships among the features, such as how lines are joined together. Perceiving Fig. 1 as a stingray requires grouping adjacent lines: line *a* with line *d* and line *b* with line *c*. Perceiving the pattern as a sail requires grouping opposite lines: line *a* with line *c* and line *b* with line *d*. Structural theories therefore provide a more complete description of patterns than feature theories.

Topological theories provide us with ways of relating patterns through global transformations of the entire pattern. For example, one could continuously transform the degree of curvature of the lines in Fig. 1 to illustrate different degrees of wind velocity pushing against the sail.

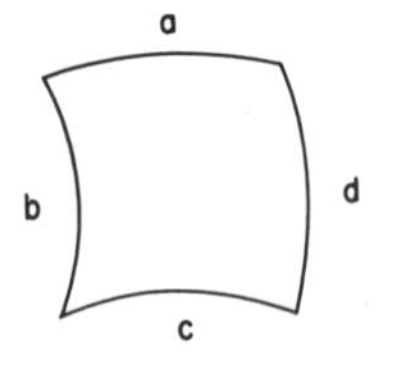

FIG. 1. An ambiguous pattern, showing a stringray or sail. (From M. Clowes, Transformational grammars and the organization of pictures. In A. Girasselli (Ed.), *Automatic interpretation & classification of images*. New York: Academic Press, 1969.)

Another example, considered in more detail later, concerns changes in facial profile that depict the age of a person. What kind of global transformations of the face would best predict perceived age?

Although having adequate techniques for describing patterns is a prerequisite for formulating theories of pattern recognition, it is also necessary to develop detailed process models of how people decide the identity of a particular pattern. How does the observer go about analyzing a pattern into its component features and synthesizing the features to arrive at its structure? The *analysis-by-synthesis model* proposes that the analysis of a pattern into its parts is guided by the requirement that the combination of those parts should form a meaningful pattern. Thus, joining lines *a* and *d* to form a unit imposes a constraint on the grouping of lines *b* and *c*. If lines *b* and *c* were not joined as a unit, the pattern would be meaningless.

Another problem for theorists of pattern recognition is that of formulating models of how people can categorize patterns when the patterns belonging to the same category (such as dogs, chairs, cars) differ in their physical attributes. One popular approach to this problem is the theory that people form category *prototypes* or patterns that are very good exemplars of each category. New patterns are then categorized on the basis of their similarity to the category prototypes.

The alternative theories are described in more detail in the following sections, and examples of recent experimental paradigms are given to illustrate how psychologists are attempting to learn more about how people recognize patterns. The different theories should not be viewed as competing against one another in a contest that will eventually produce a winner. The world contains a large variety of patterns, and psychologists can invent many different kinds of experiments to study pattern recognition. The usefulness of a particular theory may therefore depend greatly on the particular pattern and experiment. More will be said in the final section about how different theories relate to one another.

II. TEMPLATE THEORIES

Template theories propose that an unknown pattern is represented as an unanalyzed whole and is identified on the basis of its degree of overlap with various standards. A pattern-recognition device can be constructed on this principle if there is only one exemplar of each pattern, such as if all letters were of the same type font. However, one of the remarkable characteristics of humans is their ability to classify patterns that vary widely in their physical attributes. A template theory is therefore usually quickly rejected as a theory of human pattern recognition in favor of a theory that proposes that patterns are first analyzed into their parts. Although I agree that this evaluation is generally true, I think there are a few cases, such as sensory storage, in which a template theory is appropriate.

Sensory storage is generally regarded as preperceptual because it does not depend on recognition, but gives the observer some additional time after the physical termination of the stimulus to obtain more information about the stimulus. One might conceive of recognition as occurring through a feature-extraction process in which the observer uses the physical presentation of the stimulus and the sensory store to identify features (e.g., Rumelhart, 1970). The content of the sensory store might then be thought of as a template that can undergo further analysis before the sensory store decays away.

This kind of an interpretation is most clearly presented by Phillips (1974) in an article that distinguishes between sensory storage and short-term visual memory. Subjects in Phillips's experiments were asked to judge whether two sequentially presented checkerboard patterns were the same or different. Performance was highly accurate if the two patterns were separated by less than 600 msec, the first pattern was not masked, and the second pattern occurred in the same place as the first pattern. In this case, matching depended on the characteristics of a high-capacity sensory store that seemed to preserve information as a visual template. When the interval between the two patterns was longer, or the first pattern was masked, or the position of the second pattern was shifted slightly, the performance of the subjects declined and became more sensitive to the complexity of the patterns. Phillips argued that subjects, in this case, had to rely on a limited-capacity, short-term visual memory in which patterns were analyzed into their various parts.

An example in which perceptual matching seems to depend on an auditory template was demonstrated when subjects were required to judge whether two repeated sequences of sounds were the same or different (Warren, 1974). The standard pattern consisted of three or four unrelated sounds. Each sound had a duration of 200 msec, which was too rapid for untrained subjects to identify the order in which the sounds were presented. Subjects were asked to judge whether the sounds composing a comparison stimulus were in the same order as the standard. The accuracy of same–different judgments was significantly better than chance, even though subjects could not identify the order in which the sounds had occurred. Warren proposed that (*a*) the identification of order and recognition of auditory temporal patterns may represent fundamentally different processes; and (*b*) recognition may involve matching of "temporal templates."

One aspect of Warren's results makes the template interpretation particularly attractive. He found that the recognition performance was best when the duration of the sounds in the comparison stimulus exactly matched the duration of sounds in the standard stimulus. Increasing the duration (up to 600 msec) caused a decline in performance even though it should make it easier for subjects to identify the temporal order of sounds. Warren has suggested that the temporal dimension in hearing is often considered analogous to the spatial dimension in vision. The template comparison therefore depended on an exact temporal overlap in the same way that Phillips's results depended on an exact spatial overlap.

These findings suggest that there is a place in psychology for a template interpretation of pattern recognition. But these results were only obtained under special circumstances. First, the patterns were presented so rapidly that subjects did not have sufficient time (or memory capacity, in Phillips's research) to complete a feature analysis. Second, the patterns had either the same spatial or temporal form, which allowed subjects to "superimpose" the standard and comparison stimulus to check for a match. However, these conditions are not met in most pattern-recognition experiments, and subjects must use some form of feature analysis.

III. FEATURE THEORIES

A. Examples of Feature Theories

In contrast to template theories of pattern recognition, feature theories propose that a pattern is analyzed into its parts, which then form the basis for the observer's decision regarding the identity of the pattern. This is a common approach used by most pattern recognition theorists, but finding a good set of features for a given set of patterns is usually not an easy task. Gibson (1969) has suggested the following criteria as a basis for selecting a set of features for uppercase letters.

1. The features should be critical ones, present in some members of the set, but not in others so as to provide a contrast.
2. They should be relational so as to be invariant under brightness, size, and perspective transformations.
3. They should yield a unique pattern for each letter.
4. The list should be reasonably economical.

Gibson used these criteria, empirical data, and intuition to derive the set of features for uppercase letters shown in Table I.

A major empirical determinant of what constitutes a good set of features is the degree to which confusable patterns share a common set of features. A set of features is usually evaluated by its ability to predict what pairs of patterns should be difficult to discriminate. Thus if a person often confuses the capital letters *C* and *G*, these letters should theoretically share many features.

An article by Geyer and Dewald (1973) illustrates how different feature sets might be tested. The authors compared feature sets proposed by E. Gibson (1969, p. 88), Laughery (1971), and Geyer (1970) by determining how well they could predict the confusion errors that occur when adults are asked to recognize briefly exposed letters of the alphabet. Their basic assumption was that a brief exposure would sometimes result in subjects detecting only some of the features of a letter, in which case they would have to use the partial information to guess which letters had been presented. The proposed feature list would, of course, determine what kind of partial information could theoretically occur and thereby determine what kind of confusion errors should occur. Geyer's (1970) feature

TABLE I

CHART OF DISTINCTIVE FEATURES FOR A SET OF LETTERS[a]

Features	A	E	F	H	I	L	T	K	M	N	V	W	X	Y	Z	B	C	D	G	J	O	P	R	Q	S	U
Straight																										
horizontal	+	+	+	+		+	+								+				+							
vertical		+	+	+	+	+	+	+	+	+				+		+		+				+	+			
diagonal	+							+	+		+	+	+	+	+											
diagonal	+							+	+	+	+	+	+	+								+	+			
Curve																										
closed																+		+			+	+	+	+		
open vertically																				+						+
open horizontally																	+		+	+					+	
Intersection	+	+	+	+			+	+					+			+						+	+	+		
Redundancy																										
cyclic change		+							+			+				+									+	
symmetry	+	+		+	+		+	+	+		+	+	+	+		+	+	+			+					+
Discontinuity																										
vertical	+		+	+	+		+	+	+	+				+								+	+			
horizontal		+	+			+	+								+											

[a] From Eleanor J. Gibson, *Principles of perceptual learning and development*, © 1969, p. 88. Reprinted by permission of Prentice-Hall, Inc., Englewood Cliffs, New Jersey.

list, essentially a modification of Gibson's (Table I), was the best predictor of the confusion matrix. The major modification was that Geyer's list contained the number of features present in the letter (such as *two* vertical lines for the capital letter *H*) rather than simply whether that feature was present, as shown in Table I. A more detailed discussion of alternative models formulated to predict confusion matrices is contained in Chapter 11 of Reed (1973).

An example of a matching procedure based on sequential feature selection is discussed by Goldstin, Harmon, and Lesk (1972). The investigators asked 10 subjects to rate values of 21 features (such as length of hair or size of nose) for each of 255 photographs of human faces. These ratings were then given to a computer as the "official" description of each face. A different group of subjects was then asked to examine portrait photographs and describe them by sequentially selecting facial features in the order of their importance, emphasizing unusual features that might discriminate each examined face from the other, unknown faces. After the subjects rated each feature, the computer compared how similar these ratings were to the official ratings and eliminated those faces for which the ratings were too discrepant. After only seven features had been selected, less than 1% of the population of 255 faces had a better match than the selected face.

Feature theories have also played an important role in auditory perception, as is evident from Cole and Scott's (1974) review of research on the perception of speech. These authors have proposed that speech perception involves the simultaneous identification of three different kinds of cues: invariant cues that accompany a particular phoneme in any context, transitional cues that are context dependent, and cues provided by the waveform envelope. Invariant features serve to identify consonant phonemes, either uniquely or as one of several alternatives. Transitional cues, such as vowel transitions, provide information about the identity of certain phonemes and their temporal order within the syllable. The waveform envelope provides information about stress and intonation, and integrates syllables into higher-order units such as words and phrases. The three types of cues are perceived independently, but are then integrated with the others.

B. Patterns as Integrated Features

Although patterns can usually be represented in terms of their parts, there can be a wide variation in the degree to which the parts or dimensions of a stimulus combine to form a coherent pattern. Garner (1974) refers to this variable as *dimensional integrality*. According to Garner, if dimensions are integral, they are not directly perceived as dimensions at all, but are derived as a secondary process. Thus, subjects can discriminate colors which differ in hue, value, or chroma, but these dimensions combine to form a unitary stimulus. In contrast, separable dimensions are directly perceived, and the judgment of similarity depends on a secondary process in which the observer must consider the similarity of two stimuli along each dimension and combine this information into an overall impression. This would occur if a person were required to judge two stimuli that differed in size and brightness. An excellent discussion of the effect of integrality on various tasks of information processing is provided in Chapter 6 of Garner's (1974) book.

An example of two kinds of patterns that differ in their degree of integrality are schematic faces and geometric forms. Results obtained by Klatzky and Thompson (1975) indicated that the features composing a triangle or a parallelogram were easier to synthesize into a visual whole than the features composing a schematic face. If the line segments composing a geometric form were presented sequentially, observers could join the segments together, creating a geometric form that could be matched to a second form as quickly as if the first form had been presented as a visual whole. But this was not true if the features (eyes, nose, mouth) of a schematic face were presented sequentially. Observers were able to compare two schematic faces more quickly if the first face was presented as a whole, rather than as a sequence of separate components. Klatzky and Thompson (1975) concluded that people can synthesize the components of a geometric form into a visual whole, but are not able to create a visual whole from the components of a schematic face.

One question raised by these findings is whether the subjects' unfamiliarity with the schematic faces may have limited their performance. According to a theory proposed by LaBerge and Samuels (1974), the integrality of the parts of a pattern may change over time as a function of perceptual learning. LaBerge and Samuels proposed that the processing stages involved in reading become automatic and do not require attention (processing capacity) as the reader becomes more skilled. In the early trials of learning, the beginning reader must combine the separate features of a letter into a single unit; a task that requires attention. In later trials, the combination of features into a letter occurs automatically and does not require attention or processing capacity. Support for this view comes from an experiment by LaBerge (1973) in which subjects initially took longer to match unfamiliar letters (ɩɺ ɱ) than familiar letters (b d p q) when they were not expecting these letters. However, after 4 days of practice, subjects could match the unfamiliar letters as fast as the familiar letters, suggesting that the letter codes for the unfamiliar letters were automatically activated and did not require a synthesis of separate features. Sufficient practice with schematic faces may also result in the formation of visual wholes unless their greater complexity prevents the synthesis of the facial components.

IV. STRUCTURAL THEORIES

A. Visual Grammars

The feature approach to pattern recognition seems adequate when the patterns can be described as a list of independent attributes. The Goldstein *et al.* (1972) program for matching faces worked well because the faces could be characterized by a feature list specifying the length of hair, bushiness of eyebrows, size of nose, etc. However, it is less clear that a feature-list approach is adequate for characterizing patterns such as letters or numbers, in which the intersection of lines determines the structure of the pattern (see Holbrook, 1975, for a discussion of the limitations of this approach). A feature matrix like the one shown in Table I tells us what parts are present and also tells us about certain gestalt characteristics, such as whether the pattern is symmetrical or contains discontinuities. However, it does not tell us how the parts are joined together; that the uppercase letter *H*, for example, consists of two vertical lines whose midpoints are connected by a horizontal line.

In order to more directly represent the relations among features, scientists in the field of artificial intelligence developed structural or grammatical models. Figure 2 shows the features of a grammar proposed by Narasimhan and Reddy (1967) to describe uppercase letters. The features consist of different segments of letters and usually contain numbered nodes that are points at which segments can be joined. The letter *H* might be described as consisting of a horizontal line joined at Node 1 to Node 2 of a vertical line and joined at Node 3 to Node 2 of a second vertical line. A visual grammar that explicitly states how lines are joined together gives a more precise description of line patterns than the feature list shown in Table I. It also seems intuitively reasonable that we form structural descriptions of patterns. It is for these reasons that I agree with Sutherland (1968, 1973) that a visual grammar approach is superior to a feature-list approach, at least for accounting for our ability to recognize patterns in which structural relations are important.

In spite of the appeal of this theoretical approach, visual grammars have had little impact on psychology. Perhaps one reason is that their description of a pattern is too complex and precise, making it difficult to make predictions. This is particularly true for accounting for perceptual confu-

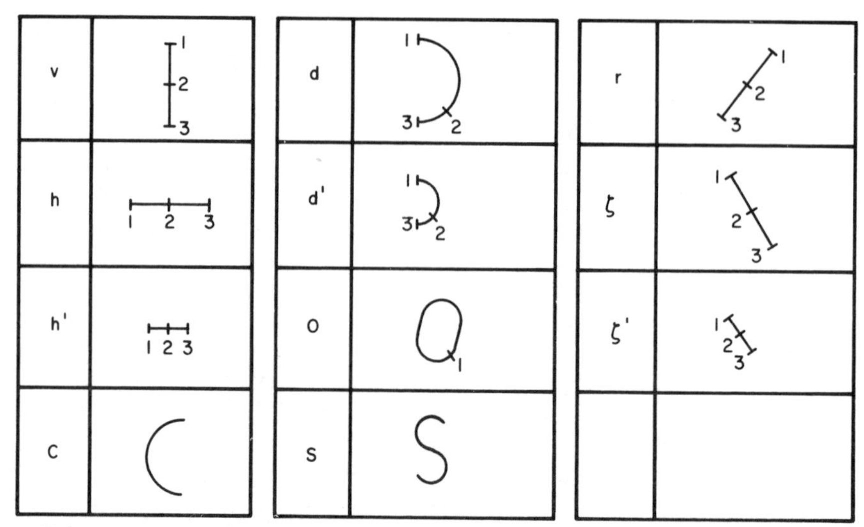

FIG. 2. Components of a visual grammar. (From R. Narasimhan & V. S. N. Reddy. A generative model for handprinted English letters and its computer implementation. *ICC Bulletin*, 1967.)

sions, which has been one of the major empirical techniques for evaluating feature models. One can use feature models to predict perceptual confusions by counting how many features two patterns share or calculating the distance between two patterns in a multidimensional space, but it is less clear how to measure the similarity of two patterns that are represented by a visual grammar. It is simply more difficult to design experiments to test a visual grammar.

However, there have appeared some promising initial attempts to construct structural models, one example being a doctoral thesis at MIT that uses 12 physical attributes to represent uppercase letters (Shillman, 1974). The attributes are shaft, leg, arm, bay, closure, weld, inlet, notch, hook, crossing, symmetry, and marker. As the names suggest, the attributes include both simple parts of letters and relational information. The attributes themselves include relational descriptors indicating their location, orientation, segmentation, and concatenation. Their selection was motivated by behavioral data, since Shillman had subjects rate the goodness of letter samples by how well each pattern represented a given letter. He was particularly interested in category boundaries, which were determined by creating ambiguous patterns that could be classified as either of two letters. Although further development and testing of this descriptive language is necessary (see Naus & Shillman, 1976, for a later application), Shillman's work appears to be a significant step toward the solution of a difficult problem.

B. Parts of Patterns

The lack of grammatical models in psychology should not imply that psychologists are uninterested in the structure of patterns. The interest in organizational phenomena expressed by the Gestalt psychologists has continued to attract attention (see Hochberg, 1974). Organizational variables can influence a variety of perceptual processes, such as detecting a feature of a pattern, recognizing a part of a pattern, and constructing a pattern from its parts.

Weisstein and Harris (1974) found that the ability of subjects to identify a briefly flashed line segment depended on the structure of the context in which it was embedded. The observer had to indicate which one of four diagonal line segments was present in a display that also contained four horizontal and four vertical lines. The arrangement of the horizontal and vertical segments determined the accuracy in identifying the diagonal line. Performance was best when the lines were arranged to form a unitary and coherent design (two overlapping squares). Weisstein and Harris concluded that although their results demonstrated the importance of context, it was still not clear how context played its role.

One technique for assessing how subjects organize patterns is to ask them questions about possible parts of a pattern. For example, subjects who have encoded the Star of David as two overlapping triangles should easily identify a triangle as a part of this pattern, but they should have difficulty in identifying a part that does not match their encoding. Reed (1974) tested this hypothesis by asking subjects to judge whether the second of two sequentially presented patterns was a part of the first pattern. He found that subjects were quite accurate in recognizing certain parts (such as a triangle in the Star of David) but were not very accurate in recognizing others (such as a parallelogram in the Star of David). The results supported Sutherland's (1968) suggestion that people form structural descriptions of patterns and find it difficult to recognize parts of a pattern that do not match these descriptions.

Palmer (1977) has studied both analysis and synthesis in order to learn more about the representation of patterns in terms of their parts. The analysis task was similar to Reed's, except that a pattern was composed of six line segments and a part was composed of three line segments. The lines could be disconnected and usually did not make a familiar geometric form. In the synthesis task, subjects were shown two three-segment parts and were asked to construct a six-segment figure by superimposing one part over the other part. Palmer measured the latency and accuracy of their constructions in order to test the hypothesis that performance would depend on the *goodness* of the component parts to be synthesized. The results supported the hypothesis. Parts based on the most obvious organization of the pattern required only about 1.5 sec to synthesize, compared to over 4 sec for parts that were judged to be of medium or low goodness. Palmer's study shows that the part structure of patterns not only influences performance in tasks requiring analysis, but can influence performance in synthesis tasks as well.

V. ANALYSIS BY SYNTHESIS

The analysis-by-synthesis model proposes that analysis and synthesis interrelate by having the attempted synthesis of features influence how the features are analyzed. Halle and Stevens (1962) proposed such a model for speech recognition by suggesting that the analysis of speech is achieved through matching acoustic information to an active internal synthesis of comparison signals. The failure to achieve a satisfactory match can result in reanalysis of the acoustic information until a satisfactory match is achieved. The model has some appeal to cognitive psychologists because it emphasizes the role of the observer as an active processor of information, rather than a passive observer (Neisser, 1967).

We should consider, however, whether the model requires too much processing, given the apparent ease with which we recognize many patterns. Posner and Boies (1971), for example, have argued that the recognition of a letter occurs relatively automatically, without requiring much processing capacity. Letters are, of course, familiar patterns, and Posner's stimuli were very legible. The analysis-by-synthesis model might be more appropriate for less familiar patterns or for familiar patterns that are very distorted. In this case, the first analysis of the features might not result in the recognition of the pattern, and a reanalysis might be necessary. Another case in which reanalysis is usually necessary is the misreading of words. This can occur either because certain features of the letters are wrongly identified, or because only some of the features of the letters are correctly identified and the wrong word is guessed—a fact that soon becomes obvious if the word does not fit the context of the sentence.

Another phenomena that can perhaps best be explained by an analysis-by-synthesis model is the creation of features that are not really there. The phonemic restoration effect can serve as an illustration. Warren and Sherman (1974) replaced a phoneme of a word by a noise meeting certain criteria. They instructed their subjects to listen to a sentence and indicate the exact place at which the noise occurred. Subjects tended to hear the deleted phoneme as being present and had great difficulty in detecting the location of the noise. Warren and Sherman indicated that phonemic restoration could produce phonemes that are perceptually indistinguishable from phonemes that are actually present. They have proposed that the produced effect is a special form of auditory induction in which linguistic rules enter into the synthesis of the restored sound, thereby facilitating the comprehension of speech.

A somewhat similar phenomena can occur for vision, as is illustrated by Figs. 8 and 9 of Gregory's (1974) chapter in Volume 1 of this series. The figures suggest two overlapping triangles, although the contours of only one triangle are physically present. There are, of course, limitations to our tendency to create missing features. The illusory contours in Gregory's example are suggested, but are not as clear as if they were physically present. Similarly, the restored phoneme effect is less likely to occur if a silent gap replaces the phoneme, rather than a burst of noise (Warren & Sherman, 1974). Other examples, in fact, would seem to contradict the analysis-by-synthesis model. Impossible patterns (such as Figs. 5 and 6 of Gregory's chapter) can be perceived, but only because the analysis is consistent for local parts of the pattern. The analysis is inconsistent with the entire pattern, so analysis by synthesis would have to be ignored in these cases.

It should be noted that the analysis-by-synthesis model is inconsistent with a template model (since analysis does not occur), but is consistent with both the feature and structural models. It is concerned not so much with the language used to describe patterns as with the temporal order of events that lead to the recognition of the pattern. It is concerned with such issues as whether the identification of one feature influences the identification of another feature and whether the analysis of features is repeated if a satisfactory match is not obtained after the first analysis. The rapid rate at which patterns are recognized makes it difficult to test the model at an empirical level, and the complexity of specifying the interactions that might occur in feature analysis makes it difficult to formulate a model at a theoretical level.

One method for slowing down the pattern-recognition process and controlling the order in which subjects analyze a pattern is to present the parts of the pattern sequentially. Chastain and Burnham (1975) used this technique to test the hypothesis that the starting segment would determine how people perceived the ambiguous rat–man figure. Subjects were shown six successive segments, with the first segment selected to suggest either a man or a rat. Their initial perception was either of the nose and eyes of the man (Segment 1 in Fig. 3) or the tail of the rat (Segment 5 in Fig. 3). The choice of the starting segment did significantly influence perception of the figure, supporting the hypothesis. People who saw Segment 1 first were more likely to perceive the figure as a man, and people who saw Segment 5 first were more likely to perceive the figure as a rat. A subsequent experiment revealed that the starting segments alone

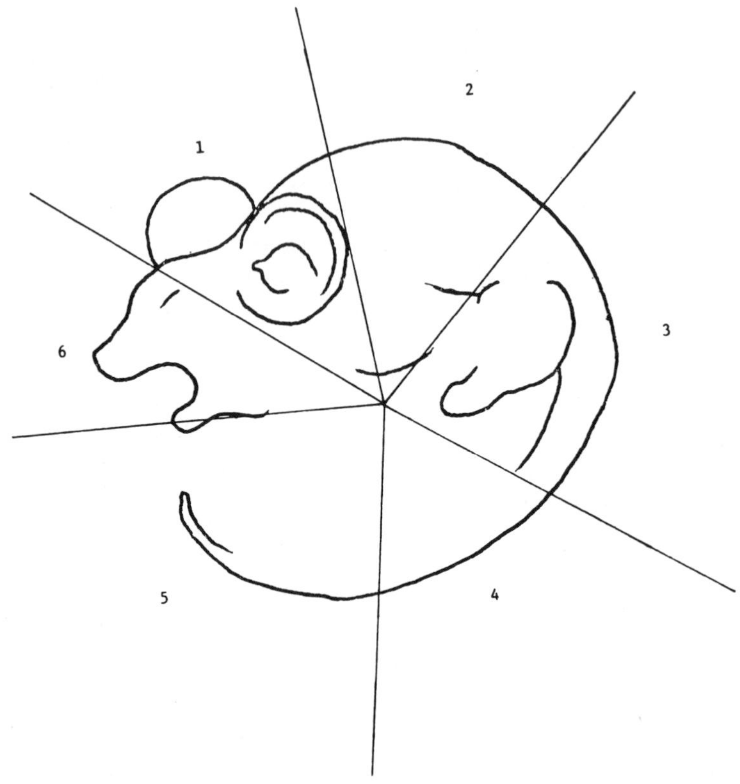

FIG. 3. Ambiguous rat–man figure with segments indicated. (From G. Chastain & C. Burnham. The first glimpse determines the perception of an ambiguous figure. *Perception & Psychophysics*, 1975, *17*, 221–224. Copyright (1978) by the American Psychological Association. Reprinted by permission.)

did not produce the effect, since it was necessary to follow the starting segment with additional parts of the figure. Chastain and Burnham proposed that the initial perception resulted in expectations that determined how subjects organized their perception of the remaining segments. Their proposal is consistent with an analysis-by-synthesis model, in which the analysis of parts of a pattern is guided by the attempt to combine the parts into a meaningful pattern.

VI. TOPOLOGICAL THEORIES

The structural theories based on visual grammars are more explicit than feature theories as to how the features or parts of a pattern relate to each other. However, visual grammars can usually be considered an extension of feature theories because they depend on a specification of features that form the primitives of the grammar. The organization of the pattern is specified in terms of the relationships among the features. The topological approaches considered in this section are attempts to quantify the overall organization of a pattern without starting with a feature analysis.

One of the most successful attempts to relate the overall organization of a pattern to perceptual performance is the series of studies reported by Uttal (1975). The stimuli used in Uttal's research were dot patterns, which made it possible to emphasize the geometric organization of the patterns, rather than their component features. The dot patterns were embedded in a noise field consisting of randomly placed dots, and the subject's task was to indicate which of two sequential bursts contained the target form, as opposed to only random noise. The major independent variables were the density of the noise dots and the geometry of the target form. Among the findings discussed by Uttal were the following:

1. The detectability of a straight dotted line was insensitive to the orientation of that line.
2. The detectability of a straight dotted line was superior to that of a curved or angled line.
3. The sides of dotted triangles contributed more to their detection than the corners.
4. The detection of dotted triangles and squares was insensitive to their orientation.
5. Distortion of a square into a parallelogram resulted in a monotonic decline in its detectability.
6. Organization of a set of dotted lines into regular polygonal forms or linear arrays enhanced detectability over that of irregular arrangements.
7. Disorganization of the corners of squares and triangles into increasingly less regular arrangements resulted in a monotonic decline in detectability.
8. Figural goodness, as defined by Garner and Clement, had little effect on the detectability of the figures.

In order to account for these findings, Uttal proposed an autocorrelation theory of form perception that is sensitive to the geometrical regularity of the pattern. The autocorrelation measures how well an identical copy of the pattern overlaps with the original pattern when the copy is displaced along the horizontal and vertical axes. Large values or peaks occur in the two-dimensional space when there is a high degree of overlap caused by regularities in the pattern. Uttal reduces this autocorrelation diagram to a single number by multiplying the amplitudes of pairs of peaks, following the formula

$$F = \frac{\sum_{n=1}^{N} \sum_{i=1}^{N} (A_n \times A_i)/d_{ni}}{N} \qquad (n \neq i), \tag{1}$$

where A_n is the amplitude of the nth peak, A_i is the amplitude of the ith peak, d_{ni} is the euclidean distance between the two peaks, and N is the total number of peaks. The higher the figure of merit (F), the easier the pattern should be to detect—a prediction supported by most of Uttal's results.

One aspect of Uttal's results that has been inconsistent with the predictions of the figure of merit expressed in Eq. (1) is the effect of pattern symmetry on detection. The model predicts that symmetrical patterns should be easier to detect—a prediction which was not supported in more recent experiments (Uttal, Eskin, & Sawyer, 1975). Since Eq. (1) is an arbitrary way of summarizing the autocorrelation, Uttal has suggested that other equations may be necessary to calculate the figure of merit, depending on the demands of the task. A more complex version of Eq. (1) gives a better prediction of the effect of pattern symmetry, but the figure of merit expressed in Eq. (1) gives a better prediction of Uttal's earlier findings.

An interesting example of how we can quantify changes in the overall shape of a pattern has been described by Pittenger and Shaw (1975). They studied the problem of how the shape of the human head changes with age without causing a loss of facial identity. For example, Mr. Smith's profile at age 50 appears different from his profile at age 30, although we recognize both profiles as Mr. Smith. Pittenger and Shaw used two different methods to transform the facial profiles in order to determine whether either transformation would result in reliable judgments of the relative ages of the profiles. The strain transformation, shown in Fig. 4, produced highly reliable judgments. Increasing strain level made the profiles appear older. The shear transformation had a small, but significant, effect on age judgments, suggesting that both transformations are influential but that strain is the more important determinant of judged age. Another experiment revealed that the strain transformation did not destroy the identity of the person, since people could distinguish two profiles differing in level of strain from an unrelated profile.

Although the strain transformation did not change the identity of a face, other kinds of transformations can result in changing how individuals interpret a form. Shepard and Cermak (1973) generated a variety of closed curves by varying two parameters. The forms could be interpreted as different kinds of objects (butterfly, animal head, the continent of Africa) depending on their location in a two-dimensional space. Shepard and Cermak suggest that the gestalt-like transformations provide a novel set of patterns that may lead to new insights when compared with patterns

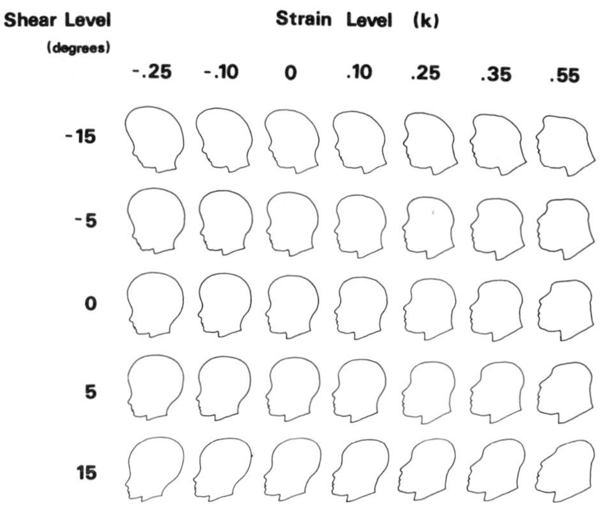

FIG. 4. Transformations of a facial profile by shear and strain. (From J. B. Pittinger & R. E. Shaw. Aging faces as viscal-elastic events: Implications for a theory of nonrigid shape & perception. *Journal of Experimental Psychology: Human Perception and Performance*, 1975, *1*, 374–382.)

composed of more verbally analyzable dimensions. If their prediction proves correct, we will have additional ways of thinking about patterns that will extend the feature-analysis theories now dominating psychology.

VII. PROTOTYPES

One important problem faced by pattern-recognition theorists is the problem of accounting for the variation in exemplars that represent a pattern. For example, there are many different kinds of patterns that represent the letter *a* or the category *dogs*. One approach to this problem is to assume that people are capable of abstracting a central tendency to represent the average member of the category. The example shown in Fig. 5 can serve as an illustration. Patterns belonging to two different categories are represented in a two-dimensional space. The prototypes (*P*) are the central tendency of the categories, and the test pattern is classified into the category that has the closest prototype.

When new patterns are classified into the category, the prototype changes to reflect the new central tendency of the category. One advantage of a prototype theory is that forming prototypes requires very little of a person's memory capacity. All that is required to change the prototype is memory of the prototype and the number of patterns previously seen. A

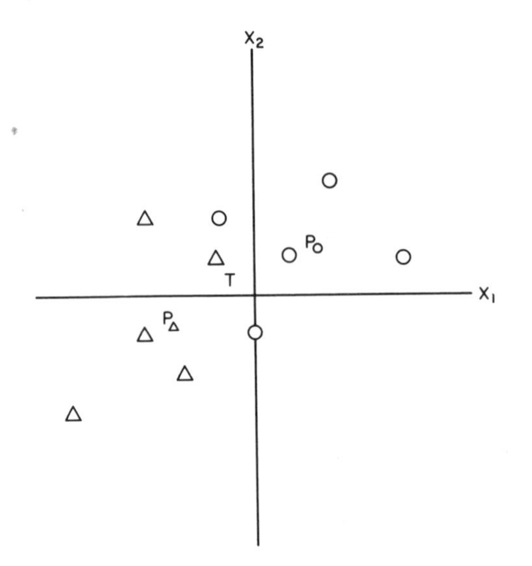

FIG. 5. Two categories of patterns represented in a two-dimensional space. (△) Category 1 patterns; (○) Category 2 patterns; ($P_\triangle$) Category 1 prototype: ($P_\bigcirc$) Category 2 prototype; (T) test pattern. [From Reed (1973). Reprinted by permission.]

change in the prototype can be expressed by Eq (2), in which P represents the feature values of the prototype, X represents the feature values of a new category pattern, and N represents the total number of patterns in the category including X.

$$\text{New Prototype} = P + (X - P)/N \tag{2}$$

Equation (2) states that when a new category pattern is learned, the prototype moves toward the new category pattern, but how far it moves is inversely proportional to the number of patterns in the category. When the prototype is based on many previously seen patterns, it should be influenced very little by a new category pattern.

A. Artificial Categories

There is, in fact, some evidence in favor of the view that subjects can abstract category prototypes and use them as a basis for classifying patterns. Posner and Keele (1968) reasoned that if prototype formation facilitates the learning and recognition of new patterns, the greatest amount of transfer should occur under conditions that make it easy for subjects to form a prototype. The prototype patterns in their experiment consisted of a triangle, the capital letters M and F, and a random pattern. Each pattern was composed of nine dots, and the degree of variation of each dot determined the amount of distortion from the prototype. Since limiting the distortion of the category patterns should have made it easier to form a prototype, the learning and recognition of new patterns should have been better following the learning of category patterns that varied little from the category prototype. But the results failed to support this hypothesis. Subjects were better able to learn and recognize highly distorted patterns when they trained on moderately distorted patterns than when they had trained on slightly distorted patterns.

In spite of this initial disconfirmation, Posner and Keele (1968) found some evidence of prototype formation. First, the subjects' ability to correctly classify new patterns depended on the degree of similarity between the new patterns and the category prototype. A rank correlation of .97 between distance from the prototype and errors indicated that the patterns most distant from the prototype were most difficult to recognize. Second, subjects who had trained on moderately distorted patterns could classify the (previously unseen) prototype as well as they could classify the original category patterns and better than they could classify new patterns that were the same distance as the prototype from the category patterns.

A greater amount of support for Posner and Keele's main hypothesis came from a study by Peterson, Meagher, Chait, and Gillie (1973). Using a procedure similar to Posner's, the experimenters confirmed the prediction that limiting the variation of category patterns would facilitate prototype formation. Accuracy in drawing the prototypes improved as the variability of the category patterns decreased. In contrast to the Posner and Keele (1968) results, low category variability also resulted in better classification of novel patterns. Although it is not certain what caused the different results, Peterson *et al.* suggest it may have been due to the fact that Posner and Keele mixed meaningful (triangle, *B*, and *F*) and random concepts while their own experiments mainly used random concepts.

Although the previous experiments provided evidence for the abstraction of prototypes, they were not designed to test a variety of models in order to investigate how subjects would classify novel patterns. Four main classes of models were compared by Reed (1972) to determine whether they could be used to make differential predictions as to which categories subjects would choose in classifying a series of novel patterns. The patterns were schematic faces that differed along four well-defined dimensions. Except for one learning experiment, the patterns composing the two categories were simultaneously presented and were available for inspection throughout the classification task (see Fig. 6).

Three of the four models tested were based on the distance between the novel patterns and the category patterns. The *average-distance model* states that subjects compare the average distance between the novel pattern and the patterns composing each category. The correct category is

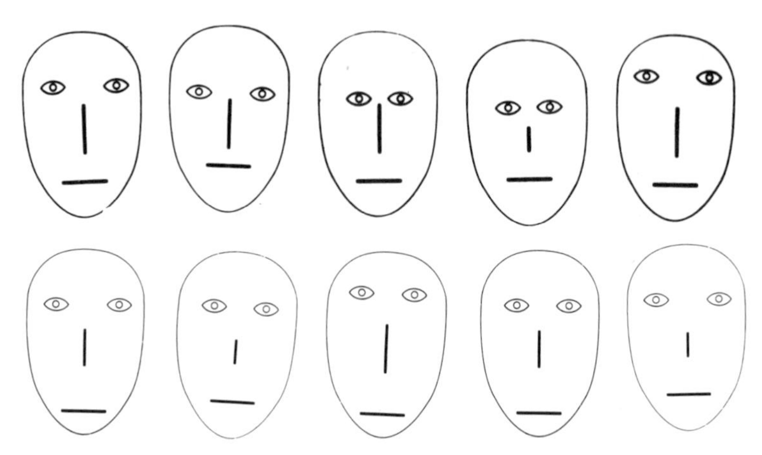

FIG. 6. Two categories of schematic faces. The upper five faces represent one category and the lower five faces represent another category. [From Reed (1972). Reprinted by permission.]

the one with the minimum average distance. The *prototype model* states that the subjects compare the distance between the novel pattern and the category prototypes, choosing the category that has the closest prototype. The prototype is defined as the central tendency of the category, and therefore has the mean category value along each pattern dimension. Because the distance metric (whether Euclidean or city-block) is a nonlinear function of the pattern dimensions, the predictions of the average distance and prototype models are not identical, but they tend to be highly correlated. The *closest-match model* states that subjects find the category pattern that is least distant from the novel pattern and choose the category containing that pattern. A fourth model, the *cue-validity model*, is a probability model, rather than a distance model, and states that subjects choose a category by using the relative frequency with which the feature values of the novel pattern appear in the two categories. The model states that subjects choose the category giving the greatest number of feature matches.

Reed (1972) found that the average-distance and prototype models made consistently better predictions as to how subjects would classify the novel patterns. Additional converging operations suggested that the prototype strategy was the strategy most often used, since subjects often indicated (in a forced-choice verbal report paradigm) that they used that strategy, and they were more accurate in recognizing the category prototypes than in recognizing equidistant control patterns. The one exception occurred in an experiment in which the category prototypes were so similar (Fig. 3, Reed, 1972) that subjects most likely used a simpler classification strategy.

Three of the previous models and one new model were compared by Hyman and Frost (1974) in a study designed to determine whether classification strategies change with learning. The experimenters used random dot patterns as stimuli, but varied the category exemplars in a systematic manner. The exemplars belonging to one category were relatively taller than they were wide, and the exemplars in the other category were relatively wider than they were tall. Reaction times were used to test the alternative models. For example, the prototype model predicts that classification times should be a function of the difference in distances of the novel pattern from the two prototypes. Similar predictions were made for the exemplar (average-distance) and nearest neighbor (closest-match) models. The fourth model, called a *rule model*, predicts that reaction time should be a function of the difference between the value on the height dimension and the value on the width dimension. This model differs from the prototype model in that it predicts that it is easier to classify extreme patterns, lying far from the boundary separating the categories, than it is to classify patterns near the central tendency of the category.

Hyman and Frost hypothesized that classification strategies change as a function of learning. During the initial learning trials, subjects would presumably store the individual exemplars and use the average-distance model to classify novel patterns. At a later stage in learning, they would abstract a category prototype and use this as a basis for responding. At a still later stage, after seeing all the category exemplars, they would be able to use a rule based on the relative heights and widths of the patterns. The results indicated that the rule model was superior for one set of patterns (first design) and that the prototype model was superior for another set of patterns (second design). There was relatively little support for the hypothesis that classification strategies change over time, except that there appeared to be a change from an average-distance strategy to a rule strategy in the first design. The prototype strategy was superior over all trials for the second design. Presumably, the obviousness of the rule will determine whether a rule strategy is used, and this may have caused the different results for the two sets of patterns.

B. Natural Categories

Although the previous studies used standard experimental techniques to investigate prototype abstraction, Rosch (1975) has directed our attention to the relevance of these ideas outside the laboratory. She begins with the proposal that the categories that make up the real world—such as plants, birds, and colors—are difficult to represent by logical rules, which characterize most of the psychological research on concept formation. One of the main points of her argument is that most natural categories are characterized by exemplars that are not all equally good members of the category. For example, some blue colors are better exemplars of the category *blue* than others, just as some birds are easier to classify as such than others. Rosch presents some evidence for her view that the more prototypical members of a category are learned first and serve as a reference for learning other members of the category.

But can subjects identify the average shape of exemplars sampled from real-world categories? Rosch, Mervis, Gray, Johnson, and Boyes-Braem (1976) have hypothesized that there is a basic level of abstraction at which subjects can form category prototypes and that this basic level includes categories such as *shirt, truck, fish,* and *table*. According to their hypothesis, there should be little gain in a subject's ability to form a prototype at a subordinate level (such as *dining-room table*) and a large loss in a subject's ability to form a prototype at a superordinate level (such as *furniture*). The experimenters tested their hypothesis by asking subjects to identify outline shapes of objects that had been formed by averaging the shapes of category exemplars, normalized for size and orientation. Subjects correctly named 25 of the 32 prototypes at the basic level and did not do significantly better at the subordinate level. Performance at the superordinate level was not significantly better than chance. These results support the view that there is a basic level of abstraction at which prototypes of real-world shapes can be recognized.

VIII. OVERVIEW

As was mentioned in the introduction, the different theories of pattern recognition should not all be viewed as mutually exclusive, competing theories. The one exception is the template theory, which proposes that patterns are not analyzed and therefore differs from the other theories. However, this theory has only limited application and will not be discussed in this section.

Three of the theories (feature, structural, and topological) are all concerned with ways of describing patterns but emphasize different aspects, such as the features or components of patterns, the relationships among the features, or the more global aspects of shape, such as pattern symmetry. The relation of feature, structural, and topological theories can be made more explicit by formulating a language for describing patterns.

Narasimhan (1969) has defined a picture language in terms of five components: a set of primitives, a set of attributes, a set of relations, a set of composition rules, and a set of transformations.

The primitives are the features or components of patterns, such as those illustrated in Fig. 2. Attributes are values that describe the features; for example, the length of a line. The primitives and their attributes define a feature theory. A structural theory includes primitives and attributes, but emphasizes relations and composition rules. The composition rules use the relations to specify how primitives are joined together. Topological theories emphasize the global relations characterizing a pattern. The research considered in Section VI quantified these relations through global transformations applied to the entire pattern.

The feature, structural, and topological theories can be considered components of a general descriptive language. Each emphasizes different aspects of the description and each has its particular advantages. The structural theory builds upon a feature theory by including rules for describing structural relations in addition to information about features and attributes. Structural theories can provide a more complete description of patterns than feature theories, but feature theories are easier to incorporate into mathematical models because they provide a less complex description than structural theories. Models based on probability theory or multidimensional scaling generally assume that patterns can be represented solely in terms of their feature attributes (e.g., Reed, 1973). Topological theories have provided us with quantitative measures of more global characteristics of patterns. Although such characteristics have long been considered important, it is only very recently that psychologists have found measures related to human performance. Since the topological transformations don't directly specify pattern features, it is unlikely that they alone will provide a complete theory of pattern recognition. But they will provide an important component of a general descriptive language.

Once we are able to represent patterns in terms of psychologically relevant descriptions, we can consider more detailed process models of performance in pattern-recognition tasks. How do people identify a pattern when they have insufficient time to identify all its features? How do we use context to aid our recognition? There are a variety of questions we can ask, but space limitations allowed consideration of only two topics, analysis by synthesis and prototype abstraction.

It is at the level of process models, rather than at the level of pattern description, that we are likely to find competing theories. But as is the case for alternative descriptive theories, it is unlikely that a particular process model will work well in all circumstances. The analysis-by-synthesis model, as one example, would seem to suggest that pattern recognition requires a rather complex decision process demanding considerable processing capacity. The model should therefore be more successful in those conditions that make pattern recognition difficult—patterns that are briefly exposed, distorted, unfamiliar, or have a complex structure.

The prototype model is a good example of a model that can be tested against competing models. The main competitor is the cue-validity model, which proposes that the observer classifies patterns on the basis of feature probabilities or the relative frequency with which the features of the pattern have appeared in each category. But there are at least two reasons suggesting that there may be some degree of truth in both models. First, most tests of the two models have used artificially constructed categories designed to maximize differences in prediction. Predictions of the models may be more closely correlated for natural categories if patterns similar to a category prototype also have high cue validity for that category. Evidence supporting this correlation has been presented by Rosch and Mervis (1975). Second, even for artificial categories, Reed (1972) found that the best predicting model was one that assumed that people form category prototypes but differentially emphasize features in comparing the similarity of a pattern to the category prototypes. Features that best discriminated between the two categories were weighted higher than features that were less predictive of category membership.

The emphasis here on the relative strengths and weaknesses of the different theories should not discourage us from attempting to develop and test pattern-recognition models. Rather, it should lead us to consider a variety of approaches when trying to understand a particular result. When we have learned which models work best in which circumstances, we will have a better understanding of how people recognize patterns. This will require a greater emphasis on a more global view in which the particular task, patterns, and proposed model for an experimental paradigm are related to other tasks, patterns, and models.

References

Chastain, G., & Burnham, C. A. The first glimpse determines the perception of an ambiguous figure. *Perception & Psychophysics*, 1975, **17,** 221–224.

Clowes, M. Transformational grammars and the organization of pictures. In A. Graselli (Ed.), *Automatic interpretation and classification of images*. New York: Academic Press, 1969.

Cole, R. A., & Scott, B. Toward a theory of speech perception. *Psychological Review*, 1974, **81,** 348–374.

Garner, W. R. *The processing of information and structure*. Potomac, Maryland: Erlbaum, 1974.

Geyer, L. H. A two-channel theory of short-term visual storage. Unpublished doctoral dissertation, State Univ. of New York at Buffalo, 1970.

Geyer, L. H., & DeWald, C. G. Feature lists and confusion matrices. *Perception & Psychophysics*, 1973, **14,** 471–482.

Gibson, E. *Principles of perceptual learning and development*. New York: Appleton, 1969.

Goldstein, A. J., Harmon, L. D., & Lesk, A. B. Man–machine interactions in human-face identification. *The Bell System Technical Journal*, 1972, **51,** 339–427.

Gregory, R. L. Choosing a paradigm for perception. In E. C. Carterette & M. P. Friedman (Eds.), *Handbook of perception* (Vol. 1), New York: Academic Press, 1974.

Halle, M., & Stevens, K. Speech recognition: A model and program for research. *IRE Transactions on Information Theory*, 1962, IT-8, 155–159.

Hochberg, J. Organization and the gestalt tradition. In E. C. Carterette & M. P. Friedman (Eds.), *Handbook of perception* (Vol. 1), New York: Academic Press, 1974.

Holbrook, M. B. A comparison of methods for measuring the interletter similarity between capital letters. *Perception & Psychophysics*, 1975, **17,** 532–536.

Hyman, R., & Frost, N. H. Gradients and schema in pattern recognition. In P. M. A. Rabbitt (Ed.), *Attention and performance V*. New York: Academic Press, 1974.

Klatzky, R., & Thompson, A. Studies of visual synthesis—Mental construction of wholes from parts. Paper presented at the 16th Meeting of the Psychonomic Society, Denver, Colorado, November 1975.

LaBerge, D. Attention and the measurement of perceptual learning. *Memory & Cognition*, 1973, **1,** 268–276.

LaBerge, D., & Samuels, S. J. Toward a theory of automatic information processing in reading. *Cognitive Psychology*, 1974, **6,** 293–323.

Laughery, K. R. Computer simulation of short-term memory: A component decay model. In G. H. Bower & J. T. Spence (Eds.), *The psychology of learning and motivation* (Vol. VI). New York: Academic Press, 1971.

Narasimhan, R. On the description, generation, and recognition of classes of pictures. In A. Grasselli (Ed.), *Automatic interpretation and classification of images*. New York: Academic Press, 1969.

Narasimhan, R., & Reddy, V. S. N. A generative model for handprinted English letters and its computer implementation. *ICC Bulletin*, 1967, **6,** 275–287.

Naus, M. J., & Shillman, R. J. Why a Y is not a V: A new look at the distinctive features of letters. *Journal of Experimental Psychology: Human Perception and Performance*, 1976, **2,** 394–400.

Neisser, U. *Cognitive psychology*. New York: Appleton, 1967.

Palmer, S. E. Hierarchical structure in perceptual representation. *Cognitive Psychology*, 1977, **9,** 441–474.

Peterson, M. J., Meagher, R. B., Chait, H., & Gillie, S. The abstraction and generalization of dot patterns. *Cognitive Psychology*, 1973, **4,** 378–398.

Phillips, W. A. On the distinction between sensory storage and short-term visual memory. *Perception & Psychophysics*, 1974, **16,** 283–290.

Pittenger, J. B., & Shaw, R. E. Ageing faces as viscal-elastic events: Implications for a theory of nonrigid shape perception. *Journal of Experimental Psychology: Human Perception and Performance*, 1975, **1,** 374–383.

Posner, M. I., & Boies, S. J. Components of attention. *Psychological Review*, 1971, **78,** 391–408.

Posner, M. I., & Keele, S. W. On the genesis of abstract ideas. *Journal of Experimental Psychology*, 1968, **77,** 3, 353–363.

Reed, S. K. Pattern recognition and categorization. *Cognitive Psychology*, 1972, **3,** 382–407.

Reed, S. K. *Psychological processes in pattern recognition*. New York: Academic Press, 1973.

Reed, S. K. Structural descriptions and the limitations of visual images. *Memory & Cognition*, 1974, **2,** 329–336.

Rosch, E. Universals and cultural specifics in human categorization. In R. Brislin, S. Bochner, & W. Lonner (Eds.), *Cross cultural perspectives in learning*. New York: Halsted, 1975.

Rosch, E., & Mervis, C. Family resemblances: Studies in the internal structure of categories. *Cognitive Psychology*, 1975, **7,** 573–605.

Rosch, E., Mervis, C. B., Gray, W., Johnson, D., & Boyes-Braem, P. Basic objects in natural categories. *Cognitive Psychology*, 1976, **8,** 382–440.

Rumelhart, D. E. A multicomponent theory of perception of briefly exposed visual displays. *Journal of Mathematical Psychology*, 1970, **7,** 191–218.

Shepard, R. N., & Cermak, G. W. Perceptual-cognitive explorations of a toroidal set of free-form stimuli. *Cognitive Psychology*, 1973, **4,** 351–377.

Shillman, R. J. Character recognition based on phenomenological attributes: Theory and methods. Unpublished doctoral dissertation, Massachusetts Institute of Technology, 1974.

Sutherland, N. S. Outlines of a theory of visual pattern recognition in animals and man. *Proceedings of the Royal Society*, 1968, **171,** 297–317.

Sutherland, N. S. Object recognition. In E. C. Carterette & M. P. Friedman (Eds.), *Handbook of perception* (Vol. 3), New York: Academic Press, 1973.

Uttal, W. R. *An autocorrelation theory of form detection*. Hillsdale, New Jersey: Erlbaum A., 1975.

Uttal, W. R., Eskin, T. E., & Sawyer, R. Symmetry and complexity effects in form detection. Paper presented at the 16th meeting of the Psychonomic Society, Denver, Colorado, November 1975.

Warren, R. M. Auditory pattern recognition by untrained listeners. *Perception & Psychophysics*, 1974, **15,** 495–500.

Warren, R. M., & Sherman, G. L. Phonemic restorations based on subsequent context. *Perception & Psychophysics*, 1974, **16,** 150–156.

Weisstein, N., & Harris, C. S. Visual detection of line segments: An object-superiority effect. *Science*, 1974, **186,** 752–755.

Chapter 5

PERCEPTUAL PROCESSING IN LETTER RECOGNITION AND READING*

W. K. ESTES

I. SOME STRATEGIC CONSIDERATIONS

At first sight there seems to be an inordinately large gap between the long-term objectives of investigators of letter and word recognition and

* Preparation of the chapter was facilitated by Grants BG 41176 and BNS 76-09959 from the National Science Foundation.

ISBN 0-12-161909-5

the experimental paradigms with which they work. The motif tying together an otherwise extremely broad range of problems and topics is the desire to uncover mechanisms and processes involved in reading. But in scanning the enormous research literature, one rarely finds an example of a study in which an experimental subject has the task of reading as much as a few lines, let alone a page, of printed material—and when one does, the outcome rarely turns out to be instructive. The reason is that the activities involved in reading ordinary printed material are so complex as to defy direct analysis. Consequently, investigators have had to resort to the strategy that has proven serviceable in so many other research areas—that of abstracting components or aspects of reading and communicative behavior for analysis, formulating principles and models on the basis of research on simplified systems, and then trying to apply these to more complex situations.

However, the actual history of research on processes basic to reading does not by any means follow the script that would be implied by the simple-to-complex strategy. Owing to the way in which the development of research methods and the growth of theory are intermeshed in any area, it is always difficult, psychologically, to begin a line of research with the logically simplest units. Thus, for example, in the early days of research on verbal learning and memory, it evidently never occurred to anyone to examine learning and memory of a single nonsense syllable or a single paired-associate item.

Similarly, in the case of letter recognition, it seemingly did not make sense to experimental psychologists at the turn of the century to look at the process of recognizing a single letter (at least not in the case of normal adults). Substantial research began with the observation that the eyes move in jumps, with several letters being simultaneously visible during a single fixation. The ensuing implementation of tachistoscopic techniques to simulate perception during a single fixation yielded striking results: on the one hand, the observation that quite a bit can be seen in an exceedingly short time, but on the other hand, the observation of definite limitations of capacity. The search for the explanation of the capacity limitations turned out to provide the main motivation for the bulk of research in this area for many decades.

One important source of capacity limitations proved to be the masking of letters by stimuli coming before or after the letters are displayed. But with procedures developed to study masking effectively, it turned out that the process of identifying a single letter was no longer a trivial matter. Thus, it was only after many decades of research on letter perception that one began to see studies of the time course of accrual of information concerning a single character, the selective effects of particular kinds of masks on individual letters, and the systematic patterns of confusion errors that occur when a single letter has to be identified in a time too brief for complete accuracy. Significant questions began to emerge as to how individual letters are perceived: Is the process one of comparing stimulus input with a template in memory, or one of abstracting dimensional information from a stimulus display and comparing it with feature lists in memory? Concerning the latter idea, how does feature information combine to yield recognition of whole letters? Can properties of response to arrays of letters be predicted from those obtained by studies of the perception of single letters in particular visual field positions, or are there emergent processes when letters occur simultaneously? If the latter, what are the emergent processes?

For purposes of a compact review of the field, I propose to make full use of advances in theory and method and to proceed in a primarily logical sequence. Thus, we shall begin by reviewing some of the basic facts concerning the perception of individual letters, then move to the question of how far properties of the perception of letter arrays can be accounted for by processes demonstrated in the case of individual letters, perhaps together with specific rules or principles of interaction. With this background we may be able to clarify some intriguing questions that have proven intractable to direct approaches in the absence of adequate theoretical preparation. Here I refer to some of the focal issues that have been responsible for much of the research in this area: the question of perceptual units in reading, the basis of capacity limitations, parallel versus serial processing, the possibility of categorical perception, and the locus of linguistic effects in letter and word recognition.

II. PROPERTIES OF THE VISUAL SYSTEM BASIC TO READING

Parameters of time, space, and energy are of central importance to the interpretation of any study of visual perception. Reference to some of these parameters may conveniently be suppressed for brevity when experiments are conducted under conditions that have become standard, but the degree of comparability of parameters must be known whenever one hopes to make meaningful comparisons of data across experiments. This point is well understood by investigators in the tradition of psychophysics. What is not widely enough appreciated is that a shift of interest to linguistically meaningful stimuli, as in work on letter and word perception, does not reduce the importance of continuing attention to the basic physical parameters. I am not advocating pedantic descriptions of every experiment in terms of absolute values of physical parameters, but it is essential to know when these are comparable across studies and when they are not.

In a similar vein, it must be a sound strategy routinely to consider the degree to which phenomena arising in studies of letter or word perception can be interpreted in terms of more basic visual processes before considering the introduction of hypotheses that entail the postulation of higher-order processes or mechanisms. Following out this strategy, in the next few paragraphs I shall point up a few of the salient properties of the visual system that are of special relevance to our present task of understanding the perception of alphanumeric displays. For more extensive discussions of the original literature bearing on my summary remarks, the reader is referred to review articles by Ganz (1975), Robson (1975), and Thomas (1975).

A. Temporal Resolution

It has often been observed that individuals can perceive letters that are exposed in a tachistoscope for intervals as brief as 2–3 msec, possibly even less. But this result is obtainable only if immediately following the exposure the field remains free of interfering inputs for a critical interval. If another stimulus follows within this interval, the change is not detected; rather, information concerning luminance relationships in the visual field is integrated in such a way that the trade-off between stimulus intensity and time is virtually perfect (Bloch's Law). The critical interval for this trade-off depends on the luminance of the target stimuli and the background, but is typically of the order of 50–100 msec for brightness discrimination and 200 msec or longer for form discrimination (including letter recognition). However, the upper limit is not sharp; the trade-off function falls off gradually with increasing exposure duration beyond the critical interval. In the case of a patterned stimulus that is changed during an exposure, the individual may see only the first, only the second, both of the patterns, or even some composite, depending on specific parameters.

For the interpretation of many tachistoscopic experiments—for example, those involving partial report—it is important to note that, from the subject's standpoint, a stimulus exposed for only a few milliseconds may be effectively present for a much longer interval—up to a second or more under favorable conditions (a relatively intense stimulus with a dark postexposure field). During this interval of what is termed *iconic memory* (Neisser, 1967), the individual can report on the contents of the preceding stimulus display almost as though it were still present. The pattern of excitation in the photoreceptors representing the stimulus pattern must presumably be transmitted to the cortex in order to produce recognition. However, Sakitt (1975) has shown that transmission need not occur immediately following the stimulus; rather, the pattern of activity is maintained in the photoreceptors for an appreciable time, even if transmission is temporarily blocked, as by flooding the visual field with bright light.

B. Spatial Resolution

When a stimulus, such as a printed letter, is presented in the visual field, an image is projected on the retina. But the eye is unlike a camera in that the work of constructing an internal representation of the letter is not done by a mosaic of receptive units corresponding point-to-point with the retinal image. Rather, the retinal ganglion cells have receptive fields that are generally large relative to the dimensions typical of letters or interletter spacings in visual displays. Because the number of ganglion cells is large and the receptive fields vary in size, information concerning the contours of a figure, such as a printed letter, can be gained by integrating luminance differences over areas surrounding the contours.

This property of the system has two consequences of particular importance for letter perception.

1. Two adjacent contours can be resolved by the system only if they are separated by a minimum distance, typically amounting to a very few minutes of visual angle for stimuli of relatively high luminance that appear near the fovea, but becoming larger in the case of stimuli that are fainter or located further toward the periphery of the visual field.
2. The perceptibility of a contour is degraded if any form of "visual noise" (e.g., random dots) appears within a distance of several degrees of visual angle; the noise stimulation reduces the luminance difference across the contour between the figure and the adjacent background and therefore, owing to the time–intensity trade-off, more time is required to reach a given level of discrimination as the amount of adjacent noise stimulation is increased.

Furthermore, as a consequence of the imperfect resolution of the visual system in the temporal as well as the spatial domain, perceptibility of a contour is affected similarly by visual noise occurring at the same or in an adjacent location either simultaneously with or shortly before or after the exposure of a target stimulus. Depending on the temporal and spatial relationships, these effects are conventionally referrred to as instances of lateral masking, paracontrast, or metacontrast (Kahneman, 1968; Lefton, 1972). However, for the sake of simplicity I shall speak only of masking by noise (following Turvey, 1973). Eriksen and his associates (Eriksen & Collins, 1965, 1968; Eriksen & Rohrbaugh, 1970a) have demonstrated the relevance of this process to the interpretation of various phenomena of forward and backward masking in research of the type that we are concerned with in this chapter.

C. Discontinuities and the Concept of Encoding

We have seen that, owing to limits of temporal resolution, the visual system typically deals with inputs in packets of information representing stimulation summed over intervals of the order of 100 msec. In reading, as a consequence of saccadic eye movements, the pattern of stimulus input to the central area of the retina changes abruptly several times per second. Consequently, to make progress in analyzing perceptual processes in reading, it is essential to understand the principles governing this "packaging" of information.

The first point to be emphasized is that effective time–intensity trade-off does not generally describe perception across a discontinuous change from one patterned stimulus to another. Consider, for example, a study reported by Potter and Levy (1969) in which subjects were shown sequences of colored pictures of scenes (for example, landscapes, still lifes) by means of film strips at rates as high as eight pictures per second, all projected on the same region of the subject's visual field. At the highest rates, the observer sometimes "missed" a picture in the sense of being unable to recognize it on a test at the end of the sequence, but all of the pictures seen remained distinct and well-organized—they were never composites of two successive scenes. How can this apparent exception to the principle of time–intensity trade-off be accounted for?

I think we can assume that there is no exception so far as the peripheral segments of the visual system are concerned. The pattern of excitation that is transmitted centrally from the photoreceptors at a point in time shortly after the change from a particular picture to its successor in the sequence must be a composite of inputs from the patterns of stimulation existing before and after the change. An aspect of the visual system that may be importantly involved in producing, nonetheless, a sharp change in the individual's sensory experience is the known fact that some input channels from photoreceptors are especially sensitive to the on-effects that are generated by increases in stimulus intensity, whereas others are primarily devoted to the transmission of steady-state or sustained inputs from continuing sources of stimulation. Furthermore, there is independent reason to assume that activation of the on-channels tends to inhibit concurrently active sustained channels (see, for example, Breitmeyer & Ganz, 1976). Thus, in the case of two successive pictures in the Potter and Levy experiment, we might expect the on-effects occurring at the onset of a new picture to inhibit input from the sustained channels transmitting information regarding the previous picture. As a consequence, within the packet of information being integrated by the system over the interval stretching from just before until just after the change of scenes, the components originating in the first picture would tend to be degraded in relative intensity by this inhibition and thus could be distinguished from the new input and suppressed or ignored by higher processing centers (perhaps much the same as in the case of the saccadic suppression that occurs with respect to the contents of the visual field during saccadic eye movements).

This interpretation seems to fit in satisfactorily with the results of a study by Eriksen and Eriksen (1971) in which simpler stimuli (single printed letters) were presented in rapid succession at the same location in the visual field. In that study it was reported that letters presented successively for exposure durations of the order of 5–10 msec with no interval between them did appear to the observer to be present all at the same time, "giving the effect of a composite puzzle to untangle [p. 309]" but apparently with the stimulus patterns that originated from the different letters having different apparent intensities, for the subjects were able to distinguish these superposed letters with considerably greater-than-chance accuracy.

If, under the conditions of either of the experiments just cited, the duration of the first of two successive displays were progressively shortened, the individual's perception of the first display would be progressively degraded, and below some threshold duration it would not be seen at all, even though at this same duration it would have been recognizable if the second display had been replaced by a uniform dark field. One speaks of *backward masking* of the first display by the second, and of the contents of the second display as a *masking stimulus* relative to the first. The interpretation of backward masking is too complex to be discussed in detail here; relatively full discussions have been presented by Ganz (1975) and Turvey (1973). For our present purposes, the principal point to be noted is that a sufficient delay of the mask permits the synthesis of an encoded representation of the preceding stimulus pattern in short-term visual memory.

This minimal interval between onset of the target stimulus and onset of the mask, termed *encoding time* by Ganz (1975), may be as short as 10 msec following a 5-msec exposure of a single letter (Kinsbourne & Warrington, 1962), but may range up to 200 msec (Eriksen & Eriksen, 1971) depending on luminance conditions. When the original stimulus display includes more than one letter, even though they are sufficiently widely spaced to avoid any known type of lateral interaction, the encoding time is substantially increased (Eriksen & Rohrbaugh, 1970a; Weisstein, 1966). The source of this effect of additional letters upon the encoding time of a given letter cannot, at present, be specified with assurance. It might represent competition for feature detectors, in which case the magnitude of the effects should depend on the particular letters that are added, but there seem to be no relevant data available. There is stronger reason to believe that the critical factor is the need to encode positional information together with item information in the multielement case, in order to permit appropriate response to a delayed indicator (Eriksen, Collins, & Greenspon, 1967).

III. PERCEPTION OF INDIVIDUAL LETTERS

A. Models for Individual-Letter Recognition

As a consequence of the historical background of research in this area, there is surprisingly little well-developed theory and even less factual material available concerning the recognition of individual letters. Three types of models have been put forward, largely on the basis of general considerations rather than specific research results (Reed, 1973). Of these, perhaps the simplest is the class of template-matching models, according to which the basis for recognition of a letter is the matching of a stimulus input to a representation of the given stimulus pattern that is maintained in the memory system. This model has been the basis for the first attempts at visual letter recognition by computers.

So long as the type font is held constant, the template model is simple and workable; at least it is for a computer, and it presumably should also be so for a human being. However, when one considers the ability of the human being to recognize letters in almost endlessly varying type fonts, it becomes a much more difficult problem to conceive a template-matching model that would adequately represent performance. Furthermore, the types of errors people make in recognizing letters do not fit well with the idea that the individual letter patterns are indivisible units. When one obtains a *confusion matrix* by presenting various individual letters to a subject at near-threshold intensities or exposure durations, the pattern of errors shows that some pairs or small sets of letters are relatively often confused with each other, but rarely with letters outside the given subgroup (see, for example, Townsend, 1971).

The varying patterns of similarities between letters can be taken into account in extensions of the template models, which have been termed *prototype* models. The assumption of these models is that the individual maintains in memory, not a template representing a letter in a particular variety of type, but rather an abstract representation of a central tendency, or prototype, of the various particular physical patterns that have represented the given letter in the individual's past experience. In any new situation, then, the sensory pattern arising from a printed letter is assessed by computing its distance from each of the possible letter prototypes in a multidimensional space, the pattern then being assigned the name of the letter prototype that generates the smallest distance measure. Little more can be said at present concerning the class of prototype models except that it cannot be ruled out on the basis of available evidence; but on the other hand, it has not been developed to the point of being rigorously testable or of generating new predictions.

By all odds, the currently most popular type of model in relation to research on letter recognition is the critical-feature model (see, for example, Massaro, 1975; Rumelhart, 1970). In this type of model, the processing system deals with a letter, not by comparing the whole pattern with some representation in memory, but rather by assigning the pattern a value on each of a small number of dimensions, *critical features*. What is maintained in memory is not a pattern or prototype but rather a set of lists of feature values, one list for each letter of the alphabet; recognition is achieved on the basis of the matching of the set of values computed for a newly occurring sensory pattern to one of the feature lists maintained in memory. Furthermore, under some task conditions, discrimination within subsets of letters can be achieved on the basis of feature differences without full identification.

There seem to have been three more or less independent sources of interest and motivation for the development of feature models. One strand originates at the neurophysiological level in the well-known work of Hubel and Wiesel (1968), which suggests the existence of detectors in the visual system, possibly single cortical cells that are selectively sensitive to stimulation arising from particular contours (for example, vertical or horizontal edges) in the visual field. It is, of course, a long step from the demonstration of these effects in the responses of cats and monkeys to simple visual stimuli to the perception of characters by human beings, but nonetheless the idea that the same type of model might apply at these different levels is an attractive one. A second strand originates in the data on confusion errors in letter recognition that led Gibson (1969) and others to propose critical-feature systems as a basis for the identification of letters. The third, and perhaps the strongest, source of support for feature models in visual letter recognition is the analogy to the well-developed theory of recognition of elementary auditory speech sounds (Jacobson, Fant, & Halle, 1969). It has been found that a rather full account of the perception of elementary vowel and consonant phonemes can be given in terms of their values on a small number of binary-valued dimensions, or critical features, and, further, that there seems to be no way at the perceptual level to resolve these features into simpler components. Thus, for example, an individual's ability to distinguish between *bill* and *vill* when these are heard by way of a tape recording turns on a particular attribute of the initial consonant (stop versus constrictive), and this same distinguishing feature provides the basis for discrimination whenever either of these consonants occurs at the beginning of a word. Experimental studies involving the perception and short-term memory of simple speech stimuli have demonstrated that the patterns of confusion errors can be predicted in impressive detail from the critical-feature model (for example, Wickelgren, 1966).

Naturally, on grounds of elegance and parsimony, one would hope to be able to develop basically similar types of models for the visual as for the auditory recognition of letters. It must be admitted, however, that specific and rigorous empirical support for the idea has been slow in appearing. Attempts have been made to predict confusion matrices for visual letters on the basis of specific critical-feature systems, but while these have shown some promise, the success has fallen far short of that achieved in the auditory case. Thus, in a detailed regression analysis of Townsend's data, Holbrook (1975) found that neither the feature system of Gibson (1969) nor that of Geyer and DeWald (1975) predicted as well as either a simple mechanical measure of interletter similarity or subjective ratings of similarity obtained from groups of human subjects. More impressive results have been obtained by Rumelhart and Siple (1974) in a study in which they predicted confusion errors among the letters of a specially constructed alphabet in which all letters were generated by combinations of 14 basic line segments. However, we have as yet no firm grounds for judging the extent to which this result can be generalized to the more general problem of the recognition of letters in varying type styles. Some tangential but mildly supportive evidence is provided by selective masking studies (Henderson, Coles, Manheim, Muirhead, & Psutka, 1971) in which, for example, a postmask vertical grid exerted the greatest effects on letters with vertical lines as constituents and a horizontal grid exerted the greatest effects on letters with horizontal lines as constituents.

Thus the present state of affairs is that the empirical evidence with regard to feature models is quite unsettled, and the popularity of feature models in this area is primarily due to the possibilities of relating psychological data on visual letter recognition to comparable work in audition and to research on underlying neurophysiological mechanisms. An additional consideration, which may, in practice, be no less important than those just mentioned, is that the concept of critical features provides the basis for a particularly tractable type of model in which the same basic filtering and combinatorial processes can be assumed to be operating at the levels of feature extraction, letter identification, and the identification of higher-order units, such as syllables and words (Estes, 1975a, b; LeBerge & Samuels, 1974).

B. The Time Course of Individual-Letter Recognition

1. Feature Extraction

In the present state of theory, it is not possible to specify with assurance the details of the process that eventuates in the encoding of a letter stimulus to the point of immunity from backward masking. An account that has at least some indirect support is based on a conception of information processing as a succession of comparisons between transformations of the stimulus input and memory structures (Estes, 1975a,b). The memory structures may be conceived as traces of previous inputs, which function as interactive filters (Anderson, 1973) or gates in the flow of information from the receptor surface to the higher cognitive centers. In the processing system that is basic to reading, ensembles of these traces are assumed to be organized at the levels of critical features, letters, and frequently occurring letter groups.

During the encoding interval following onset of a letter stimulus, the pattern of excitation consistituting the substrate of the icon is transmitted to the first level of the filter system. If attributes of the input pattern match the subset of traces at this level corresponding to critical features of the given letter (feature detectors), then the output to the next level activates the trace corresponding to the letter; in this event, transmission continues to the appropriate response mechanism, and we say that recognition has occurred. Further, outputs of the letter detectors are transmitted to the next higher level, where an appropriate combination will activate, in turn, a representation of a letter group such as a syllable or word. If the matching process fails at either the feature or the letter level, then the stimulus is not recognized.

The conception of the feature detection and encoding process just outlined leads naturally to a simple model that may serve to interrelate some of the properties of individual letter recognition. Among the first-order facts to be accounted for are the approximately ogival growth of the probability of letter identification as a function of exposure duration and the effect of variables related to acuity (for example, line orientation and retinal location).

In a provisional formalization, I shall assume that the time t from onset of a letter stimulus to its termination by a postexposure mask can be divided into an initial segment t_0 during which a pattern of excitation is established in the peripheral visual system, and a second segment $t - t_0$ during which attributes of this input pattern are compared with feature representations in memory. During any brief interval Δt within the second segment, there is probability $\lambda \Delta t$ that any one feature detector of the set corresponding to the stimulus letter will be activated by the input. Consequently, the probability distribution of activation times for a feature detector has the exponential density

$$f_t = \lambda \exp[-\lambda(t-t_0)], \tag{1}$$

where "exp" is the base of natural logarithms, and the probability that the detector has been activated by the end of the interval is

$$F_t = 1-\exp[-\lambda(t-t_0)]. \tag{2}$$

Finally, since the comparison processes for different detectors are assumed to proceed independently, if the stimulus letter comprises N features, the probability that all are activated is given by

$$P_t = \{1 - \exp[-\lambda(t-t_0)]\}^N. \tag{3}$$

Within the successive filter model, once all of the detectors associated with a letter have been activated within a critical interval, the representation of the letter in the memory system is activated in turn. In this event, we say that the letter has been encoded, with the dual implication that recognition will not be impaired by backward masking and that an overt identification response will be made if called for by the task.

It will be easy to demonstrate that the family of curves generated by Eq. (3) with various combinations of values of λ and N is descriptive of the time course of individual letter identification. First, however, let us consider the variables that would be expected to determine the value of the rate parameter λ. This parameter, reflecting the probability of activation of a feature detector within any short interval of stimulus exposure, should depend to a major extent on the conditions determining visual acuity for the relevant types of contours.* In particular, resolvability of

* The relevance of a number of aspects of acuity to the interpretation of feature detection was pointed out to me by C. W. Eriksen.

vertical and horizontal line segments is known to be better than that of oblique segments (Thomas, 1975). Furthermore, length of a straight contour favors discriminability; hence curves should be least readily detectable, since they are in effect composed of concatenations of short rectilinear and oblique segments. As a consequence of these differential acuity requirements, we should expect that feature detectors associated with different types of contours would have different energy thresholds and, therefore, that sufficiently precise measurements would show that different aspects or components of a letter become visible at different points during the encoding interval. Some possibly relevant evidence is available in a study by Johnson and Uhlarik (1974). Their subjects were presented with very brief exposures of various geometric forms and were asked immediately following each exposure simply to draw what they believed they had seen. The forms included such items as a square, a triangle, and a composite of uppercase letters *H* and *B*. Initially, a given form was exposed for 2 msec, then, if the subject's drawing did not completely identify the form, another exposure was given with a 1-msec increment in duration, and this procedure was continued until identification was complete. The data yielded an orderly pattern in the emergence of different components with increasing exposure time; in general, vertical and horizontal line segments appeared first, then oblique segments, then curves.

Taking this last result, together with the feature model for identification, we might expect that different letters of the alphabet would differ systematically in their encoding times. Some relevant data are available from an unpublished experiment by David H. Allmeyer and myself. In a part of a study conducted for another purpose, we displayed single uppercase letters on a CRT screen for exposure durations ranging from 25 to 85 msec, with each exposure preceded and terminated by a pattern mask. The subject's task was to identify the letter displayed on each trial. For an analysis bearing on this problem, we can categorize the consonants used into three groups: a rectilinear subset, with constituents including only horizontal and vertical lines (*F, H, L, T*); an oblique subset, in which each letter includes oblique constituents (*K, M, N, V, W, X, Y, Z*); and a curved subset (*C, D, G, J, P, Q, R, S*). Mean proportions of correct identifications of letters in the three subsets, representing pooled data for four highly experienced observers, are shown in the lower panels of Fig. 1, plotted as a function of exposure duration. It will be seen that the curves differ appreciably and exhibit precisely the ordering expected on the supposition that encoding of a letter depends on the activation of all of its feature detectors, and that detectors for rectilinear, oblique, and curvilinear components become available in that sequence. The theoretical

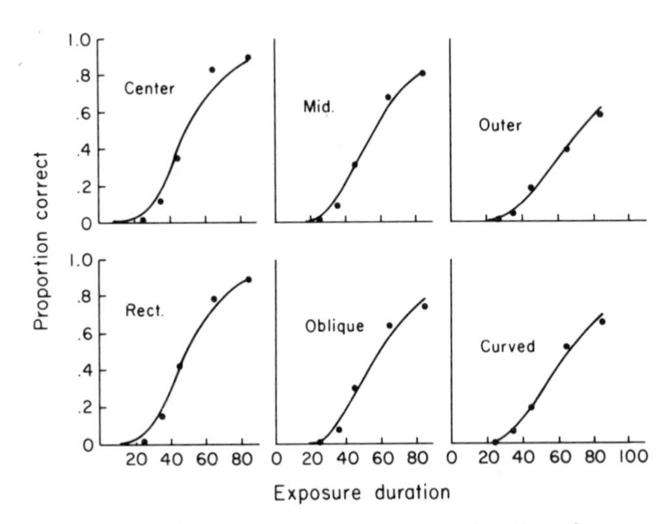

FIG. 1. Identification of single consonant letters as a function of exposure duration. Upper panels represent data grouped by position of the letter array in the visual field; lower panels represent data grouped by letter type. Solid curves are theoretical functions (see text).

functions were computed from Eq. (3). A computer search of parameter values with a least-squares criterion yielded estimates of 10 msec for t_0, the minimum encoding time, and 5 for N, the number of features per letter, both being constant over the three letter types. Values of λ were .052, .042, and .036 for rectilinear, oblique, and curved letters, respectively.

Of course these data must be interpreted with due caution, since other properties of letters may be confounded with differences in geometric form. For example, one might immediately ask whether there are systematic differences among these letter groups with respect to the frequencies with which the given letters occur in English text. Indeed there are; however, the ordering of the three letter groups on the basis of frequencies in English is not the same as that exhibited in Fig. 1. The vertical–horizontal subset is highest in average frequency, but the other two groups are interchanged. Furthermore, vowels are much higher in average frequency than any of the consonant groups (this being the reason why they were not included in the analysis), but the identification function for vowels falls appreciably below that for the rectilinear consonants. It might be added that data reported by Mayzner (1972) for a single, intensively studied subject, using a somewhat different style of type and a range of exposure durations from 12 to 20 msec, yielded the same ordering of letter groups.

In addition to the contours making up a letter and the intensity and durational properties of an individual display, the principal additional parameter of individual letter identification is spatial position of the letter in the visual field. It is well known that at photopic levels of illumination the resolving power of the visual system is highest when a simple target stimulus (for example, a grating) is viewed with the center of the fovea; resolution decreases as the image of the target moves toward the periphery (Thomas, 1975). Thus, one might anticipate that the same would be true for the identification of single letters.

Some relevant data are at hand in the study by Allmeyer and myself, cited earlier. In our experiment, the stimulus letters subtended viewing distances of approximately .37° wide and .53° high, and appeared in random sequence at any one of 14 possible letter positions spaced in a horizontal row across the visual field, extending approximately 4° on each side of the central fixation point. It would be interesting to examine the perceptibility of individual letters at different distances from the fixation point, but our data are not adequate for this purpose. However, pooling all of the letters used, we were able to construct functions representing probability of identification of a letter as a function of position in the visual field for each of the exposure durations represented in Fig. 1. These functions proved to be virtually symmetrical and concave downward, and, of course, ordered one above the other in relation to exposure duration. The time course of letter identification at different positions is illustrated in the upper panels of Fig. 1 in terms of pooled data for the central four letter positions, the intermediate positions, and the outermost four positions. The theoretical functions again represent least-squares fits of Eq. (3). Estimates of t_0 and N are as before; λ values are .050, .044, and .032 for the center, middle, and outer positions, respectively. It is clear that, as anticipated, the rate of feature extraction from letter stimuli is greatest near the fixation point and decreases systematically as the target moves toward the periphery. In a later section we shall consider the implications of these findings for the recognition process when letters varying in both constituent features and spatial positions are presented simultaneously in multiple-letter displays.

2. Processing Stages and Short-Term Memory

Once the encoding of a letter is accomplished, the coded representation remains in an active state for only a short time, the level of activity evidently decaying exponentially to zero over the course of a few hundred milliseconds (Averbach & Sperling, 1961; Ganz, 1975). However, during this interval, neural messages from the coded representations of letters

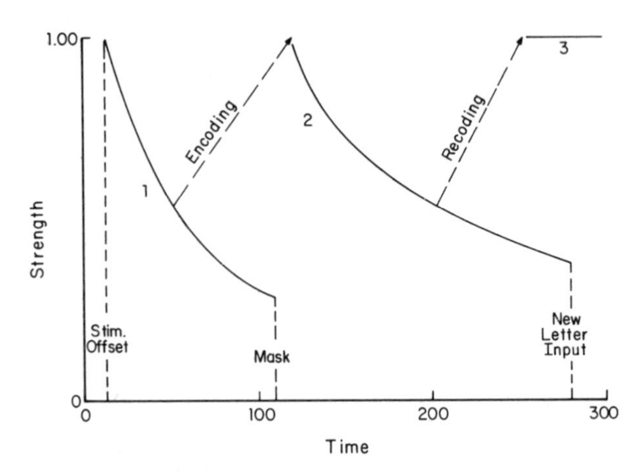

FIG. 2. Hypothetical functions illustrating the assumed course of transformations of information following display offset. Curve 1 shows decay of visual persistence terminated by a postmask. Curve 2 represents retention loss of an encoded representation of the target letter, terminated by new letter input. Curve 3 denotes a recoded representation of the letter in the auditory–articulatory system, maintainable indefinitely by rehearsal.

may be transmitted over associative paths to excite representations of the same letters in the auditory–articulatory system, commonly termed *short-term auditory memory*, where the encoded information can be maintained indefinitely by verbal rehearsal (Estes, 1973, 1975a; Sperling, 1967). A resumé of the overlapping stages of processing is illustrated in Fig. 2. Curve 1 denotes the course of visual persistence, Curve 2 the decay of an encoded visual representation, and Curve 3 the maintenance of a recoded auditory representation. The values illustrated for rates of decay and durations of the different representations should not be taken as hard-and-fast figures; rather, they represent summary impressions arising from a review of the substantial relevant current literature.

It should be emphasized that since the hypothetical curves in Fig. 2 are idealized representations of processes that overlap in time, an observed retention curve, as might be obtained in a partial report experiment with a delayed indicator (see Section IV,A,2), must constitute a mixture. A bit of notation is needed in order to show how the mixture might arise: Let x_t denote the magnitude of visual persistence, y_t the probability that an encoding of the stimulus letter exists in an active state in memory, z_t the probability that there is a recoded representation of the letter in auditory short-term memory, and p_t probability of correct report if an indicator is presented at time t following a brief stimulus exposure. Curve 1 in Fig. 2 is described by the function

$$x_t = ax_{t-1},$$

and the descending portion of Curve 2, representing decay of the visual code, is described by

$$y_t = \beta y_{t-1},$$

where α and β are constants with values between 0 and 1. However, the letter code may also be active at time t if encoding occurred during the preceding interval, this event occurring with some probability c if visual persistence is still available. Hence the full recursion for y_t is

$$y_t = \beta y_{t-1} + (1 - \beta y_{t-1})cx_{t-1}.$$

Similarly, an auditory code will be available during interval t if it existed during the preceding interval or if it did not but recoding occurred,

$$z_t = z_{t-1} + (1 - z_{t-1})\acute{c}\beta y_{t-1}.$$

Finally, probability of correct report at time t is the sum of the probabilities of report based on (a) visual persistence; (b) an active letter code in the absence of visual persistence; or (c) an auditory code in the absence of a visual code:

$$P_t = x_t + (1 - x_t)[y_t + (1 - y_t)z_t]. \tag{4}$$

Illustrative curves computed from Eq. (4) are presented in Fig. 3, assuming minimum coding times of 1 and 2 units for the visual and auditory codes, respectively. Also, just to show that the predicted form does arise in practice, empirical points are included, read from individual retention curves reported by Averbach and Sperling (1961, Fig. 7). In order to accommodate the large difference between retention functions

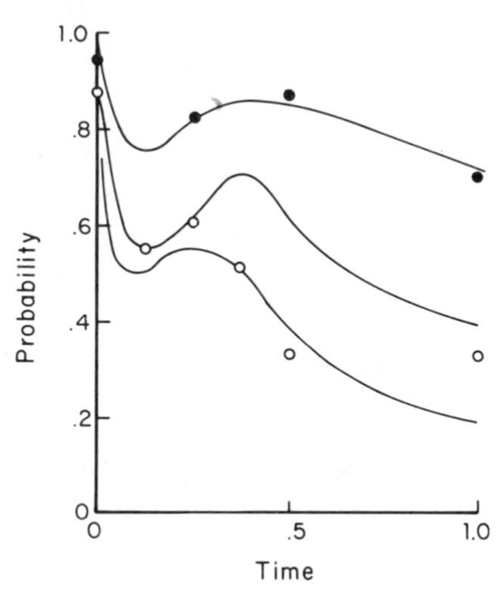

FIG. 3. Theoretical functions for probability of report of a letter at varying intervals between display offset and indicator. (● = dark field; ○ = light field.) [Empirical points after Averbach & Sperling (1961).]

obtained with dark versus bright postexposure fields in this model, it is necessary to assume that both the rate of decay of visual persistence and the rate of decay of the visual code are greater (i.e., α and β, respectively, are lower) with the bright field. The nonmonotonicity of the curves might well not be apparent, of course, in the case of group data, since in general the maxima and minima for subjects with different parameter values would not coincide.

IV. PERCEPTION OF MULTIPLE-LETTER DISPLAYS

A. Problems of Methodology

To proceed from the perception of individual letters toward perceptual aspects of reading, our next step must be the consideration of the changes that occur in the perception of letters when they appear in the context of other letters, as in the words of a printed message. We shall be primarily concerned with interactions that may occur at the level of stimulus processing; but before reviewing these, we need to mention some of the considerations of methodology that must be kept in mind when interpreting relevant research.

Surrounding a target letter with other letters in a display necessarily introduces other factors that may influence an observer's performance beyond possible interactions at the level of visual processing. There is no doubt that in much of the reading that occurs in everyday life, individuals gain considerable information from context and are able to identify words and larger passages without necessarily perceiving every individual letter (as in reading a signboard at a distance or a restaurant menu in dim light, etc.). Consequently, research on perceptual interactions must generally be conducted with displays of unrelated letters and exposure conditions that render negligible the effects of other aspects of context. However, even unrelated letters introduce possible complicating factors: confusability between letters, increased memory load, and response competition are among the most conspicuous.

1. Decision Problems Arising from Confusability

Even if the letters surrounding a target letter do not produce a familiar or meaningful unit, they may produce effects at a decision level, rather than a perceptual level, owing to the confusability of letters that have features in common. Suppose that a subject's task is to report which of two target letters, L_1 or L_2, occurs in a briefly presented display of unrelated letters, with exposure conditions such that the stimulus input is somewhat degraded, making perfect accuracy impossible. If the input from L_1 is sufficient to activate some but not all of its constituent features, and if there are no other letters simultaneously present that share features with L_2, then identification of L_1 may occur without error. But if there are one or more other letters present whose degraded inputs include features overlapping with those of L_2, then the observer has a decision problem and is likely to conclude erroneously that L_2 rather than L_1 was included in the display. These effects of confusability at the decision level have been investigated extensively, and the results have been incorporated in quantitative models (Estes, 1974; Gardner, 1973; Shiffrin & Geisler, 1973). Consequently, the role of confusability can—and should—be adequately allowed for in interpreting the results of research on multiple letter displays.

2. Positional Uncertainty and Response Competition

In one of the most widely used techniques for exploring recognition and identification in multiple-letter displays, a row of characters is displayed briefly, following which an indicator (for example, an arrow under one of the character positions) signifies to the subject the location whose contents he or she is to try to report. In order to interpret the results of experiments using this paradigm, it is necessary to know something concerning the precision with which information concerning positions of display characters is entered and maintained in short-term visual memory. I have not been able to find in the literature any reports of direct parametric attacks on this problem; hence, to give an idea of the positional uncertainty that is to be expected under standard tachistoscopic conditions, I shall have to use some data from an unpublished study conducted in my laboratory.

Except for the obtaining of positional information, the apparatus and procedures of this study were identical to those described by Estes (1972). Single rows of letters were displayed on a CRT screen, with the subject's task being to determine which of two uppercase target letters, *A* or *T*, was included in each display. Each trial began with the display of a row of six dot matrices in the positions in which letters would appear, then a 100-msec display of one of the target letters embedded in a row of heterogeneous noise letters, then again the row of dot matrices, which remained in view until the subject operated a response key to indicate the target he thought he had seen. Immediately following the identification response, the subject operated another key, which was in a row of six, to indicate the position in which he believed the target had appeared. The letters, as displayed, each subtended a visual angle of about .75° in width with interletter spacing of approximately .45°, so that the six-letter display, centered with respect to the fixation point, spanned about 6.75° of visual angle.

Data obtained for 18 subjects are the basis for the functions plotted in Fig. 4. These curves, representing approximately 300 observations per point, show the proportions of instances in which a correctly identified target letter occurring in the location signified by the Arabic numeral on the curve was assigned by the subject to each of the six possible display locations. The most conspicuous trends are the sharply decreasing accuracy of localization from the center of the field toward the periphery (as evidenced by the progressive reduction in proportion of correct placements and increasing dispersion of positional responses around the correct location) and the increasing skewness of the functions as the target moves from center to periphery. Analyses of the distributions of placement responses on trials when the target letter was not correctly identified indicate that these functions cannot be accounted for on the basis of response bias and may be taken to provide information concerning the uncertainty attaching to the spatial position of the encoded representations of letters in the individual's short-term visual memory system. In my study, determinations were made only immediately following display offset, but evidence from other experiments makes it clear that memory for position decays rapidly over a few hundred msec following

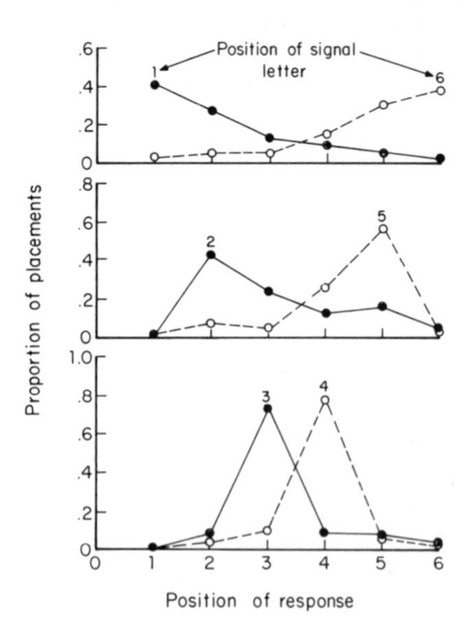

FIG. 4. Positional uncertainty functions. The Arabic number on each curve indicates the position of the target letter, and the plotted points represent proportions of instances in which a target presented in the given position and correctly detected was assigned by the subject to the various serial positions.

offset of a display (Townsend, 1973). This process is of considerable importance for the interpretation of experiments based on indicator methods (see, for example, Eriksen *et al.*, 1967; Lowe, 1975).

Although the facts presently available do not provide a basis for a definitive theory of the way in which indicator information is processed, we can at least offer an account that is compatible with the information (summarized in preceding sections) regarding the processing of individual letters and with the available information concerning spatial uncertainty. A basic assumption is that the result of the encoding stage of letter processing is a set of letter representations in the visual short-term memory system, each including feature information that identifies a letter and positional information that specifies its location in the visual field relative to the fovea and to nearby anchor points, such as other characters or spaces. The positional information is, however, sufficient only to locate the letters subject to an uncertainty gradient of the form illustrated in Fig. 4.

If a positional indicator follows a display, the indicator stimulus is similarly encoded in terms of both feature and positional information. The positional component of the encoded indicator representation is then compared to that of each of the active letter codes, and the closest match determines the letter position that will be the basis for response. Over an interval of time following display offset, the level of activation of both feature and positional components of the letter representation decays, and consequently, response to an indicator becomes less accurate.

As a consequence of positional uncertainty, we must expect an apparent interaction between simultaneously presented letters (even if they are sufficiently dissimilar to preclude effects of confusability and spaced widely enough to make lateral masking negligible), provided that the letters share the same response mechanism. In studies of Eriksen and Eriksen (1974) and Eriksen and Hoffman (1972), subsets of letters were assigned to different indicator responses. For example, the subject might be instructed that whenever a letter belonging to Subset 1 occurs at a target location, he is to operate one response switch, and whenever a member of Subset 2 appears at the target location, he is to operate a second response switch. Under these conditions, it is found that the subject's speed in responding correctly to a letter at the target location is slowed if a member of the other subset is present simultaneously, with the effect increasing as the distance between the two letters is reduced. Since this effect is independent of the degree of similarity between the stimulus patterns of the letters, it must be attributed to a process of response competition.

A tentative interpretation of this interference effect has been offered in terms of the observer's problem in collating information concerning the identities and positions of letters (Eriksen & Eriksen, 1974; Estes, Allmeyer, & Reder, 1976). This provisional interpretation assumes that the processes of feature extraction and encoding of letter representations occurring within a small area of the visual field take place simultaneously and in parallel, or nearly so. But if the observer's task is to report the one letter of a group that is pointed out by an indicator or that occurs at a predesignated target location, then it is necessary also to encode information concerning the spatial locations of the letters relative to the indicator or the target position. This encoded positional information concerning any two letters becomes increasingly similar as the two letters are brought closer together in the visual field, consequently, more processing time is required to resolve the uncertainty and determine which should be the basis for the required response. Because the sometimes substantial effects of response competition have only become evident very recently, it is impossible to adequately take these effects into account in interpreting much of the earlier work addressed to the problem of stimulus interactions in multiple-letter displays. Hence, our understanding of stimulus interactions is much more limited than one might have supposed on the basis of the rather extensive literature.

B. Lateral Masking

On the basis of what we know about the course of processing, what should we expect to be the consequences of a display of another stimulus at the same or a closely neighboring location, either just prior to, simultaneously with, or just subsequent to the display of a target letter? First of all, we certainly shall have to take account of effects arising from the imperfect temporal and spatial resolution of the visual system (see Section II). The feature-extraction or encoding process operates on a pattern of excitation maintained in the photoreceptors for an interval following a brief letter display; consequently, if another stimulus, whether it be a patterned stimulus or simply a random noise field, is displayed in the same location within a few tens of milliseconds before or after the display of the letter, then the pattern of receptor-cell activity transmitted to the cortex will include a composite of that arising from the letter and that arising from the mask stimulus. Furthermore, since the receptive fields of individual receptor cells overlap, the same will be true if the mask stimulus is not superposed on the letter stimulus but occurs in a sufficiently close, neighboring location.

Experimental studies of this type of masking show just the properties that would be expected if we assume an integration of input from patterns of receptor activity that occur within the same short interval of time, regardless of the sources (Eriksen, 1966; Turvey, 1973): The degree of the masking effect is approximately symmetrical for mask stimuli occurring before and after the test stimulus, decreases with lateral separation of the target and mask stimulus in the visual field, increases with intensity of the mask, and decreases with intensity of the target.

The critical separation between adjacent letters needed to preclude lateral noise masking can be quite closely specified on the basis of research with simpler stimuli. It has been known for well over a century that adjacent line segments must be separated by a visual angle of 1–5 min of arc at retinal eccentricities of .5 to 8° in order to be resolved by the eye (for example, data of Hueck, cited by Ruediger, 1907). Similarly, the gap in a Landolt ring can be detected by the normal eye in the presence of a tangential adjacent line segment in the field only if the ring and the line are separated by a space of about 5 min of visual angle (Flom, Weymouth, & Kahneman, 1963). At very small separations, then, we must expect a letter to suffer noise masking from adjacent stimuli, whether these are letters or unpatterned stimuli. Thus, we are prepared for the finding of Eriksen and Rohrbaugh (1970b) that perception of a target letter is equally impaired by an adjacent letter or an adjacent disc stimulus at a separation of 5 min of arc, the degree of masking decreasing to become negligible beyond about .5°.

Of greater import for the perception of groups of letters is the pattern masking that must be expected if two letters (or a letter and another stimulus including letter-like contours) occur in close temporal or spatial juxtaposition. When the pattern of excitation initiated by a letter stimulus has been transmitted to the cortex, the central processor has the task of determining which aspects are to be encoded together as a representation of an object—in the present example, a letter. But signals coming from the periphery over an interval of the order of 100 msec are processed together. Consequently, if a letter and another patterned stimulus occur either successively and in the same location or simultaneously and spatially close together during this interval, then the central processor has the additional problem of determining which of the contours should be encoded together to form a representation of the target, and which should be interpreted as constituents of another stimulus or dismissed as part of the background. Two of the guiding principles in this process are intensity of the traces and spatial proximity (Ganz, 1975). Among coexistent traces, the more intense ones tend to be interpreted as part of the figure, and the weaker ones as part of the ground; those occurring in spatial proximity tend to be encoded together.

We should anticipate that this pattern masking would be stronger in the backward than the forward direction (that is, of two successively presented stimuli in the same location, the second would tend to mask the first), since, among coexistent traces of two successive stimuli, those of the second will normally be more intense than those of the first. This is indeed the case (Turvey, 1973). More importantly for our present interests, we should expect that lateral pattern masking will extend over somewhat greater spatial separations than noise masking and that the masking effect on a letter target will be greater if the adjacent stimulus is made up of letter-like contours than if it is not. Though no single studies have systematically varied all of the relevant parameters, on the whole, results of a number of related studies seem to fall quite well in line with expectations. At separations of less than about .1° of visual angle, masking effects presumably represent a mix of noise and pattern masking; as separation increases, the effects decrease, but more rapidly for masking stimuli such as solid disks or rectangles, which do not include letter-like contours (Eriksen & Rohrbaugh, 1970b; Shaw, 1969; Strangert & Brännström, 1975). Furthermore, the minimal separation required to preclude lateral masking of a target letter increases approximately linearly with retinal eccentricity of the target (Bouma, 1970). This pattern of relationships was the basis for our conception of feature extraction in terms of input channels to feature detectors, with the channels being distributed over the visual field, decreasing in density from center to periphery, and exhibiting lateral inhibitory interactions that decrease with increasing separation (Estes, 1972).

A parameter that should be expected to be of critical importance with respect to lateral inhibition is the relative onset time of adjacent characters. It is well known that on-effects, as distinguished from steady-state effects, are an important constituent of visual stimulation, especially in peripheral vision (Breitmeyer & Ganz, 1976), and consequently we should anticipate that maximal lateral masking of a letter will occur if an adjacent letter occurs simultaneously with or shortly after its onset. This expectation is confirmed in part by results showing that a noise character that appears simultaneously with a letter at a separation of .1–.5° exerts a very much larger masking effect than the same character at the same separation if its onset appreciably precedes that of the target letter (Estes *et al.*, 1976; Estes, Bjork, & Skaar, 1974). Less is known about the specific kinds of contours of mask stimuli that exert the greatest effects on adjacent letters. The local character of the lateral interactions is, however, further indicated by the results of Wolford and Hollingsworth (1974a) showing that for right-hand letters (that is, letters such as an uppercase *B* or *R*, whose distinguishing contours are on the right-hand side) a greater masking effect occurs if an adjacent character appears on the right than if it appears on the left.

C. Serial Position Effects

One of the most frequently reported empirical functions from tachistoscopic studies is the serial position curve, that is, a function exhibiting the proportion of letters detected or identified from each of the letter positions in a multiletter array. Scores, possibly hundreds, of these serial position curves have appeared in the literature, and they take on a bewildering variety of forms—concave upward, concave downward, symmetrical, skewed, *W*-shaped. A number of reviewers have attempted to make sense of this heterogeneity, proceeding from analogies to serial position effects in verbal learning (Harcum, 1967) or from various hypotheses concerning the habits that individuals develop with respect to scanning arrays of letters in various orders—left to right (Geyer, 1970) and outside to inside (Coltheart, 1972), among others. None of these efforts seems up to the task of interpreting the specific effects of various parameters, and none of them has won wide acceptance. However, by proceeding from the information we have already assembled concerning the local interactions that occur between adjacent letters, I think we can arrive at a sensible organization of a considerable range of findings.

1. Single-Letter Displays

We have noted in a previous section that if letters are presented singly at various locations across the visual field, the serial position function is smoothly concave downward, as illustrated by the solid curve (no mask) in Fig. 5(a). This and the other curves in the figure represent patterns that seem to be well established for the given conditions. The curve just referred to closely follows one reported by Merikle, Coltheart, and Lowe (1971), but could just as well represent one of those obtained by Estes and Allmeyer (Section III,B,1) with exposure duration as a parameter. This function for single letters represents just what one would expect on the basis of the well-known systematic decrease in resolving power of the visual system as the stimulus moves from the center toward the periphery.

I have found no published serial position curves for the case of single-letter presentations with a forward mask, but the study of Merikle *et al.* (1971) includes a backward mask condition, illustrated in the dashed curve in Fig. 5(a). Since there can be no lateral interactions in this

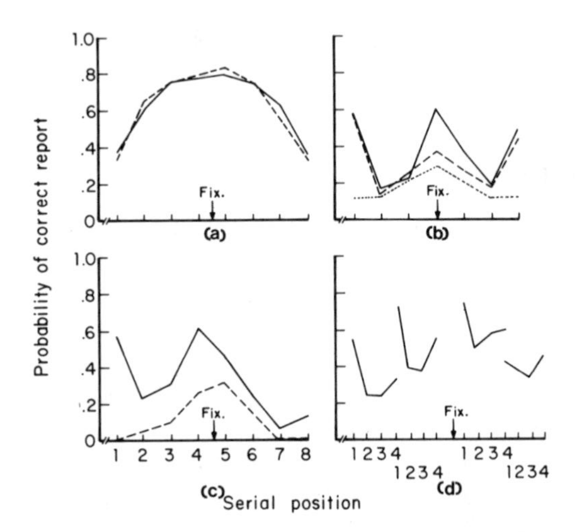

FIG. 5. Representative forms of serial position curves for (a) single letters (— = no mask; --- = back mask); (b) a centered multiletter display with a single-letter probe procedure (— = no mask; --- = back mask; --- = for mask); (c) a centered multiletter display with full report procedure (— = 100 msec; --- = 25 msec); and (d) four-letter displays appearing in different locations in the left and right visual fields with a full-report procedure.

paradigm, we observe a uniformly concave downward serial position curve.*

2. MULTIELEMENT DISPLAYS WITH SINGLE LETTER PROBES

Suppose, however, that a row of letters were presented simultaneously at the same positions for a brief exposure, with the subject being asked after each display to report the letter that he or she had seen at some one of the display locations. Then we would have to expect the central–peripheral gradient of acuity to combine with the effects of lateral masking between neighboring letters (for a similar analysis, see Wolford, 1975). The lateral masking effects would necessarily be greatest in the interior of the display, where all letters have neighbors, and least at the ends, where each terminal letter has only a single neighbor and blank space on the other side. Combining these considerations, we are prepared for the W-shape represented in the upper curve (no mask) in part (b) of Fig. 5; this function closely resembles published curves reported by Winnick and Bruder (1968), Henderson and Park (1973), and Merikle and Coltheart (1972).

* The mask curve does not run below the no mask curve only because the exposure durations were adjusted (30 and 15 msec, respectively) to equate the mean heights.

If the multielement display is preceded by a noise masking stimulus, which, as we have seen, would be expected to exert its effects at an early stage of processing, then we should anticipate an overall lowering of the serial position curve, since the patterned inputs from all display locations would be degraded. Further, there would be a reduction in lateral masking effects since the forward mask would tend to blur contours that might otherwise enter into lateral interactions. The forward mask curve in Fig. 5(b), closely following one reported by Merikle and Coltheart (1972), shows much the anticipated effect.

In the case of a multiletter display we are on less firm ground in attempting to predict the effect of a backward mask, since we have to deal with the concurrent lateral masking effects arising from interactions between traces of the simultaneously presented letters of the display and between these and the contours of the subsequent pattern mask. The result, which seems to be quite well established by several studies, is illustrated by the backward masking curve in part (b) of Fig. 5. This curve closely follows one reported by Henderson and Park (1973) and appears to agree in essentials with the results of Merikle and Coltheart (1972). The sharp difference between the single- and multiple-letter cases is the occurrence in the latter case of a substantial effect of the backward mask at central locations but no effect at peripheral locations. Most likely, the simplest interpretation of this result is that the encoding time for a letter is increased by the occurrence of closely adjacent patterned stimuli as a consequence of the additional processing needed to overcome the effects of lateral masking. Letters toward the center of the display have more neighboring letters to generate masking, and hence would have longer encoding times and higher probabilities of being still unencoded at the time of onset of the postmask.

3. Supraspan Full Report

The single-letter probe technique used to obtain curves of the type illustrated in Fig. 5(b) is perhaps the best technique available for investigating many aspects of the serial position function, since it minimizes complications having to do with an individual's limited capacity for maintaining short-term memory for many letters simultaneously. However, the technique suffers the limitation of not bringing into play processes involved in constructing an ordered perceptual representation of a group of letters, which must be an important constituent of the reading process. Hence, substantial interest continues to attach to experiments in which an array of letters is displayed, with the subject's task being simply to report as many of the letters as he can.

When a full-report procedure is used with arrays and viewing conditions otherwise similar to those in the studies just referred to, we find a drastic change in the appearance of the serial position curve, as illustrated in the upper, solid curve in part (c) of Fig. 5. This curve closely resembles one reported by White (1969) but could almost equally well have been taken from studies of Bryden (1966), Merikle, Lowe, and Coltheart (1971), or many others. It is characteristic of the results obtained for full reports when the number of letters presented is greater than four or five and with exposure duration of the order of 100–200 msec. If, however, with the same material, exposure duration is shortened to about 25 msec, then one observes a curve such as the lower, dashed curve of Fig. 5(c), which closely resembles one from the same study of White (1969) just referred to.

As usual, we shall attempt to account for these results in terms of simpler processes with which we are already familiar. Let us first consider the upper curve. We must assume that if White's subjects had been tested by single-letter probes following exposure of the eight-letter display, they would have yielded a symmetrical serial position curve much like the upper curve of Fig. 5(b). The skewness that appears in the case of the full-report procedure I would attribute to limitations of short-term memory. The visually encoded representations of letters that are established during the display and the interval of visual persistence decay rapidly (see Fig. 2), and thus must be recoded in the auditory–articulatory system in order to be reported. But we have already seen that this recoding requires time of the order of 200 msec per letter, and the time of the stimulus exposure plus the interval of visual persistence does not suffice for more than four or five letters to be so recoded. If the recoding and reporting occur in a left-to-right order, owing to well-established reading habits and the requirements of the experimental task, then the first few letters at the left can be entered in short-term memory and maintained for the time needed to report them, but those farther to the right will have progressively less chance of becoming encoded in short-term memory and, therefore, progressively greater chance of failing to appear in the report.

Some investigators—for example, Wolford and Hollingsworth (1974b)—have raised an objection to this interpretation on the ground that reports from tachistoscopic displays do not reflect the auditory confusion errors that have been taken to index auditory coding of letters in other short-term memory experiments (as by Conrad, 1967; Estes, 1973). However, I am not satisfied that the objection is cogent. In the short-term recall experiments, conditions of stimulus presentation are such that the subject can clearly perceive and even pronounce all of the letters as they appear, and hence we must assume that they are all entered in auditory short-term memory. It is during a subsequent retention interval, when the subject is required to shadow digits or perform some other interfering task, that this memory becomes degraded and gives rise to confusion errors reflecting auditory properties of the letters. But in the tachistoscopic situation, the source of the incomplete reports comes earlier. The problem for the subject is that there is insufficient time to accomplish the recoding into the auditory system of all of the letters for which visual codes were established during and immediately following the display. Errors in report do not reflect auditory confusions because the source of these errors is degraded or forgotten visual codes (rather than auditory codes that have become degraded as a function of interference following display offset).

The effect of reducing exposure time in the full-report situation can be interpreted on the assumption that the effect of this change is primarily one of reducing the amount of information entering the feature-processing system. With only a 25-msec exposure, the peripheral sensory system is unable to integrate the luminance differences within the visual field over a sufficiently long time to resolve all of the contour differences, especially toward the periphery, and consequently fewer letters can be encoded. Since, on this reasoning, the effect of a reduction in exposure time, per se, is, like that of noise masking, primarily one of degrading the input, we should expect the resulting serial position curve to be similar to that produced by forward masking. Indeed, this is the result observed (Fig. 5(c).

4. Subspan Full Report

If our interpretation of the full-report functions is correct, then the marked skewness and asymmetry in the visual field observed at the longer exposure durations should be eliminated if we simply reduce the number of letters displayed simultaneously so as to minimize the role of the bottleneck imposed by the necessity for recoding the letters in short-term memory. With displays of up to four or five letters, we should expect the forms of the serial position functions to be predictable almost exclusively from the overall decrease in acuity from center to periphery of the field and from the lateral masking effects that tend to reduce the recognizability of the interior letters of a string, regardless of where the string is located.

Serial position curves for strings of four unrelated letters show precisely the properties anticipated, as illustrated in part (d) of Figure 5, the curves following closely in form those reported by Estes *et al.* (1976). In their experiment, the subjects were required to fixate at the center of the display field, and strings of four letters, with patterned pre- and post-masks, were displayed at any one of four locations left or right of the fixation point for 150 msec. In the same experiment, data were also obtained for a much longer exposure duration of 2400 msec (but with the subjects' eye movements monitored to ensure that fixation was maintained at the center of the field) and yielded serial position functions very similar in form to those shown in Fig. 5(d), except for an elevation in overall level. These results, together with those of Townsend, Taylor, and Brown (1971), which were obtained with constant fixation but unlimited viewing time, make it clear that the principal features of serial position functions result from lateral interactions that are not limited to brief displays. Further, Taylor and Brown (1972) have shown that the forms of the functions are virtually identical under monocular, binocular, and dichoptic viewing, thus confirming our supposition that the major interactive effects are central, rather than peripheral, in nature.

5. Transposition Errors

Not only lateral masking, but also processes having to do with positional uncertainty of encoded letter representations should have predictable effects on data obtained under the full-report procedure. Referring back to our consideration of positional uncertainty in Section III,A (and, especially, to Fig. 4), we note that in experiments of the kind represented in part (d) of Fig. 5, the positional uncertainty gradients of letters adjacent in a display will appreciably overlap. Consequently, we must expect that even when subjects can identify letters, they will not always be able to report them in their correct relative positions. The earliest study I have found in the literature addressed specifically to this point was reported by Koestler and Jenkins (1965), and revealed what, at the time, seemed a surprisingly large incidence of "transposition errors." These errors, reflecting incorrect ordering of correctly reported letters, occurred with substantial frequency in the 150-msec data of the Estes, Allmeyer, and Reder study, reaching levels as high as 40% at interior positions of displays for some subjects. That these errors reflect an imprecision in the encoding of information about letter position, rather than solely retention loss following display offset, is evidenced by the fact that transposition errors occurred with only slightly reduced frequency at a 2400-msec exposure duration. At all exposure durations, the frequency of transpositions tended to increase from center toward periphery of the visual field (as we should expect from consideration of the uncertainty gradients in Fig. 4), and was greater for interior than for end positions of the four-letter displays.

Another implication we might draw from the forms of the uncertainty gradients of Fig. 4 arises from the increasing skewness of these gradients as the target moves toward the periphery. If we consider any one letter in a multiple-letter display as a target, then we must expect that the uncertainty gradient of a letter that is a given distance peripheral to the target will overlap the target location to a greater extent than the gradient around a letter that is the same distance central to the target; consequently, we should expect that replacing the peripheral neighbor by a blank space would facilitate perception of the target letter to a greater extent than replacement of the central neighbor by a blank space. This effect has been amply documented in a number of studies (Estes & Wolford, 1971; Shaw, 1969; Townsend *et al.* 1971; Wolford & Hollingsworth, 1974a) and is apparent also in part (d) of Fig. 5, where we see that a letter in a given absolute position four letter spaces to the left or right of the fixation point is better reported if it is the outermost, rather than innermost, member of a four-letter display.

D. Hemispheric Laterality

Since, with fixation maintained at the center of a display, letters presented a few degrees to either the left or right side of the fixation point project entirely to the opposite hemisphere, it may seem, at first thought, quite easy to compare various properties of letter recognition when the letters are presented in the left or right visual field and to infer from the results differences in hemispheric functioning.

Unfortunately, there are many possible confoundings in experiments that have been contrived for this purpose and definitive findings are scarce. If a row of more than a very few letters is centered in the visual field, then, as we have observed previously, the result is a conspicuous asymmetry in the serial position curve, with accuracy of report greatest at the left extreme and decreasing across the visual field from left to right. However, one cannot draw any inference as to the superiority of the right hemisphere in view of the likelihood that the asymmetry is almost entirely attributable to a combination of report order and short-term memory effects (see Section III,D).

When smaller displays are presented either entirely in the left visual field or entirely in the right visual field, the result is near equality of performance, but usually with some slight right-field advantage (Heron, 1957; Winnick & Bruder, 1968). Estes *et al.* (1976) obtained a similar result, with a slightly greater right-field advantage for short than for long exposure durations, but the asymmetry disappeared when the subjects were asked to give clarity ratings of letters, rather than verbal reports. Thus, here again there is reason to suspect that effects of memory and report order may be implicated. One may note, for example, that when subjects report in a normal left-to-right order from an array of four letters presented just to the right of the fixation point, the first letter reported is nearer to the center of the visual field, and therefore less subject to masking, than the first letter of a similar array displayed just to the left of the fixation point. There does not seem any way of completely cancelling out other factors to arrive at any pure differences in hemispheric processing at the *perceptual* level in experiments that present only full-report data.

In contrast to the results from full-report studies, measures of visual acuity have been found to decline quite symmetrically on the left and right sides of a central fixation point (Ruediger, 1907), and single-letter probes by means of indicators following multiletter displays have yielded serial position curves that are very nearly symmetrical in the left and right visual fields (Merikle *et al.* 1971; Winnick & Bruder, 1968). However, these studies may lack sensitivity, in that the displays have not been followed by pattern masks, so that any differences in speed of processing between the two hemispheres might not have manifested themselves.

McKeever and Suberi (1974) have reviewed a number of studies indicating, in fact, that processing of characters presented only in the left visual field takes 30–50 msec longer than characters presented only in the right visual field. However, when the stimulus is nonverbal, the difference may appear only if the response is vocal—as distinguished from a motor response, such as pressing a key (Filbey & Gazzaniga, 1969).

In an additional study, McKeever and Suberi evaluated the masking effects of an annulus surrounding the position of a randomly selected target letter at varying delays following the display of an array of letters and found that metacontrast functions (i.e., probability of correct recognition as a function of the interval between display and mask) were roughly parallel for displays presented in the left and right visual fields, but with the curve for the left field displaced toward longer asynchronies. That is, over most of the range of delays, to attain any given level of recognition accuracy, a letter presented in a given position in the left visual field had to be followed by a 10–15 msec longer delay of the postexposure mask than a letter in the corresponding position of the right visual field. This result suggests that encoding to the point of enabling a naming response is accomplished more quickly for letters processed initially by the left hemisphere than letters processed by the right hemisphere. However, this difference in processing speed can be expected to yield observable performance differences in studies of letter recognition only under conditions that sharply restrict stimulus encoding time—a qualification that probably excludes studies utilizing full-report procedures.

V. TARGET–BACKGROUND RELATIONSHIPS AND CATEGORY EFFECTS

A. Detection as a Function of Display Size and Redundancy

On a common-sense basis, the full-report procedure would seem to be the way of studying letter recognition that is closest to ordinary reading. This view may be justified for some instances of reading, as when, for example, individuals read difficult material or foreign language text, with the attendant necessity of constantly spelling out words and looking them up in a dictionary. But in more rapid reading, we must consider the possibility that there is insufficient time, and perhaps no necessity, to identify each individual letter in a passage of text. It may often suffice only to identify enough attributes of letters to permit the recognition of familiar letter groups and words.

The detection of differences in attributes between letters, as distinguished from full identification, has been investigated in two principal experimental paradigms: visual search and forced-choice detection. The search paradigm was first extensively used in research aimed at the issues of serial-versus-parallel processing and template-versus-critical-feature models (Neisser, 1967). Because stimulus factors cannot be closely controlled when the subject is searching rows or columns of material for a target letter, the search experiment in its original form has not turned out to be a method of choice for distinguishing perceptual from other factors determining performance, but it has been extensively used in the investigation of such factors as familiarity (Krueger, 1975). Some recent variations on the search paradigm, in which the subject is not free to scan a passage of displayed material, but rather is exposed to a series of frames by means of a modified tachistoscope or CRT screen, provide better control of stimulus and temporal variables (Shiffrin & Schneider, 1977; Sperling, Budiansky, Spivak, & Johnson, 1971), but little literature is yet available on these variations.

The forced-choice detection procedure was introduced by Estes and Taylor (1964), originally with the purpose of supplementing Sperling's (1960) partial-report method for assessing processing capacity with a method that provides even sharper control of memory factors. In this procedure, the subject knows in advance the set of alternative targets (usually two) that may appear during a series of trials. Each display of the series includes one of the two target letters embedded in an array of other (noise) letters. Following each display, the subject chooses a response, usually by operating one of two alternative response keys, to indicate which of the two target letters he or she believes was present. Thus, unlike the report procedure, both accuracy and speed of response can conveniently be monitored simultaneously.

The original experiments, utilizing exposure durations of the order of 50 msec, yielded estimates of letters identified that agreed quite well with those obtained with partial-report procedures. Thus, it appears that the detection task can, under some circumstances, be performed on the same basis as report, with the individual identifying the letters in the briefly seen display, and then responding on the basis of the contents of short-term memory. However, continuing research has shown that detection need not be based on identification. It appears that, at least for subjects experienced in the experimental situation and thoroughly familiar with the particular set of targets, the response decision on each trial can be made on the basis of detection only of some feature or attribute that differentiates the two possible targets (Estes, 1975b). Thus, it is found that detection with accuracy much above chance can be accomplished even with very short exposure durations (of the order of 5–15 msec) and postexposure pattern masking such that the subject is rarely able to report any of the letters of a display if asked. Furthermore, under these circumstances detection performance proves to be unaffected by variables (such as embedding the target letter in a word) that normally influence identification (Bjork & Estes, 1973; Estes, 1975b; Thompson & Massaro, 1973).

The detection method has proven particularly useful in the investigation of the dependence or independence of processing of letters in different display locations. In previous sections, we have considered some specific types of interactions between letters—response interference, lateral masking, etc. We might ask whether, once the processes responsible for these interactions have exerted their effects, there are other sources of nonindependence in the processing of multiple-letter displays. For displays of unrelated letters the answer appears to be no (Eriksen & Lappin, 1967; Estes & Taylor, 1966; Wolford, Wessel, & Estes, 1968). In the latter study, for example, the subject's task was to detect which of two targets was included in a square matrix display of letters presented tachistoscopically. On trials run under the usual procedure, with only one target letter present in each display, it was possible to determine, for the sample of subjects used, the probability that a target would be detected in any one of the possible display locations. From these data, predictions could then be computed as to the probability of a correct detection if two duplicate targets were included in a display at any two locations. These predictions, computed on the basis of a simple probability model that assumed independence of detection at different locations, closely described the observed data—a result that has been replicated in a number of variations.

In consequence of these findings, it would seem that if displays are constructed with locations and separations of letters selected so as to eliminate any substantial effects of lateral masking, then detection of targets in multiletter arrays should be almost perfectly predictable on the basis of information concerning detectability at particular visual field locations and confusability between letters. This expectation proves to be well founded, for when such parameters as display size and number of redundant signals are varied, one obtains exceedingly orderly sets of empirical functions that can be well accounted for in terms of the confusability relationships and decision processes that are assumed in current feature models (Estes, 1975b; Gardner, 1973; Shiffrin & Geisler, 1973).

This point can be illustrated in terms of the data reproduced in Fig. 6 (Estes, 1972). In that study, the subjects were presented with linear arrays of four, six, or eight letters, each of which might contain one, two, three, or four instances of whichever of the two possible target letters had been selected for representation on a given trial. The probability of correct detection is seen to vary as a function of both set size and number of redundant signals, in much the manner that one might intuitively expect. But we can readily go considerably beyond intuitive expectations. If we assume (Estes, 1975b) that on any trial there is some probability, β, that stimulus input from any one target letter will activate a feature that differentiates it from the alternative target, and a probability β' that any

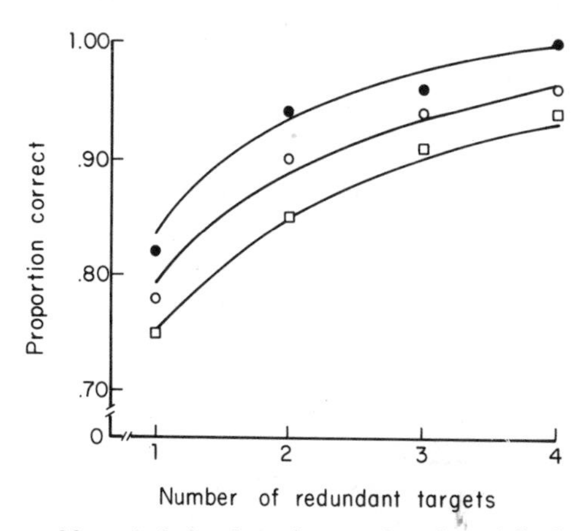

FIG. 6. Accuracy of forced-choice detection as a function of display size and number of redundant targets included in the display (Display size: ● = 4; ○ = 6; □ = 8.) Theoretical functions are computed from an equation given in the text. [After Estes (1972).]

noise letter will activate a feature compatible with one of the targets, and if we further assume that on the conclusion of stimulus processing on any trial, the probability that the subject will decide that a given target was present is equal to the proportion of activated features compatible with that target, then a straightforward probabilistic argument based the assumption of independent processing of the different display locations yields the following function:

$$P(c) = (1-\beta)^M + \sum_{i=1}^{M} \sum_{j=0}^{N-M} \binom{M}{i} \beta^i (1-\beta)^{M-i} \binom{N-M}{j} \beta'^j (1-\beta')^{N-M-j} \frac{2i+j}{2(i+j)},$$

where N denotes display size and M the number of duplicate target letters included in the display.

Any one term in the summation represents the probability that i target letters and j noise letters activate their relevant features. This probability is multiplied by the expected proportion of these letters whose activated features would trigger the correct response (all i of the target letters plus half of the j noise letters, divided by $i + j$). With estimates of .86 and .16 for the parameters β and β', this function generates the theoretical curves plotted along with the observed values in Fig. 6. The closeness of fit of the function lends some support to the conclusion that under these conditions, effects of set size and target redundancy are well accounted for by the assumptions of feature detection and confusability. Further, Shiffrin and Geisler (1973) have shown that essentially the same assumptions about confusability, taken together with a simple memory search model, generate a good account of reaction-time data for the same experimental conditions.

In another variation on the same theme, with the number of targets and the number of noise elements fixed, one may vary the homogeneity or heterogeneity of the noise elements. The empirical result is that, with other factors controlled, increasing heterogeneity of the noise background reduces detectability of a target (McIntyre, Fox, & Neale, 1970). Why should this result obtain? It appears that again an adequate answer can be given in terms of confusability (Estes, 1974). In a set of possible noise letters, different individual letters will have different numbers of features in common with the target letters used in a particular experiment. This being the case, increasing heterogeneity of the noise letters that appear in individual displays will increase the likelihood that each display will include one or more of the noise letters that are most confusable with the targets and, therefore, most detrimental to correct performance. More elaborate hypotheses have been proposed for effects of heterogeneity, but this very parsimonious one appears sufficient to account for the available data (Estes, 1974).

B. The Question of Categorical Detection

The set-size function, that is, the negative relation between detectability of a target letter and the size of a set of letters in which it is embedded, has proven extremely robust, being manifest in much the same form over a considerable range of display parameters and for such variations in procedure as full and partial report (Sperling, 1960), forced-choice detection, (Estes & Taylor, 1966), and even successive presentation of items (Eriksen & Spencer, 1969). However, one major qualification needs to be mentioned. Namely, for the function to hold, the confusability between a target letter and the noise characters constituting the background must be such that, with some degradation of input, features activated by the noise characters may be indistinguishable from those activated by the target. In the most common variations of both the detection and search experiments, this qualification is obviously satisfied, since both the target and noise characters are letters. However, when the noise characters are made highly nonconfusable with the targets, the slope of the set-size function is reduced, sometimes virtually to zero (Estes, 1972; Gardner, 1973). Thus, the set-size function provides a tool for investigating relationships between the encoded representations of different types of characters.

If the target and noise characters in a given situation differ in some dimension or category, then the significance of this categorical difference with respect to the individual's processing system can be investigated by way of the set-size effect. Of the many possible categorical differences, the one most studied to date is that of digits versus letters. It has quite uniformly been found that for a target letter in a background of letters or a target digit in a background of digits, the set-size function has a substantial slope, but for a target letter in a background of digits or a target digit in a background of letters, the slope of the function is reduced, sometimes to zero (Egeth, Atkinson, Gilmore, & Marcus, 1973; Jonides & Gleitman 1972). Most strikingly, this result was obtained by Jonides and Gleitman when they made the digit zero physically identical to an uppercase letter *O*, but nonetheless observed a flattening of the set-size function when this character was viewed as a letter but embedded in digits, and when the same character was viewed as a digit but embedded in letters. However, this last result has not, to my knowledge, yet been replicated.

One interpretation of the vanishing slope of the set-size function when the target and background characters differ in category, suggested by Jonides and Gleitman (1972), is that an individual is able to detect the category of characters much more quicly than he can identify them. According to this idea, if a target letter is embedded among digits, it is supposed that the individual can very rapidly detect higher-order categorical features that differentiate these as a class from letters or other characters. Therefore, the digits need not be identified individually, and the number of them present in the display is inconsequential to total processing time.

However, independent sources of evidence bearing on the concept of categorical features are both fragmentary and indecisive. Nickerson (1973) asked subjects both to categorize tachistoscopically presented characters as letters or digits and to identify them; he found virtually chance accuracy of categorization when the characters were not correctly identified, which he took to be a negative finding for the concept of categorical features. In another approach, LaBerge and Brownston (1974) presented tachistoscopic displays including a mixture of red and black letters to their subjects, who were in some cases precued and in some cases not precued as to the color of the target; they found higher accuracy of detection in the precued condition but almost no difference in slopes of the set-size functions.

At least two lines of explanation of the effect of same versus different background on slope of the set-size function have been put forward that do not depend on the postulation of higher-order categorical features. One suggestion is that whenever the instructions or context of a task indicate to the subject in advance of a display that some subset of characters is likely to be sampled, then the feature detectors responsible for discrimination within the subset are partially activated and ready to respond more rapidly or more sensitively than other detectors at display onset (Estes, 1972; 1975a). If, for example, a task involves only letters as targets, then the level of excitation of detectors that discriminate among letters would be maintained at a high level. As a consequence, detection of a target letter would be facilitated; but there would be similar facilitation of input channels to letters in the background that share features in common with either the target presented or alternative targets not included in the display. Therefore the probability of confusion errors would increase with the number of background letters. In contrast, detectors that discriminate among digits would not be preactivated, and therefore if a target letter were embedded in a background of digits, increasing the number of digits would not increase the likelihood of confusion errors.

A second, not unrelated, suggestion advanced by Shiffrin and Schneider (1977) is that as a consequence of practice with a particular collection of target and background characters, subjects shift from an initial, controlled processing mode to an automatic mode. Typically, subjects begin a new task in the controlled mode, in which they compare the elements of the display serially with memorial representations of the targets and therefore are subject to confusion errors. In the automatic mode, a difference in salience has developed between the set of possible targets and the set of background characters, and only the former enter into comparison with memorial representations of targets. The basis for the difference in salience has not been specified in more molecular terms, but might be similar to that assumed in the feature-activation hypothesis. In any event, Shiffrin and Schneider have provided substantial evidence for the important role of practice, showing that even when the target and background characters are of the same type (e.g., all letters), extended practice with a particular combination of target and background sets leads to a flattening of the set-size function.

VI. PARALLEL VERSUS SERIAL PROCESSING

A considerable segment of research on letter recognition during the past 10 years has been organized in terms of efforts to determine whether the items of multiletter displays are processed serially or in parallel (see, for example, Neisser, 1967; Sperling, 1967, 1970). The issue seems to have had two quite distinct origins in the general, informal models assumed tacitly or explicitly by investigators of reading and tachistoscopic perception. Because reading (at least by American and European readers) is closely tied to left-to-right scanning of lines of print, it has seemed natural to many investigators to conjecture that perception of text, even within a single fixation, involves a successive scanning of letters, presumably in a left-to-right order, with a word being identified only after all of its constituent letters have been recognized. The most thorough assemblage of evidence for a model of this type was presented by Gough (1972) and included some fairly specific supporting evidence from reaction-time studies that seemed to show a direct relationship between the time required to recognize or read a word and the length of the word.

On the other hand, for investigators working with tachistoscopic perception of letter displays at exposure durations brief enough to preclude eye movements, it has generally seemed more natural to assume that information from all display locations is processed in parallel. This view was preferred, for example, by Woodworth (1938). But even within this tradition, some considerations arose that suggested the possibility of serial processing of a rather special kind; specifically, that the sensory information in a display is registered in parallel in the peripheral sensory apparatus but that the input from the periphery is then scanned serially by a central processor. This conception was suggested especially by the early findings on backward masking. The observation (Sperling, 1960) that the number of letters reported from a brief exposure increases linearly for small display sizes if the exposure is terminated by a masking field seemed, at the time, to be nicely explained by the assumption that a decaying trace of the display pattern in a short-term visual memory system is encoded a letter at a time until the process is disrupted by a masking pattern. Furthermore, evidence from several types of partial-report and detection studies was taken to indicate specifically a serial scan of the trace at a rate of 10–40 msec per character (Averbach & Sperling, 1961; Sperling, 1963). This assumption was embodied in models that were momentarily successful in accounting for a fair range of data (Estes & Taylor, 1964; Sperling, 1963).

These models did have the merit of engendering research that bore sharply on the assumptions—so sharply, in fact, that the models were swiftly refuted by new evidence (Estes & Taylor, 1966; Sperling, 1967). In addition, mathematical investigations by Townsend (1974) demonstrated that one could not expect to decide in a general way between the class of parallel and serial models. In view of Townsend's demonstration, we can put aside as misguided any attempts to decide once and for all in a general, model-free way whether the perceptual processor is a serial or parallel mechanism. One can, however, expect to gain evidence with regard to specific, well-defined, serial or parallel processes, and here some progress can be reported.

Within the theoretical framework of this chapter it is possible to narrow down the question of serial versus parallel processing to the point where a fairly specific answer begins to emerge. First of all, I think it is clear that the general question is not *whether* but *where* in the processing of letter arrays serialization appears. Certainly, upon the onset of a display the impinging stimulus energies from the various constituent letters activate photoreceptors in the retina essentially simultaneously, and these energies initiate the peripheral processing of the sensory information from different locations in parallel. Similarly, it is clear that when an individual's task is to report what he has seen in the display, there is a final stage at which the encoded visual information is recoded within the auditory–articulatory system, leading to the activation of motor patterns that produce the observed responses in reporting individual letters. We can neither speak nor write more than one letter at a time, so beyond some point in the processing sequence the display information is clearly dealt with on a serial, letter-by-letter basis. The area of uncertainty is the intermediate stage of processing at which feature information is extracted from individual display locations and organized into encoded representations of letters that are immune to pattern masking. We can meaningfully ask whether this process proceeds in parallel or in some specific (presumably, left-to-right) serial order.

A number of kinds of evidence have been taken to be relevant to the issue of parallel versus serial encoding. The first type to receive major attention was reaction-time data. On a strict serial model, with the assumption that encoding time is constant over the various display locations in an array, the time required to detect a target letter in a multiletter array should be an increasing linear function of display size. This expectation is indeed well realized at the level of observables (Estes & Taylor, 1966; Estes & Wessel, 1966). However, reaction time must be assumed to depend only in part on encoding time, being influenced also by the requirements of the motor apparatus and the time required to make decisions on the basis of ambiguous inputs. Continuing and progressively more analytical examination of reaction times, leading to corrections of the observed data for decision factors, led to the conclusion that the observed functions actually lend no special support to a serial scanning conception (Bjork & Estes, 1971; Estes & Wessel, 1966; Estes, 1972).

Another aspect of reaction-time data that has been adduced in this context involves comparisons of reaction times to recognize or name single letters versus words of varying lengths. For both recognition and naming latencies, the findings, in general, are that latency increases with word length. This increase is not, in general, linear, and has a slope considerably lower than that implied by the estimates of processing time per letter coming from other sources (Cattell, 1886; Cosky, 1976; Gough, 1972; Theios & Muise, 1972). In view of the inevitable confounding of other factors with word length (for example, number of syllables—shown to be related to voicing latency by Eriksen, Pollack, & Montague, 1970, and Klapp, 1974—and the extension of the display to more peripheral locations in the visual field with longer words), this type of data can not be the basis of conclusions regarding the encoding process.

Johnson (1975) found that recognition latency was shorter for a short familiar word than for the first letter of the word or for an underlined letter included within the word, and argued on this basis for an extreme form of parallel processing in which the word as a whole is processed as a unit. But again, there are worrisome confoundings. We have observed in a previous section the important role of response competition with respect to recognition or naming of individual letters. This process would inevitably be operating in Johnson's study, but since only one word was displayed at a time, there would be no similar source of response competition at the word level. Also, the amount of previous experience subjects have had in the decision processes involved in generating responses to words as compared to their constituent letters is, of necessity, uncontrolled in this type of study. Once again, the data, viewed critically, do not seem germane to theoretical questions regarding serial or parallel encoding.

Somewhat sharper evidence is available in studies that vary the interval between a display and a postmask or that include such manipulations as changing one or more letters of a display at some point during the exposure. With regard to the former, if serial processing involved an orderly left-to-right scan of the letters of a display, then a plot of the probability of identification versus delay of the postmask for letters in different serial positions of a display should yield a family of curves that leave the baseline in a regular order. Sperling (1967) found, however, that these functions have nearly the same intercept, thus providing no support for the serial conception. Carterette and Friedman (1973) carried out investigations in which they changed a single letter of a tachistoscopic display part way through the exposure interval. They found that the probabilities of identification of the first letter of the two occurring in a changed location were highest, and nearly equal, at the two ends of a display and lower in the middle, rather than following the monotone trend that would be expected on the basis of a left-to-right serial scan. Finally, Shiffrin and Gardner (1972) found detection of a target letter in a four-element display to be at least as high when all four characters were presented simultaneously for 50 msec as when they were exposed successively for 50 msec each.

On the whole, I think it must be concluded that the kinds of data that have been taken to implicate a serial encoding process can be accounted for on the basis of other factors. Furthermore, some relatively cogent evidence is negative with respect to the idea of a left-to-right serial encoding mechanism and is more easily accommodated by the idea of a parallel process. However, this conclusion is meant to imply only that the processes of feature extraction and letter encoding are initiated in parallel by the inputs from different display locations. As we have seen in previous sections, there is considerable reason to believe that *rates* of encoding may vary systematically over locations. It is almost certain that these rates have an overall central-to-peripheral gradient, and it is quite possible that the encoding time for input from any given location is modifiable by lateral inhibitory processes. With regard to the temporal course of encoding, there seems to be some support for (and no cogent evidence against) the Poisson process assumed by Rumelhart (1970), with the proviso that the rate parameter of the process should depend both on location in the visual field and on the separation of the target letter from adjacent characters.

VII. LINGUISTIC FACTORS IN LETTER RECOGNITION

A. Problems of Definition

It has been well known for nearly a century on the basis of empirical experimental studies that letters are more quickly and accurately recognized when they appear as constituents of words or other familiar linguistic units than when they appear alone or as constituents of meaningless letter strings. For much of this period, investigators have been intrigued with the question of whether a letter is actually more clearly perceived in a familiar linguistic context or whether the contribution of context is simply one of compensating for deficiencies of perception by supplying auxiliary information.

Although the question may seem simple and straightforward at first glance, an answer has proven elusive, despite major advances in our understanding of letter recognition. The reason is to be found, I think, in the fact that the term *perception* does not refer to a unitary process. Throughout this chapter we have been dealing with phenomena that are conventionally classified under the rubric *letter perception*. In every case we have found that there must be involved a mixture of sensory processes, memory processes, and decision processes, among others. Which of these should be considered perceptual? It may be that "all" and "none" are equally good answers.

As nearly as I can circumscribe the term, *percept* has the status of a hypothetical construct referring to a reaction pattern of the organism that arises as a consequence of interacting sensory and mnemonic processes. The heuristic and theoretical value of the concept lies in the fact that an organism's behavior appears to be more simply related to the environment as represented in its percept than to the environment as it would be described in physical terms by an external observer. But since only the behavior can be directly observed, the effects on a percept of any variable, linguistic or otherwise, can only be determined by means of inferences carried out within the framework of a theory or model that identifies the constituent processes and the manner of their interaction.

Since we do not, as yet, have a body of perceptual theory that is adequate to justify the necessary inferences, we cannot hope to answer definitively the original question concerning the effect of linguistic variables on the perception of letters. We can, however, hope to illuminate the problem considerably by considering the range of processes that have been identified in the task of recognizing letters and the points at which effects of linguistic factors can be identified on the basis of available evidence. As a first step, let us consider what we should expect on the basis of both general theory and the factual information concerning perception of individual letters that has been reviewed in previous sections.

B. Constraints Imposed by Visual Processing Theory

To begin with the earlier, or more peripheral, end of the sequence of processing stages, I think we can find no firm ground in either theory or fact for the supposition that linguistic factors influence the registration of sensory information or the establishment of the icon. Some writers have entertained the idea of a feedback from semantic to sensory levels (for example, Osgood & Hoosain, 1974), but at present, hypotheses of this sort must be considered highly conjectural.

It is doubtful that we can expect any model-free answer to the question of whether linguistic context could influence feature extraction. Within the framework advanced here, the possibility of an effect of simultaneously displayed context would virtually be ruled out by definition. This view seems to be in agreement with the position of Turvey (1973) who speaks of *context-independent* features at this level of processing. Like Turvey, I conceive the initial stage of feature extraction to be a process of comparing attributes of the sensory input with feature representations in the memory system. With respect to the extraction of features from a target letter, the two ways in which information from concurrently displayed letters could affect the process are by increasing the accessibility of the feature representations in memory and by modifying the criterion for a match between input and memory. But these processes could only be influenced by events occurring earlier. Therefore, it seems that at this level we should expect no effect of properties of context letters if conditions are controlled so that input from the context is strictly simultaneous with that from the target letter. Context would, on the other hand, be expected to affect feature extraction if available in advance (so that it could activate memory representations prior to sensory input from the target, Estes, 1975a).

Essentially the same analysis is applicable to the encoding of feature combinations to synthesize representations of letters. The process here is conceived to be one of comparing the combined outputs of the feature detectors activated by the input from a letter location with letter representations in memory. The argument is not quite so tight, however, since information from letters at widely separated locations in the visual field might become available to the central processor at appreciably different times, even though the inputs were initiated simultaneously. Consequently, we must allow for the possibility that information from some subset of letters in a display could conceivably affect the individual's criterion for a match between memory and input from another location.

Furthermore, it would seem that few tasks occurring outside of the laboratory can depend solely on the encoding of feature information without regard to location. Certainly any activity having to do with reading requires that a feature combination be encoded along with information regarding the location of the input relative to other stimuli in the visual field, as this is necessary in order to establish short-term memory of a letter in its position in a display. This requirement opens new possibilities of interaction between a target letter and context. To the extent that information from the letters in a concurrent context makes the occurrence of a given letter more or less probable in a particular location, less or more information, respectively, must be encoded together with a representation of the letter in order to specify its position. Thus, we must anticipate that any regularity in a letter display that bears on the relative probabilities of different letters in particular locations or sequences will influence any performance that depends on memory for letters in positions or in letter groups constituting ordered sequences.

Extensive research on short-term memory (e.g., Murdock, 1974) makes it amply clear that the recoding of letter representations in the auditory–articulatory system and their maintenance via rehearsal processes must be greatly facilitated if letters occur as constituents of pronounceable letter groups. However, when interpreting research bearing on this type of interaction, one must keep in mind the difficulty, if not impossibility, of separating pronounceability of letter groups from other attributes having to do with orthographic regularity. One step toward the needed analysis has been taken in studies showing that letters are better identified in tachistoscopic displays if they are embedded in pronounceable nonwords than if they are embedded in strings that conform to spelling rules but are unpronounceable (Baron & Thurston, 1973; Spoehr & Smith, 1975). However, there seem to be no studies addressed to the problem of whether either overt or covert vocalization of the pronounceable letter strings is actually involved in these effects.

Several aspects of the recognition of letter groups such as syllables and words require separate consideration. It has been documented beyond question that recognition at this level depends strongly upon both the familiarity of particular letter groups and the conformity of the statistical structure of particular displays to the characteristic structures of language. We should expect, of course, that these effects would be mediated to a major extent by variations in the accessibility or availability of memorial representations of the letter groups involved in the matching between memory and input that is basic to the recognition process. There seems to be no specific ground in theory to expect either the familiarity or the statistical structure of letter arrays to produce, in addition, interactions at a level that could be termed perceptual. Nonetheless, hypotheses that presume such interactions have been perennially revived and, as will be seen later, ingenious controls have been needed to permit their evaluation.

C. Linguistic Effects in Relation to Task Requirements and Processing Levels

The remainder of this section will constitute a necessarily selective review of the large literature bearing on effects of linguistic factors at the various levels of processing. For more extensive surveys, the reader is referred to Baron (1978), Massaro (1975), Smith and Spoehr (1974), Theios and Muise (1977). I shall pay special attention to the methods that have been developed to rule out the effects of higher-level processes when effects at lower levels are at issue.

1. Predictability: The Role of Familiarity and Redundancy

If conditions are such that an individual perceives some letters of a display, but fails to perceive others, he must be better able to guess correctly the contents of the missed locations if these are predictable on a rational or statistical basis from attributes of the perceived letters. Predictability is enhanced if the letters to be identified constitute part of a familiar or meaningful sequence or even in the absence of these attributes, if their statistical structure is close to that of ordinary prose.

Familiarity is doubtless of great importance in ordinary reading and appears as a significant determiner of speed of recognition in relatively uncontrolled letter-search tasks (Krueger, 1975). However, in the perception of brief displays in tachistoscopic experiments, the role of familiarity turns out to be secondary, when appreciable at all. Thus, Baron and Thurston (1973), using a forced-choice task with the alternatives presented following the display, found that the recognition of target letters embedded in nonwords that conformed to the orthographic and phonetic rules of English was as good as that of targets embedded in familiar words (and in both cases significantly superior to that of letters embedded in nonwords that violate orthographic rules); these results were confirmed and extended by Baron (1975). These results are not universal, for Spoehr and Smith (1975) found a small but significant advantage for letters embedded in words rather than pronounceable nonwords, thus implicating some combination of meaningfulness and familiarity. At the same time, they found that letters embedded in strings that conformed to spelling rules but were unpronounceable were better recognized than letters included in strings of unrelated consonants.

Sequential letter redundancy, like familiarity, is of major importance in search or reading tasks that do not involve close control of stimulus variables (Krueger, 1975; Lefton, 1973; Miller, Bruner, & Postman, 1954; Neisser, 1967). However, these factors become secondary or insignificant in the case of brief tachistoscopic displays. Thus, Spoehr and Smith (1975) found a near-zero correlation between recognition accuracy and bigram frequencies of the letters in their displays, in agreement with the results of a still more searching study reported by Johnston (1975).

The nature of this disappearance of the redundancy effect at short exposures was elucidated by Estes by means of a procedure that permitted the exposure duration of a target letter and the surrounding context letters to be varied independently. His results showed that when a display is presented in a forced-choice detection situation at durations insufficient for perfect detection, and the exposure duration of the context letters is the same as that of the target letter, there is no measurable effect of a word versus a nonword context. If, however, the context letters remain in view for several hundred milliseconds following the offset of the target, then the advantage for a letter embedded in a redundant context reappears. Furthermore, the advantage grows as a function of reaction time within a trial, being absent when responses occur with latencies less than 500 msec and becoming substantial when latencies are longer than 1000 msec (Estes, 1977). But since a plot of accuracy of detection versus latency reaches a maximum at about 500 msec and then decreases, there is no apparent support for the idea that context facilitates the earlier stages of recognition (feature extraction and encoding). Rather, it seems that a redundant context retards the loss of information from very short-term visual memory.

Work by Mason (1975) suggests that some of the relationships involving letter redundancy may have to be further investigated. Whereas previous work has considered only the frequency with which letters occur either singly or in particular order in the written language, Mason introduced a measure of spatial, or more descriptively, *positional*, redundancy (that is, the frequency with which a given letter occurs at a particular position in words of a given length in the language). Both in a relatively uncontrolled search situation and in one involving restricted exposures (2 sec) of individual letter strings, Mason found a strong relationship between positional redundancy and the speed and accuracy of detecting target letters. In fact, for relatively skilled readers, performance proved even better on nonword letter strings constructed so as to represent extremely high average positional redundancy than on words of the same length.

2. Possible Effects at the Level of Feature Extraction

All aspects of a linguistic context would be expected to influence the more cognitive levels of processing, especially the recognition of letter groups such as syllables and words. A principal question at issue is whether any of the differences between word and nonword contexts influences processing at the levels that would be considered most characteristically perceptual, that is, the abstraction of critical features and the encoding of feature information into representations of letters in short-term visual memory.

Our ability to deal with this question has been substantially furthered during the past decade by a series of advances in methodology. One of the first, and thus most critical, of these is associated with the work of Reicher (1969). I think it is fair to say that prior to Reicher's study, all effects of linguistic factors on letter perception that had been demonstrated could have involved the higher-level, more cognitive stages, and thus no inferences were possible regarding any effects at the level of feature extraction or encoding. The experimental design introduced by Reicher, which I shall denote the WW–NN control, did appear to permit such inferences and thus gave rise to an enormous burst of new research in this area.

In Reicher's experiment, the subject was presented on each trial with a brief display that comprised either a single letter, a four-letter word, or a string of four unrelated letters. Immediately after the display, two target letters were shown, and the subject responded by indicating which of the two he believed he had seen in a designated location in the display. The word and the unrelated letter displays were made up in such a way that in the former case, either of the two possible target letters would complete a word, whereas in the latter case, neither target letter would produce a word. For example, if the target letters for a trial were *m* and *r*, an acceptable word display would be *came* and an acceptable nonword display would be *amce*. It will be seen that with this design the subject cannot improve his performance by utilizing redundant information from the context of a word to improve his guesses on trials when he perceives the other letters but fails to perceive the target letter.

Using this design, Reicher found that target letters embedded in words were identified significantly more often than the same letters embedded in nonword strings or presented alone, the latter two conditions being approximately equivalent. Later, Bjork and Estes (1973) added the logical extension of Reicher's paradigm to a W–N factorial design by including also displays of the form WN and NW (that is, cases in which a word is presented but choice of the alternative target letter would convert it into a nonword, or in which a nonword is presented but choice of the alternative target letter would convert it into a word). With the full design, one can rule out effects of redundant contextual information by considering the WW versus NN displays, or one can assess the magnitude of such effects by comparing WN versus NW displays (Estes *et al.*, 1974; Estes, 1975b).

A series of variations on the Reicher paradigm by Smith and Haviland (1972) and Wheeler (1970) seemed to dispose of interpretations of Reicher's finding in terms of guessing on the basis of redundant information or of short-term memory. Thus, by exclusion, these experiments pointed to some effect at the level of feature extraction or encoding.

With respect to feature extraction, more direct evidence appears to be negative. The most senstitive experimental paradigm that has been applied to this problem is a forced-choice detection task in which the subject knows the alternative targets for an entire series of trials in advance and can thus respond to each display in terms of the presence or absence of critical features that differentiate the two targets. Studies using this procedure (with the WW–NN control) have uniformly failed to show any trace of a word advantage (Bjork & Estes, 1973; Estes *et al.*, 1974; Massaro, 1973; Thompson & Massaro, 1973).

3. Possible Effects on Encoding and Identification

The question of the effect of linguistic context on the encoding of feature information is more difficult to anser. Johnston (1975) has suggested an affirmative answer, primarily on the basis of substantial word–nonword differences in experiments that seem to exclude most alternative interpretations. However, arguments by exclusion cannot be entirely satisfying, since we are still at a stage of research in which new variables of significance (for example, positional redundancy) continue to emerge, and we cannot be sure that we have yet identified all of the processes involved in letter recognition.

On the positive side, a series of studies reported by Estes (1975b), with a W–N factorial design and a postexposure, single-letter probe procedure that required actual letter identification (rather than simply response to feature differences), yielded evidence of a word advantage that could apparently be localized at the encoding stage of processing. However, the detailed error patterns suggested that only the encoding of positional information with item information, rather than the encoding of critical feature combinations, might be implicated.

At the present time, in my estimation, the question of whether linguistic context can influence the encoding of feature information remains open. But although this question is intriguing from a theoretical standpoint, and remains a challenge, our present inability to deal with the issue decisively should not be overemphasized, especially from the standpoint of understanding the contribution of linguistic factors to proficiency in reading. I think there is little doubt that the principal effects of linguistic context are localized, not in the perception of individual letters, but rather in recognition and short-term memory for letter groups.

For a graphic example of the way in which context exerts its facilitating effects by assimilating a target letter into a group or sequence that is relatively accessible in memory, consider the data portrayed in Fig. 7, which derives from an unpublished study by David H. Allmeyer and myself (cited by Estes, 1977). In this experiment, subjects were presented tachistoscopically with four-letter words, pronounceable nonwords, or unrelated letter strings at varying exposure durations in a full-report task. The functions presented in Fig. 7 exhibit the probabilities of reporting at

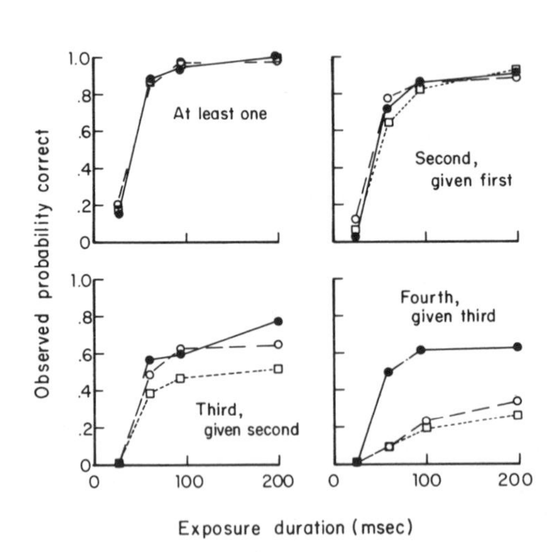

FIG. 7. Proportion of cases in which a first, second, third, or fourth letter was correctly reported from a word (W), pronounceable nonword (P), or nonpronounceable (N) display of four letters, without regard to position (● = W; ○ = P; □ = N).

least one letter (upper left panel), a second letter, given that at least one was reported (upper right panel), a third letter, given that at least two were reported (lower left panel), and a fourth letter, given that at least three were reported (lower right panel).

Plainly, the linguistic context has no effect whatever, at any exposure duration, on the probability of reporting at least one letter nor even on the probability of a second given the first. But surely one would expect that, if global features of words were perceived first and facilitated the recognition of letters, the probability of obtaining no information from a word would be less than for a nonword. Thus, these data lend no encouragement to the idea that linguistic context operates primarily, if at all, by directly activating units or features of higher order than those associated with letters.

Further work is required to specify precisely the attributes of linguistic context responsible for the divergence of the curves in the lower two panels of Fig. 7. The pattern observed fits quite well with the idea that the process is one in which perception of some of the components of a familiar letter group activates a representation of the group in the memory system (as hypothesized in the interactive filter model of Estes, 1975a,b or the related model of LaBerge and Samuels, 1974) and thus permits recognition of the group even under conditions such that partial information extracted from some of the individual letter locations would be insufficient to permit synthesis of a full representation of an unrelated letter string.

VIII. AFTERTHOUGHTS

Although research on perceptual processes in reading has a long history, a great part of what we know is a result of work within the past 10 years. An important reason is to be found in the advances in methodology, both instrumental and theoretical, that have led to the emergence of new systematic variables (for example, spatial redundancy, positional uncertainty, response competition, and orthographic regularity) that either were not appreciated or were not seen at all in the earlier literature. Among the major advances identified in the present review are: (*a*) the working out of the implications of the limited spatial and temporal resolution of the visual processing system for a variety of phenomena, especially forward, backward, and lateral masking; (*b*) the experimental and conceptual separation of decision problems entailed by letter confusability from perceptual interactions; and (*c*) the emergence of a conception of stages and levels of processing, together with some progress toward the ordering of experimental tasks in a theoretically significant way with respect to the levels of processing that provide the basis for response.

One might ask whether substantial progress on some fronts has been gained at the cost of neglecting phenomena and ideas that may ultimately prove equally important. One possible blind spot has to do with the effects of practice. In the realm of tachistoscopic research, there is perhaps no other phenomenon that is at the same time so conspicuous and yet so little studied as the effect of practice on performance. Only a few very recent studies, for example, LaBerge (1973) on unitization in selective attention and Shiffrin and Schneider (1977) on modes of visual search, have begun to offer some promise of a break with tradition and perhaps a beginning of more analytical treatments of the learning that occurs within studies intended to deal only with perception.

A topic that has been much more conspicuously represented in general expository literature than in specific research reports is that of global principles of form and grouping. At the more specific level, one continues to see mainly suggestions that particular phenomena lend themselves to description in terms of perceptual grouping or organization, for example, Banks, Bodinger, and Illige (1974) in connection with spacing effects in visual detection, or Kahneman (1973) with regard to a *visual suffix effect*. But simply hypothesizing global principles accomplishes little. Further progress will surely require careful analyses of the phenomena that have suggested these principles, with meticulous determination of what can (and what, if anything, cannot) be handled via principles already established by independent lines of investigation. Prospects may improve as we begin to see a body of theory taking shape in this area that, although not yet highly formalized, is proving increasingly useful in the guidance and direction of analytic research.

Acknowledgments

I am indebted to Charles W. Eriksen and Richard M. Shiffrin for incisive reviews of an earlier version of this chapter that have helped materially to reduce the noise level in my treatment of many topics.

References

Anderson, J. A. A theory for the recognition of items from short memorized lists. *Psychological Review,* 1973, **80,** 417–438.

Averbach, E., & Sperling, G. Short-term storage of information in vision. In C. Cherry (Ed.), *Information theory*. London and Washington, D.C.: Butterworth, 1961. Pp. 196–211.

Banks, W. P., Bodinger, D., & Illige, M. Visual detection accuracy and target-noise proximity. *Bulletin of the Psychonomic Society,* 1974, **2,** 411–414.

Baron, J. Successive stages in word recognition. In S. Dornic & P. M. A. Rabbitt (Eds.), *Attention and performance V.* New York: Academic Press, 1975. Pp. 563–574.

Baron, J. The word-superiority effect. In W. K. Estes (Ed.), *Handbook of Learning and Cognitive Processes* (Vol. 6). Hillsdale, New Jersey: Erlbaum, 1978.

Baron, J., & Thurston, I. An analysis of the word superiority effect. *Cognitive Psychology,* 1973, **4,** 207–228.

Bjork, E. L., & Estes, W. K. Detection and placement of redundant signal elements in tachistoscopic displays of letters. *Perception & Psychophysics,* 1971, **9,** 439–442.

Bjork, E. L., & Estes, W. K. Letter identification in relation to linguistic context and masking conditions. *Memory & Cognition,* 1973, **1,** 217–223.

Bouma, H. Interaction effects in parafoveal letter recognition. *Nature,* 1970, **226,** 177–178.

Breitmeyer, B. G., & Ganz, L. Implications of sustained and transient channels for theories of visual pattern masking, saccadic suppression, and information processing. *Psychological Review,* 1976, **83,** 1–36.

Bryden, M. P. Accuracy and order of report in tachistoscopic recognition. *Canadian Journal of Psychology*, 1966, **20,** 262–272.

Carterette, E. C., & Friedman, M. P. *Are spatial visual arrays processed in serial or in parallel?* Paper presented at the Psychonomic Society Meetings, St. Louis, November 1973.

Cattell, J. M. The time taken up by cerebral operations. *Mind,* 1886, **11,** 220–242, 377–387, 524–538.

Coltheart, M. Visual information processing. In P. C. Dodwell (Ed.), *New horizons in psychology II.* Baltimore, Maryland: Penguin, 1972. Pp. 62–85.

Conrad, R. Interference or decay over short retention intervals. *Journal of Verbal Learning and Verbal Behavior,* 1967, **6,** 49–54.

Cosky, M. J. The role of letter recognition in word recognition. *Memory & Cognition,* 1976, **4,** 207–214.

Egeth, H., Atkinson, J., Gilmore, G., & Marcus, N. Factors affecting processing mode in visual search. *Perception & Psychophysics, 1973,* **13,** 394–402.

Eriksen, B. A., & Eriksen, C. W. Effects of noise letters upon the identification of a target letter in a nonsearch task. *Perception & Psychophysics,* 1974, **16,** 143–149.

Eriksen, C. W. Temporal luminance summation effects in backward and forward masking. *Perception & Psychophysics,* 1966, **1,** 87–92.

Eriksen, C. W., & Collins, J. F. Reinterpretation of one form of backward and forward masking in visual perception. *Journal of Experimental Psychology,* 1965, **70,** 343–351.

Eriksen, C. W., & Collins, J. F. Sensory traces versus the psychological moment in the temporal organization of form. *Journal of Experimental Psychology,* 1968, **77,** 376–382.

Eriksen, C. W., Collins, J. F., & Greenspon, T. S. An analysis of certain factors responsible for nonmonotonic backward masking functions. *Journal of Experimental Psychology*, 1967, **75,** 500–507.

Eriksen, C. W., & Eriksen, B. A. Visual perceptual processing rates and backward and forward masking. *Journal of Experimental Psychology,* 1971, **89,** 306–313.

Eriksen, C. W., & Hoffman, J. E. Some characteristics of selective attention in visual perception determined by vocal reaction time. *Perception & Psychophysics,* 1972, **11,** 169–171.

Eriksen, C. W., & Lappin, J. S. Independence in the perception of simultaneously presented forms at brief durations. *Journal of Experimental Psychology,* 1967, **73,** 468–472.

Eriksen, C. W., Pollack, M. D., & Montague, W. E. Implicit speech: Mechanism in perceptual encoding? *Journal of Experimental Psychology,* 1970, **84,** 502–507.

Eriksen, C. W., & Rohrbaugh, J. Visual masking in multielement displays. *Journal of Experimental Psychology,* 1970, **83,** 147–154. (a)

Eriksen, C. W., & Rohrbaugh, J. W. Some factors determining efficiency of selective attention. *American Journal of Psychology,* 1970, **83,** 330–342. (b)

Eriksen, C. W., & Spencer, T. Rate of information processing in visual perception: Some results and methodological considerations. *Journal of Experimental Psychology Monograph,* 1969, **79,** No. 2.

Estes, W. K. Interactions of signal and background variables in visual processing. *Perception & Psychophysics,* 1972, **12,** 278–286.

Estes, W. K. Phonemic coding and rehearsal in short-term memory for letter strings. *Journal of Verbal Learning and Verbal Behavior,* 1973, **12,** 360–372.

Estes, W. K. Redundancy of noise elements and signals in visual detection of letters. *Perception & Psychophysics,* 1974, **16,** 53–60.

Estes, W. K. Memory, perception, and decision in letter identification. In R. L. Solso (Ed.), *Information processing and cognition: The Loyola symposium.* Hillsdale, New Jersey: Erlbaum, 1975. Pp. 3–30. (a)

Estes, W. K. The locus of inferential and perceptual processes in letter identification. *Journal of Experimental Psychology: General,* 1975, **104,** 122–145. (b)

Estes, W. K. On the interaction of memory and perception in reading. In D. LaBerge & S. J. Samuels (Eds.), *Basic processes in reading: Perception and comprehension.* Hillsdale, New Jersey: Erlbaum, 1977.

Estes, W. K., Allmeyer, D. H., & Reder, S. M. Serial position functions for letter identification at brief and extended exposure durations. *Perception & Psychophysics,* 1976, **19,** 1–15.

Estes, W. K., Bjork, E. L., & Skaar, E. Detection of single letters and letters in words with changing versus unchanging mask characters. *Bulletin of the Psychonomic Society,* 1974, **3,** 201–203.

Estes, W. K., & Taylor, H. A. A detection method and probabilistic models for assessing information processing from brief visual displays. *Proceedings of the National Academy of Sciences,* 1964, **52,** 446–454.

Estes, W. K., & Taylor, H. A. Visual detection in relation to display size and redundancy of critical elements. *Perception & Psychophysics,* 1966, **1,** 9–16.

Estes, W. K., & Wessel, D. L. Reaction time in relation to display size and correctness of response in forced-choice visual signal detection. *Perception & Psychophysics,* 1966, **1,** 369–373.

Estes, W. K., & Wolford, G. L. Effects of spaces on report from tachistoscopically presented letter strings. *Psychonomic Science,* 1971, **25,** 77–80.

Filbey, R. A., & Gazzaniga, M. S. Splitting the normal brain with reaction time. *Psychonomic Science,* 1969, **17,** 335–336.

Flom, M. C., Weymouth, F. W., & Kahneman, D. Visual resolution and contour interaction. *Journal of the Optical Society of America,* 1963, **53,** 1026–1032.

Ganz, L. Temporal factors in visual perception. In E. C. Carterette & M. P. Friedman (Eds.), *Handbook of Perception* (Vol. V). New York: Academic Press, 1975. Pp. 169–231.

Gardner, G. T. Evidence for independent parallel channels in tachistoscopic perception. *Cognitive Psychology,* 1973, **4,** 130–155.

Geyer, L. H. A two-channel theory of short-term visual storage. (Doctoral dissertation, SUNY at Buffalo, Buffalo, New York, 1970). (University Microfilms, No. 71–7165).

Geyer, L. H., & DeWald, C. G. Feature lists and confusion matrices. *Perception & Psychophysics,* 1973, **14,** 471–482.

Gibson, E. J. *Principles of perceptual learning and development.* New York: Appleton, 1969.

Gough, P. B. One second of reading. In J. F. Kavanaugh & I. G. Mattingly (Eds.), *Language by ear and by eye.* Cambridge, Massachusetts: MIT Press, 1972. Pp. 331–358.

Harcum, E. R. Parallel functions of serial learning and tachistoscopic pattern perception. *Psychological Review,* 1967, **74,** 51–62.

Henderson, L., Coles, S. H., Manheim, M., Muirhead, J. E., & Psutka, P. M. Orientation-specific masking of letter features. *Nature,* 1971, **233,** 498–499.

Henderson, L., & Park, N. Are the ends of tachistoscopic arrays processed first? *Canadian Journal of Psychology,* 1973, **27,** 178–183.

Heron, W. Perception as a function of retinal locus and attention. *American Journal of Psychology,* 1957, **70,** 38–48.

Holbrook, M. B. A comparison of methods for measuring the interletter similarity between capital letters. *Perception & Psychophysics,* 1975, **17,** 532–536.

Hubel, D. H., & Wiesel, T. N. Receptive fields and functional architecture of monkey striate cortex. *Journal of Physiology,* 1968, **195,** 215–243.

Jacobson, R., Fant, C. G. M., & Halle, M. *Preliminaries to speech analysis: The distinctive features and their correlates.* Cambridge, Massachusetts: MIT Press, 1969.

Johnson, N. F. On the function of letters in word identification: Some data and a preliminary model. *Journal of Verbal Learning and Verbal Behavior,* 1975, **14,** 17–29.

Johnson, R. M., & Uhlarik, J. J. Fragmentation and identifiability of repeatedly presented brief visual stimuli. *Perception & Psychophysics,* 1974, **15,** 533–538.

Johnston, J. C. The role of contextual constraint in the perception of letters in words. Paper presented at Eastern Psychological Association Meeting, New York City, April, 1975.

Jonides, J., & Gleitman, H. A conceptual category effect in visual search. O as letter or as digit. *Perception & Psychophysics*, 1972, **12,** 457–460.

Kahneman, D. Method, findings, and theory in studies of visual masking. *Psychological Bulletin,* 1968, **70,** 404–425.

Kahneman, D. *Attention and Effort.* Englewood Cliffs, New Jersey: Prentice-Hall, 1973.

Kinsbourne, M., & Warrington, E. K. The effect of an aftercoming random pattern on the perception of brief visual stimuli. *Quarterly Journal of Experimental Psychology,* 1962, **14,** 223–234.

Klapp, S. T. Syllable-dependent pronunciation latencies in number naming: A replication. *Journal of Experimental Psychology,* 1974, **102,** 1138–1140.

Koestler, A., & Jenkins, J. J. Inversion effects in the tachistoscopic perception of number sequences. *Psychonomic Science,* 1965, **3,** 75–76.

Krueger, L. E. Familiarity effects in visual information processing. *Psychological Bulletin,* 1975, **82,** 949–974.

LaBerge, D. Attention and the measurement of perceptual learning. *Memory & Cognition,* 1973, **1,** 268–276.

LaBerge, D., & Brownston, L. S. Control of visual processing by color cueing. *Bulletin of the Psychonomic Society,* 1974, **4,** 417–418.

LaBerge, D., & Samuels, S. J. Toward a theory of automatic information processing in reading. *Cognitive Psychology,* 1974, **6,** 293–323.

Lefton, L. A. Metacontrast: A review. *Psychonomic Monograph Supplements,* 1972, **4,** 245–255.

Lefton, L. A. Guessing and the order of approximation effect. *Journal of Experimental Psychology,* 1973, **101,** 401–403.

Lowe, D. G. Processing of information about location in brief visual displays. *Perception & Psychophysics,* 1975, **18,** 309–316.

Mason, M. Reading ability and letter search time: Effects of orthographic structure defined by single-letter positional frequency. *Journal of Experimental Psychology: General,* 1975, **104,** 146–166.

Massaro, D. W. Perception of letters, words, and nonwords. *Journal of Experimental Psychology,* 1973, **100,** 349–353.

Massaro, D. W. Primary and secondary recognition in reading. In D. W. Massaro (Ed.), *Understanding language: An information processing analysis of speech perception, reading, and psycholinguistics.* New York: Academic Press, 1975. Pp. 241–289.

Mayzner, M. S. Visual information processing of alphabetic inputs. *Psychonomic Monograph Supplements,* 1972, **4,** 239–243.

McIntyre, C., Fox, R., & Neale, J. Effects of noise similarity and redundancy on the information processed from brief visual displays. *Perception & Psychophysics,* 1970, **7,** 328–332.

McKeever, W. F., & Suberi, M. Parallel but temporally displaced visual half-field metacontrast functions. *Quarterly Journal of Experimental Psychology,* 1974, **26,** 258–265.

Merikle, P. M., & Coltheart, M. Selective forward masking. *Canadian Journal of Psychology,* 1972, **26,** 296–302.

Merikle, P. M., Coltheart, M., & Lowe, D. G. On the selective effects of a patterned masking stimulus. *Canadian Journal of Psychology,* 1971, **25,** 264–279.

Merikle, P. M., Lowe, D. G., & Coltheart, M. Familiarity and method of report as determinants of tachistoscopic performance. *Canadian Journal of Psychology,* 1971, **25,** 167–174.

Miller, G. A., Bruner, J. S., & Postman, L. Familiarity of letter sequences and tachistoscopic identification. *Journal of General Psychology,* 1954, **50,** 129–139.

Murdock, B. B., Jr. *Human memory: Theory and data.* Hillsdale, New Jersey: Erlbaum, 1974.

Neisser, U. *Cognitive psychology.* New York: Appleton, 1967.

Nickerson, R. S. Can characters be classified directly as digits vs. letters or must they be identified first? *Memory & Cognition,* 1973, **1,** 477–484.

Osgood, C. E., & Hoosain, R. Salience of the word as a unit in the perception of language. *Perception & Psychophysics,* 1974, **15,** 168–192.

Potter, M. C., & Levy, E. I. Recognition memory for a rapid sequence of pictures. *Journal of Experimental Psychology,* 1969, **81,** 10–15.

Reed, S. K. *Psychological processes in pattern recognition.* New York: Academic Press, 1973.

Reicher, G. M. Perceptual recognition as a function of meaningfulness of the stimulus material. *Journal of Experimental Psychology,* 1969, **81,** 275–280.

Robson, J. G. Receptive fields: Neural representation of the spatial and intensive attributes of the visual image. In E. C. Carterette & M. P. Friedman (Eds.), *Handbook of Perception* (Vol. V). New York: Academic Press, 1975. Pp. 81–116.

Ruediger, W. C. The field of distinct vision. *Archives of Psychology,* 1907 (Whole No. 5).

Rumelhart, D. E. A multicomponent theory of the perception of briefly exposed visual displays. *Journal of Mathematical Psychology,* 1970, **7,** 191–218.

Rumelhart, D. E., & Siple, P. Process of recognizing tachistoscopically presented words. *Psychological Review,* 1974, **81,** 99–118.

Sakitt, B. Locus of short-term visual storage. *Science,* 1975, **190,** 1318–1319.

Shaw, P. Processing of tachistoscopic displays with controlled order of characters and spaces. *Perception & Psychophysics,* 1969, **6,** 257–266.

Shiffrin, R. M., & Gardner, G. T. Visual processing capacity and attentional control. *Journal of Experimental Psychology,* 1972, **93,** 72–82.

Shiffrin, R. M., & Geisler, W. S. Visual recognition in a theory of information processing. In R. L. Solso (Ed.), *Contemporary issues in cognitive psychology: The Loyola symposium*. Washington, D.C.: V. H. Winston, 1973. Pp. 53–101.

Shiffrin, R. M., & Schneider, W. Controlled and automatic human information processing: II. Perceptual learning, automatic attending, and a general theory. *Psychological Review,* 1977, **84,** 127–190.

Smith, E. E., & Haviland, S. E. Why words are perceived more accurately than nonwords: Inference versus unitization. *Journal of Experimental Psychology*, 1972, **92,** 59–64.

Smith, E. E., & Spoehr, K. T. The perception of printed English: A theoretical perspective. In B. H. Kantowitz (Ed.), *Human information processing: Tutorials in performance and cognition*. Hillsdale, New Jersey: Erlbaum, 1974. Pp. 231–275.

Sperling, G. The information available in brief visual presentations. *Psychological Monographs,* 1960, **74,** No. 489, 1–29.

Sperling, G. A model for visual memory tasks. *Human Factors,* 1963, **5,** 19–31.

Sperling, G. Successive approximations to a model for short-term memory. *Acta Psychologica,* 1967, **27,** 285–292.

Sperling, G. Short-term memory, long-term memory, and scanning in the processing of visual information. In F. A. Young and D. B. Lindsley (Eds.), *Early experience and visual information processing in perceptual and reading disorders*. Washington, D.C.: National Academy of Sciences, 1970. Pp. 198–218.

Sperling, G., Budiansky, J., Spivak, J. G., & Johnson, M. C. Extremely rapid visual search: The maximum rate of scanning letters for the presence of a numeral. *Science,* 1971, **174,** 307–311.

Spoehr, K. T., & Smith, E. E. The role of orthographic and phonotactic rules in perceiving letter patterns. *Journal of Experimental Psychology: Human Perception and Performance,* 1975, **1,** 21–34.

Strangert, B., & Brännström, L. Spatial interaction effects in letter processing. *Perception & Psychophysics,* 1975, **17,** 268–272.

Taylor, S. G., & Brown, D. R. Lateral visual masking: Supraretinal effects when viewing linear arrays with unlimited viewing time. *Perception & Psychophysics*, 1972, **12,** 97–99.

Theios, J., & Muise, G. The word identification process in reading. In N. J. Castellan & D. Pisoni (Eds.), *Cognitive theory* (Vol. 2). Hillsdale, New Jersey: Erlbaum Associates, 1977.

Thomas, J. P. Spatial resolution and spatial interaction. In E. C. Carterette & M. P. Friedman (Eds.), *Handbook of Perception* (Vol. V). New York: Academic Press, 1975. Pp. 233–264.

Thompson, M. C., & Massaro, D. W. Visual information and redundancy in reading. *Journal of Experimental Psychology,* 1973, **98,** 49–54.

Townsend, J. T. A note on the identifiability of parallel and serial processes. *Perception & Psychophysics,* 1971, **10,** 161–163.

Townsend, J. T. Issues and models concerning the processing of a finite number of inputs. In B. H. Kantowitz (Ed.), *Human information processing. Tutorials in performance and cognition*. Hillsdale, New Jersey: Erlbaum, 1974. Pp. 133–185.

Townsend, J. T., Taylor, S. G., & Brown, D. R. Lateral masking for letters with unlimited viewing time. *Perception & Psychophysics,* 1971, **10,** 375–378.

Townsend, V. M. Loss of spatial and identity information following a tachistoscopic exposure. *Journal of Experimental Psychology,* 1973, **98,** 113–118.

Turvey, M. T. On peripheral and central processes in vision. Inferences from an information processing analysis of masking with patterned stimuli. *Psychological Review,* 1973, **80,** 1–52.

Weisstein, N. Backward masking and models of perceptual processing. *Journal of Experimental Psychology,* 1966, **72,** 232–240.

Wheeler, D. D. Processes in word recognition. *Cognitive Psychology,* 1970, **1,** 59–85.

White, M. J. Laterality differences in perception. *Psychological Bulletin,* 1969, **72,** 387–405.

Wickelgren, W. A. Distinctive features and errors in short-term memory for English consonants. *Journal of the Acoustical Society of America,* 1966, **39,** 388–398.

Winnick, W. A., & Bruder, G. E. Signal detection approach to the study of retinal locus in tachistoscopic recognition. *Journal of Experimental Psychology,* 1968, **78,** 528–531.

Wolford, G. Perturbation model for letter identification. *Psychological Review,* 1975, **82,** 184–199.

Wolford, G. & Hollingsworth, S. Lateral masking in visual information processing. *Perception & Psychophysics*, 1974, **16,** 315–320. (a)

Wolford, G. & Hollingsworth, S. Evidence that short-term memory is not the limiting factor in the tachistoscopic full report procedure. *Memory & Cognition*, 1974, **2,** 796–800. (b)

Wolford, G. L., Wessel, D. L., & Estes, W. K. Further evidence concerning scanning and sampling assumptions of visual detection models. *Perception & Psychophysics,* 1968, **3,** 439–444.

Woodworth, R. S. *Experimental psychology.* New York: Holt, 1938.

Chapter 6

EYE MOVEMENTS AND VISUAL PERCEPTION

GEOFF D. CUMMING

I. INTRODUCTION

Look carefully at the eyes of an observer engaged in reading or some other visual task. The eyes make frequent, very fast flicks and between flicks appear to remain quite steady. Occasionally the eyes may appear to move together, smoothly and slowly, but only when the observer follows

ISBN 0-12-161909-5

the movement of some object in his field of view. It seems easy to follow quite closely what the observer is doing.

Eye movements do indeed betray very conveniently a great deal about the distribution of a person's visual attention and, indirectly, about his deeper mental processes also. Growing realization of this fact, together with continuing refinement of techniques for monitoring eye position, largely explain the recent increase in interest in eye movements.

Eye movements are no longer mainly studied in their own right in specialized laboratories; eye-movement measurement is becoming a routine and valuable contributor to studies of many perceptual and cognitive problems. Two books marked the change in emphasis: Ditchburn (1973) gave an authoritative summary of the past, with a detailed description of the study of eye movements themselves and of such closely related perceptual topics as the stabilized image and saccadic suppression; Monty and Senders (1976), on the other hand, in reporting the proceedings of the first scientific meeting to deal with the subject of eye movements and psychological processes, covered a very wide range of topics and gave a foretaste of likely future work on eye movement.

This brief review concentrates on the role of eye movements in perception, rather than the study of eye movements themselves. But monitoring eye movements in order to study perception and cognition requires some knowledge of types of eye movement and also of measurement technique. Section II, therefore, contains a classification of types of eye movement, and in Section III there is a description of the more useful methods of monitoring eye position. The study of the physiological control of eye movements is omitted, save for mention of the symposium on the topic edited by Bach-y-Rita and Collins (1971) and the reviews included in Monty and Senders (1976).

Sections IV and V consider, respectively, the benefits and possible costs of having a perceptual system which is very reliant on eye movements: benefits in the sense that eye movements ensure the maintenance of vision and allow the effective deployment of the high-acuity foveal channel; costs in the sense that perception of a stable world is required despite the gross changes in retinal images produced by large and frequent eye movements.

II. TYPES OF EYE MOVEMENT

A more detailed discussion of types of eye movement is given by Alpern (1970, 1972). He also discusses torsional movements, which are omitted here as being of lesser importance for the topics covered in Sections IV and V.

A. Large Movements

The prime function of large eye movements is to deploy in space and time the observer's tiny volume of very highly acute vision. Very high acuity is achieved within a cone less than 1° across, by the concentration of retinal receptors in the fovea and by their relatively rich representation in the visual cortex. This cone is restricted in depth to a small volume of accurate vision by the accommodation of the lens of each eye to bring stimuli at a particular distance into focus on the retina and by convergence of the eyes at the same distance. Three types of eye movement are available for moving this volume of fine vision, which will be referred to as the *fixation point*. In each case the eyes are constrained to move in close coordination. *Saccades* are the sudden changes in fixation position between points at the same viewing distance; *smooth*, or *tracking*, movements occur when the eyes move smoothly, maintaining fixation on a moving object or on a stationary point while the head moves; and *vergence* movements allow the degree of convergence of the eyes to change and so move the fixation point in depth.

Saccades occur continually in normal vision, several per second—several billion in a lifetime. They bring fixation to a succession of points in the visual field under voluntary, but usually unnoticed, control. Saccades smaller than 5° take about 30 msec to execute; the duration of larger saccades increases linearly to about 100 msec for 40° movements. During a saccade, velocity increases smoothly to a maximum, which can be over 1000° sec^{-1} for large movements, then decreases smoothly to zero. The great majority of saccades we make are smaller than 10°, and for such movements the eyes move closely together. But coordination is not very good during larger movements: Note the diplopia (double images) briefly observable at the end of a deliberately very long saccade. The latency of a saccade to a sudden change in stimulus position is quite long, 200 msec or more, and the movement itself is ballistic: Once execution has commenced the end point cannot be changed. If a change in fixation of more than 10° is attempted, the initial saccade is frequently observed to fall short of the target and be followed quickly by a second, corrective saccade.

Smooth movements are made in response to a difference between eye and target velocity. Latency can be 150 msec or less, and moving targets may be tracked accurately up to about 30° sec^{-1}. Saccades usually occur during the early part of smooth pursuit movement as the observer attempts to match the fixation point to the target in both position and speed.

Saccadic and smooth movements are different types of movement, serving quite different functions, produced by the same extraocular muscles, but by different control systems. Saccades are under voluntary control, whereas relative movement between the head and a stimulus is usually necessary for smooth movements. Heywood (1972), however, discusses reports that subjects can sometimes voluntarily produce smooth movements in the absence of both target and head movement.

Binocular coordination is good during smooth movements as well as saccades, but Berger (1969) and Berger and Walker (1972) argue that constant practice is required to establish and maintain good binocular coordination and that one function of rapid eye movement (REM) in paradoxical sleep is to provide such practice.

Vergence movements allow movement of the fixation point in depth. They are evoked by disparity, most commonly horizontal disparity, between the two retinal images. They may also be evoked by accommodation changes to viewing distance (for example, if one eye is covered). Vergence movements have a latency of about 150 msec and are quite slow, taking up to 800 msec to complete. They differ from saccades in that the course of the movement may be altered if disparity changes unexpectedly during the movement itself. In normal viewing, many shifts of the fixation point require a movement both laterally and in depth; in such cases a saccade and a vergence movement are both made, the saccade occurring during the vergence movement, and the characteristics of each may be seen clearly in an eye-movement record.

B. Fixational Movements

Even during careful attempts to hold the eyes still, unavoidable small movements persist. These eye movements of fixation are conventionally classified as *tremor, microsaccades*, and *drifts* and are illustrated in Fig. 1.

1. Tremor is a small, irregular lateral oscillation with frequency components up to 100 Hz at least and an amplitude equivalent to a few foveal cone diameters.
2. Microsaccades are small, fast, conjugate flicks, taking some 20 msec and moving the eyes a few minutes of arc.
3. Between microsaccades the eyes drift haphazardly at about 5 min of arc sec^{-1}, with tremor superimposed on the drifting motion. The two eyes drift independently and probably also undergo tremor independently, and so these movements have usually been attributed to unavoidable residual instability in the oculomotor system.

Microsaccades typically occur every few hundred milliseconds, but they become less frequent when fixating in the presence of an afterimage

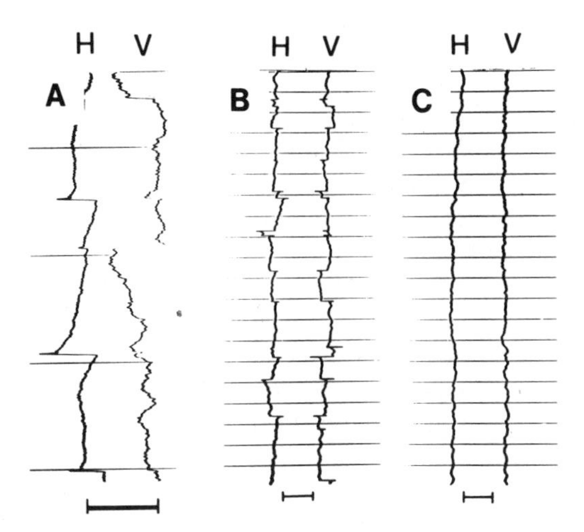

FIG. 1. Two-dimensional records of the horizontal (H) and vertical (V) components of eye movements during maintained fixation of a bright red point of light viewed in the dark. The records begin at the bottom and the horizontal lines mark seconds. The horizontal calibration bar represents 15 min of arc. Parts (A) and (B) show, on different time scales, the eye movements of fixation: The abrupt changes are microsaccades, between these saccades the eyes drift, and superimposed on the drift are the small, high-frequency tremor movements. For (C) the observer voluntarily suppressed microsaccades: The record shows that stable fixation could, nevertheless, be maintained accurately. [Reproduced with permission of author and publisher from Steinman, Haddad, Skavenski, & Wyman, *Science*, 31 August 1973, **181,** 810–819. Copyright © 1973 by the American Association for the Advancement of Science.]

(Steinbach & Pearce, 1972) and may be suppressed for 10 sec or more by practiced observers (Fiorentini & Ercoles, 1966; Steinman, Cunitz, Timberlake, & Herman, 1967) without any reduction in the accuracy of fixation, as can be seen in Fig. 1(c). The remaining movements—mainly drifts—must therefore be able to maintain steady eye position and so must be under some central control (St-Cyr, 1973); they cannot be attributed simply to system noise.

Microsaccades have usually been classed as involuntary and therefore different from larger, voluntary saccades. But any distinction seems to be unwarranted. Zuber, Stark, and Cook (1965) found that the time course of the velocity of all saccades is similar, and Cunitz and Steinman (1969) obtained similar distributions of intersaccadic intervals for saccades made during fixation and during reading. In addition, Wyman and Steinman (1973) showed that appropriate microsaccades can indeed be made voluntarily even to stimulus movements of 3.4 min of arc or smaller.

III. MEASUREMENT OF EYE POSITION

The monitoring of eye movements is being applied more widely and usefully (see Sections IV and V) now that a range of adequate measurement devices is widely available and small computers may easily be used to record and analyze the large amounts of eye-position data gathered during even brief experiments. The standard objective methods continue to be made more accurate and less obtrusive and uncomfortable for the observer; further refinements will certainly come, particularly with the exploitation of recent advances in electronics.

Lévy-Schoen (1969) has given a comprehensive survey of methods that have been used to measure eye position. Young and Sheena (1975) have given a detailed report of techniques for measuring the position of the eyes relative to the head (eye-in-head position) and for measuring the position of the head in space. If head position is stabilized with a bite bar, then measurement of eye-in-head position specifies the line-of-sight in space, but if the head is free to move, then head position must also be measured if absolute eye-in-space position is required. Young and Sheena give many more references than are mentioned in the following text and also describe some of the instruments that are available commercially.

The main objective methods for measuring eye position and lateral eye movements are briefly described in the following text; the measurement of torsional eye movements (see Howard & Evans, 1963) and pupil size (see Hess, 1972; Tryon, 1975) are not specifically considered. But first a description will be given of some subjective methods for observing eye movements and demonstrating some of their characteristics.

A. Subjective Methods

Once again, look carefully at the eyes of an observer engaged in some visual task. Besides the characteristics of the different types of eye movements, note also the presence of lid movements, blinks, and small head movements, all of which complicate the measurement of eye position. Direct observation of the eyes, which are usually observed reflected in a mirror or through a peephole in the observer's field, can be used to note the occurrence of large eye movements and to make a crude estimation of viewing direction.

Alternatively, strong foveal afterimages allow an observer in a dimly lit room to observe his own saccades, smooth movements, and convergent and divergent eye movements. Saccades past a rapidly flashing light (Lamanski, 1869) or over an oscilloscope screen on which an intense spot is moving rapidly back and forth (Hendry, 1975) give percepts corresponding to the path of the image of the light or spot across the rapidly moving retina. These percepts allow the observer to estimate the duration, path, and velocity characteristics of his own saccades.

Lord and Wright (1950) provide a good review of subjective methods. More recently, Kaufman and Richards (1969) have described a procedure that lends itself to group demonstration and that allows at least most observers to see their own instantaneous fixation points. A scene is presented in plane-polarized blue light and the plane of polarization is rotated. This causes an observer to see at his fixation point a revolving "propeller"—Haidinger's brush formed by crystals in the fovea sensitive to polarized light—seemingly superimposed on the scene and moving over it with eye movements.

B. Electrooculography

Surface electrodes placed around the eye sockets can record small changes in potential produced by eye-in-head movements. With appropriate placement and electronic decoding, horizontal and vertical recording from either or both eyes is possible, but only the best systems achieve a sensitivity of 1°, so only large eye movements may be recorded. Some disadvantages of this technique are that the potential changes are typically less than 10 μV for 1° of eye movement and are immersed in variations attributable to changes in muscle action, luminance level, and state of adaptation of the observer. Its advantages are that eye movements may be made over a very large field, recordings may be made during sleep, and there is no interference with the field of view. In addition, radiotelemetry from small head-mounted units may now be feasible.

C. Contact Lenses

The most accurate measurements of small eye movements have been made by using an accurately ground, individually fitted contact lens. Suction is used to hold the lens tightly against the sclera to minimize slippage between lens and eye. Head position is carefully stabilized, and eye rotation is sensed by measuring the rotation of light reflected from one or more small plane mirrors incorporated in the contact lens. Alternatively, the visual stimulus may be mounted on a short stalk on the lens itself if the aim is to study stabilized vision (see Section IV,A). Contact lens systems have never been in use in more than a handful of laboratories. They can offer a sensitivity of 5 sec of arc but only over a field of view of some 5° and at the cost of observer discomfort. The technique is appropriate for use in the specialized study of fixational eye movements rather than for general monitoring of eye position.

Robinson (1963) introduced a contact lens method with a larger useful field of view and no requirement for accurate head restraint. The position of the lens, and thus the eye, is sensed by measuring the voltages in two wire search coils embedded in the lens that are induced by an alternating magnetic field surrounding the head. Collewijn, Van der Mark, and Jansen (1975) reported a promising development of Robinson's method in which an annulus of flexible silicone rubber, fitting tightly against the sclera but not touching the cornea, was used to carry the wire coils. They claimed to have retained the accuracy and flexibility of Robinson's method while removing the requirement that lenses be individually fitted and while also greatly reducing observer discomfort.

D. Limbic Sensing

The white sclera reflects a larger proportion of incident light than does the colored iris. If the eye is irradiated with infrared radiation and an appropriate photodetector is aligned so that the iris–scleral boundary, the limbus, falls across its acceptance field, its output will vary as the eye moves. Both vertical and horizontal movements can be sensed with suitably positioned detectors, within the limits set by lid movements. The unit may be head-mounted for eye-in-head measurement or may be bench-mounted to give eye-in-space monitoring if head position is held fixed. A sensitivity considerably better than 1° is attainable and the useful field is fairly large, but attachments are required close to the eyes. Limbic sensors are available commercially, and the technique is becoming widely used; an advantage is that the output is electrical and in a form suitable for input to a computer.

E. Corneal Reflection

Look into the eye of an observer. A small, erect virtual image of any bright feature in the visual field is formed by the cornea and is easily seen. This image, the corneal reflex, moves with eye movements, but because the radius of the cornea is about half that of the eye, the reflex moves only about half as far as the pupil. If a field containing well-lit stimuli is used, the experimenter or camera looking through a peephole in the display can record which of the stimuli the observer is fixating, since the image of the fixated stimulus appears at the center of the pupil. This procedure requires no attachments to the head and is now wisely used to determine fixation direction in human newborns. An accuracy of a few degrees may be achieved, and head position need not be closely stabilized. There is, however, an error introduced by the difference in orientation between the visual axis and the line through the center of the pupil that is perpendicular to the corneal surface. In adults, the angle between these axes is about 5° and may easily be allowed for by calibration. But in newborns, with whom use of the method is particularly attractive, the angle is larger (Slater & Findlay, 1975a) and satisfactory correction is more difficult.

Alternatively, a point source may be fixed relative to the head, and movement of the corneal reflex may be used as a measure of eye movement. Mackworth and Mackworth (1958) extended this technique by using a half-silvered mirror to superimpose an image of the reflex on the scene recorded by a head-mounted camera. The image of the reflex indicates where in the visual field the observer is fixating. The accuracy of the system is limited to about 2°. Commercially made versions of the Mackworth "eye-mark" system are now available.

F. The Oculometer

When the eyes move, both the pupil and the corneal reflex move, and the position of either may be used an an index of eye-in-head position. But the relative movement of reflex and pupil contains more information: It provides an index of eye orientation in space, and so may be used to indicate the line of sight, independently of head orientation and position. Cowey (1963) and Slater and Findlay (1972) have discussed the principles of the method. Merchant, Morrissette, and Porterfield (1974) describe the Honeywell Oculometer, a device requiring no attachments to the head and allowing some head movement. This device uses a single infrared source to provide the corneal reflex and to light the pupil from behind by reflection from the interior of the eye. A television camera that is sensitive to infrared then gives a sufficiently clear picture of the bright pupil and the very bright reflex for the vector joining the centroids of the pupil and the reflex to be determined by a computer monitoring the television scan signal. This displacement vector gives, after calibration, the observer's line of sight.

The field of view of such systems is usefully large, but sensitivity is about 1° at best, and speed is limited by the 50- or 60-Hz television scan rate. Monty (1975) has exploited the freedom from head restraint to monitor an obsever's eye movements covertly.

G. The Double-Purkinje Eye Tracker

Purkinje images are the images formed by reflection at the optical interfaces of the eye; the corneal reflex is the first Purkinje image. The fourth Purkinje image is formed by reflection at the rear surface of the lens and, just as the relative positions of reflex and pupil indicate eye orientation, the relative positions of the first and fourth Purkinje images may also be used to give the line of sight independently of head position and movement. A double-Purkinje eye tracker was developed by Cornsweet and Crane (1973), and later improvements have been described by Clark (1975). This very promising new development in eye movement measurement is available commercially and is already in fairly widespread use. Infrared first and fourth Purkinje images are directed by movable mirrors onto two photodetectors. A servosystem controlled by the photodetectors rotates the mirrors to keep the images centered. The signals to the servomotors give an analogue measurement of the separation of the Purkinje images and thereby indicate the observer's line of sight. The head may move up to 1 cm in any direction, so chin-rest stabilization of head position is adequate. The useful field of view is 30° high and wide, and both sensitivity (about 1 min of arc) and speed of response (a lag of some 4 msec is introduced by the mechanical movements) are much better than the corresponding values for the Oculometer.

H. Stimulus Presentation Contingent on Eye Position

A recurrent theme in the remainder of this chapter is that a great contribution will be made to future eye-movement work by contingent systems. A contingent system allows a subject to view the consequences of his own eye movements. The stimulus may be an oscilloscope spot controlled by eye-position signals (Rashbass, 1960), or the link between eye and stimulus may be optical (Wallach & Lewis, 1966). In either case, transformations are introduced in order to study vision for stimuli that move in chosen ways when the eye moves. Very flexible contingencies between eye position and stimulus presentation are now obtainable by using a computer to accept the eye-position signal and then control stimulus presentation (see Fig. 2). Reder (1973) has explored these possibilities and describes one system using limbic sensing of eye position and an oscilloscope display for stimulus presentation. Some of the contact-lens devices and the double-Purkinje eye tracker may also give eye-position information sufficiently quickly to provide the input to a contingent system. A limbic sensing device operating on direct current should even allow presentation of stimuli well within 1 msec of the

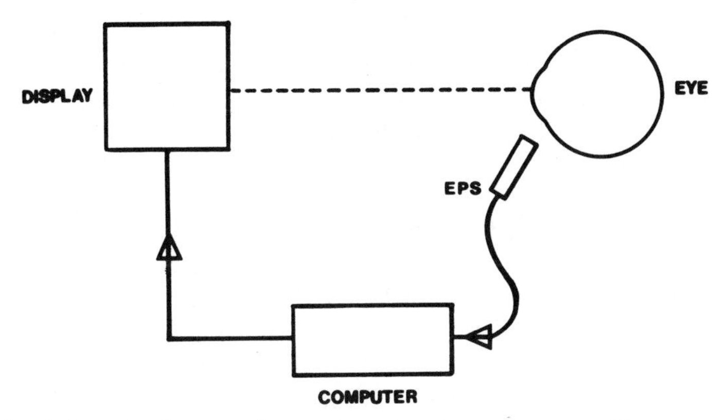

FIG. 2. Schematic diagram of a contingent system. An eye-position sensor (EPS) monitors the instantaneous orientation of the eye. A computer both accepts eye orientation information and controls the presentation of stimuli on a display viewed by the subject.

determination of eye position, and so permit contingent stimulus presentations during saccades.

Contingent systems, by allowing stimuli to be presented, moved, or changed in any chosen way as a function of eye movements, will allow rich experimental attacks on a great variety of problems, including some of the great perceptual questions raised by eye movements: What may be seen during saccades? Is any use made of input during saccades? What use is made of input from the periphery in perception and in guiding eye movements?

IV. THE VALUE OF EYE MOVEMENTS FOR PERCEPTION

A. Fixational Eye Movements and Perception

1. Stabilized Retinal Images

The contact-lens method of measuring fixational eye movements may be adapted to present an image that moves just as the eye moves, so that a stabilized retinal image (SRI) is seen by the observer. The light beam reflected from the lens is passed through an optical system and projected on a screen visible to the observer. Ditchburn and Ginsborg (1952) and Riggs, Ratliff, Cornsweet, and Cornsweet (1953) independently developed such optical lever systems and reported the now familiar rapid fading of perception of a stabilized image. Tiny movements of the image, as well as any blink, disturbance of the lens, or change in stimulus illumination, were found to give immediate regeneration. Spontaneous reappearance, both partial and full, of stabilized images was also observed, but Barlow (1963) provided evidence that the reappearances could be attributed to slight movement of the contact lens across the eye.

Yarbus (1967) achieved excellent stabilization using a fitted cap held to the eye by suction. The stimulus object and a converging lens were mounted on the cap itself. Yarbus reported never having observed the reappearance of a well-stabilized image, and described the appearance of the *empty field*, a homogeneous field left after fading, which contained (at most) a slight trace of the brightness and hue initially seen in the stimulus.

Heckenmueller (1965) reviews early SRI work, including observations of the stabilized images that we all fail to see all the time: the shadows of the retinal blood vessels and other entoptic images that are only visible in particular lighting conditions.

The results of studies of stabilized images are relevant to a consideration of perception because they demonstrate that the eye movements of fixation, condemned at first glance as merely nuisance movements arising from noise in the oculomotor system, are vitally necessary for all normal visual perception: If a retinal image is not continually moved over the retinal mosaic it does not give rise to perception. In fact, the retina requires continual change both in space (to avoid the fading seen in a Ganzfeld) and time (to avoid SRI fading). These results fit well with conceptions of the nervous system as signaling change. Arend (1973) gives one of the very few attempts to press the argument further and examine the implications for visual theory.

Pritchard, Heron, and Hebb (1960) found that observers describe SRI fading to be structured and to consist of the disappearance and reappearance of meaningful units. Most SRI work since then has been directed to the study of structured fading, with hopes of identifying pattern-perception mechanisms in the visual system corresponding to those identified by neurophysiologists in animals. The idea is attractive—Weisstein (1969) has explored the logic of the method—but early work is unconvincing because the requirement for verbal description can exaggerate the amount of structure attributed to the fading, as Schuck (1973) has demonstrated by using steady fixation and an objectively changing stimulus. But experiments (e.g., Atkinson, 1973; Brown *et al.*, 1975; Wade, 1974) using simpler stimuli and less subjective responses have established the description of SRI fading as a promising method for investigating visual perception. An important advance will be the introduction of apparatus that allows flexible objective variation of the stimulus during stabilized viewing. Gerrits and Vendrik (1974) have described a suitable system, although they have not reported using it for this purpose (see Fig. 3).

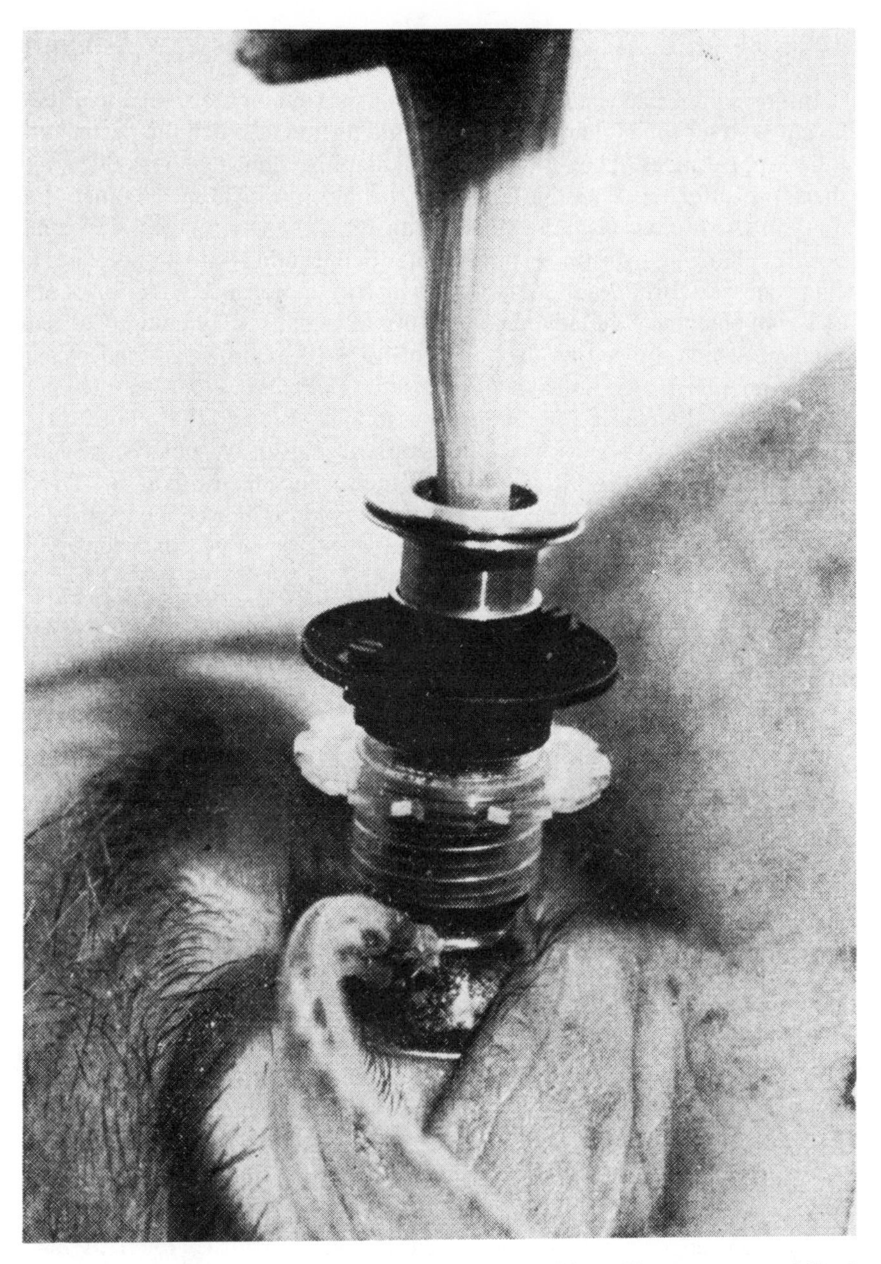

FIG. 3. Part of the apparatus used by Gerrits and Vendrik (1974) to present a stabilized retinal image that may be varied by the experimenter. An image of the stimulus is transmitted along a flexible bundle of optic fibers, which is visible at the top of the figure. The lower end of the fiber bundle is attached to a contact lens worn by the subject and so moves freely with the subject's fixational eye movements. [Reproduced from H. J. Gerrits & A. J. H. Vendrick. The influence of stimulus movements on perception in parafoveal stabilized vision. *Vision Research*, 1974, *14*, 175–180, with permission.]

Intense afterimages are certainly stabilized well on the retina, and they do show, over time, characteristic SRI fading and also definite partial and full reappearances. They are the easiest stabilized images to produce and therefore offer an attractive approach to the investigation of functional units in the human visual system (Evans, 1967). Afterimages, however, besides changing over time as the result of continuing retinal photochemical processes (Brindley, 1962), also change in appearance with blinks and eye movements. Even eye movements as small as 10 min of arc or smaller, which are within the range of sizes of fixational saccades, can cause an afterimage to disappear briefly (Kennard *et al.*, 1970). So, ironically, although an afterimage is a stabilized image that is produced easily without the need for accurate monitoring of eye position, its appearance can depend critically on the microsaccades of fixation. These eye movements therefore need to be suppressed, or at least measured and allowed for, if the results of afterimage work are to be consistent with results found by other approaches to SRI fading (Wade, 1974).

2. Retinal Image Movement Necessary for Vision

Fixational eye movements provide sufficient movement of the retina relative to the image for normal vision. What minimum relative movement is necessary for vision? The answer certainly depends on the stimulus and the viewing conditions, for the speed and pattern of fading of a stabilized image depend on the stimulus and its position in the visual field. High-contrast, sharply defined foveal stimuli fade most slowly, and so could be expected to require the least movement for vision to be maintained. For low-intensity, low-contrast targets, particularly those seen in the parafovea, even fixational eye movements do not suffice for vision, since fading is observed with voluntary fixation (Schuck, 1973). On the other hand, Steinman *et al.* (1967) found that for a small white fixation dot normal vision is not impaired when microsaccades are suppressed, so, in this case at least, drifts and tremor provide sufficient image movement.

The adequacy of various movements for the maintenance of vision can be studied by imposing motion on a stabilized image. The early work of Ditchburn, Fender, and Mayne (1959) used high-contrast foveal stimuli and suggested that movements over quite a wide range of frequencies were required to maintain vision. Gerrits and Vendrik (1974) used an ingenious fiber optic device, part of which is shown in Fig. 3, to present a television picture as a stabilized image. They imposed a variety of types of movement on a bright square viewed in the near periphery and obtained results suggesting that movements similar to drifts and the lower frequencies of tremor were sufficient to maintain vision. They found that movements must not be too regular, or their regenerative effect habituated. They also found that microsaccades may play a role in the maintenance of vision for nonfoveal stimuli, which in general fade more easily and require larger movements for regeneration.

An incidental question briefly mentioned by a few of these papers is of greater interest than finding the minimum movement needed for vision: Are such small movements perceived, and is some compensation for normal fixational movements necessary to hold the world still? Gerrits and Vendrik (1974) mentioned that their imposed random image motion was seen as stimulus movement, whereas in normal vision the similar motion of the retinal image produced by fixational eye movements is certainly not. However, Clowes (1961) found that if an optical lever system is used to exaggerate rather than to cancel image movement produced by eye movements, the subject does not report any perception of motion. More work is required, but an attractive possibility is that image motion synchronous with eye movement is ignored, even if, within limits, the extent or direction of the motion is inappropriate. This possibility is certainly more parsimonious than the compensation system based on movement-detecting neurones suggested by Arend (1973).

3. Acuity and Fixational Eye Movements

Weymouth, Anderson, and Averill (1923) and Marshall and Talbot (1942) suggested that tiny, fast movements of the retina may effectively "fill in" the receptor mosaic and improve vision for stimulus details that give images smaller than the foveal cone separation. The possibilities may be analyzed theoretically in detail, but it suffices to say that Keesey (1960), amongst others (see Riggs, 1965), has found very similar acuity for briefly visible targets whether seen in normal viewing or as stabilized images, suggesting that fixational movements serve neither to degrade nor enhance fine vision. Westheimer and McKee (1975) have recently provided further support for this conclusion by showing that acuity is not changed even when the briefly visible target is presented in motion at $1^\circ \sec^{-1}$.

B. Large Eye Movements and Perception

Walls (1962) argued that eye movements first appeared in phylogenesis to compensate for head and body movements, thus holding the image of the world stationary on the retina. Humans undoubtedly still make many more smooth eye movements to compensate for head and body motion than to track objects moving in the visual field. By contrast, voluntary saccadic eye movements serve to distribute the fixation point in space and time: They only appeared in evolution when a small part of the retina acquired superior resolving power. Foveas of high acuity, together with a fast and accurate saccadic system, imply a two-stage model of vision: Initial peripheral analysis locates objects and events of interest; this then leads to fixation and thus to more detailed analysis. The tendency to fixate is so strong that it is the one component of directional response that possesses such speed and reliability that it is considered part of the generally nonspecific orienting reflex (Sokolov, 1963). Indeed, a very weak stimulus that is just below foveal threshold and can only be seen when viewed in the near periphery, disappears periodically because the observer cannot suppress reflexive drifts that insist on aligning the image on the fovea, where it is invisible (Steinman & Cunitz, 1968). But usually, however, the observer sees a stable world and can coordinate beautifully within that world his fixation point movements, body movements, and manipulative movements of objects. The fixation point may be moved purposively throughout a large space surrounding the body, with the sequence of movements being well matched to the observer's current task or interest. But at any time the first stage of analysis can enable a sufficiently conspicuous peripheral event to interrupt and capture fixation temporarily.

Two-stage models of vision have obvious parallels in models of general human performance. These, too, seek to describe wide input channels that are followed by detailed analysis applied only to a selected tiny portion of all impinging stimulation. An enormous range of possibly important stimuli is surveyed by the wide input channels, but detailed analysis can only be applied to selected stimuli, because the capacity of the central mechanisms is limited. Neisser (1967) and Broadbent and Treisman (see Broadbent, 1971) have developed such models.

A two-stage visual model has been advanced by Trevarthen (1968). Trevarthen's model distinguishes between ambient and focal vision: Ambient vision relies heavily on peripheral visual input and visual exploration by gross head and eye movements to define space around the body; focal vision employs the foveal channel to examine in great detail, with the aid of saccades, a small part of space. The model is strongly based on neurophysiological findings, with the midbrain subserving ambient vision, while focal vision is largely cortical. Links with motor control centers are emphasized, as they must be if the model is to handle the observed richness of interaction between vision and action.

Breitmeyer and Ganz (1976) have developed in detail a two-channel model of vision in which the role of the cortex in processing for each channel is emphasized. They have proposed that *transient channels* carry out initial peripheral analysis, whereas *sustained channels* are needed for the perception of fine detail. Transient channels, formed by transient cells both in the geniculostriate system and in the (subcortical) superior colliculus, have short latencies and brief responses, are particularly sensitive to movement, but have poor spatial resolution. Sustained channels, on the other hand, are formed by cells in the retina, geniculate, and visual cortex that have fine spatial resolution, but are relatively insensitive to movement and have long latencies and long integration times. In the retina, sustained cells are heavily concentrated near the fovea, whereas transient cells are numerous in the periphery. A novel stimulus moving into the peripheral visual field may therefore stimulate transient channels, which call for an eye movement to bring fixation to the stimulus and so allow the sustained channels to make a detailed examination of it.

Breitmeyer and Ganz have presented a major synthesis in the field of visual research. They draw heavily on physiological results to justify the assumptions of their model and are able to explain an impressively wide range of findings in the fields of visual masking, visual information processing, and saccadic suppression (see Section V). Their analysis makes many contributions to the subject of this chapter. For example, it would be easy to interpret the results discussed in Section IV,A,2 in their terms: Sustained channels should require less image movement for the maintenance of vision than should transient channels. Indeed, least movement is required for highly detailed stimuli presented to the fovea, and these are precisely the stimuli that are assumed to be analyzed by sustained channels.

A two-stage model of vision implies that fixation implies attention and that vision in the periphery is important. These two points will be considered in the next three sections.

C. Scanning

The use of fixation position as an index of current visual attention is particularly attractive with infants, since controlled looking appears early in life (Slater & Findlay, 1975b; Trevarthen, Volume III of *Handbook of Perception*) and appears to be used in a flexible and discriminating way. The richness of infants' looking remains to be explored, but the number and duration of fixations are now widely used to provide operational definitions of both visual preference and the ability to discriminate among stimuli, and so to help in the study of the development of perception and cognitive abilities (McCall, 1971).

In adults, observation of fixation sequences gives a basis for the description, modeling, and teaching of complex skills, such as driving (Mourant & Rockwell, 1972), which require rapid and wide deployment of visual attention. In more spatially limited cognitive tasks, fixation patterns have been used as indications of the sequence, organization, and speed of the processing chosen by the subject attempting the task. Examples are reading (Judd & Buswell, 1922), visual search (Gould, 1973) and the analysis of chess positions (Simon & Gilmartin, 1973). The study of large eye movements can even be necessary for a full understanding of tasks that are concerned in some way with space, yet are not visual—such as judging the position of a sound in the dark (Jones & Kabanoff, 1975), using mental images (Hebb, 1968), or paying attention to the input to one ear while ignoring simultaneous input to the other (Gopher, 1973). Subjects engaged in such tasks do often make appropriate eye movements, even though the movements cannot lead to the gaining of relevant visual information; rather, they must serve some function of spatial orientation or organization.

1. Looking at Pictures

The scanning pattern observed when we look at a picture gives tantalizing suggestions about how we build up detailed knowledge of a picture over time. Early work by Buswell (1935) found that fixation points are not uniformly distributed over a picture; Mackworth and Morandi (1967) showed that fixation points cluster about angles, junctions, and other parts that are informative. But of more interest than just the points fixated are the sequences followed during scanning, which must be closely linked to the processes of perception, recognition, or memory invoked by the observer. Noton and Stark (1971) found that an observer viewing any particular picture tends to make repeated sequences of fixations. Extra viewing time is spent on further fixations of the same important parts of the picture, often even with similar sequences of fixations recurring. They termed such sequences *scanpaths* and found that a subject often exhibited the same scanpath in recognizing a picture as he had when initially examining it for memory. Mackworth and Bruner (1970) analyzed picture scanning by adults and 6-year-old children. Their records showed that adults fixated predominantly on informative parts of a picture; large areas of a picture might never be fixated, and sharp contours or particularly bright or dark parts might only be fixated if they contributed to the meaning or interest of the whole picture. Mackworth and Bruner found that adults scanned in a highly skilled and appropriate fashion, whereas children could not pick out important features so well and sometimes stopped short of surveying the whole picture, seemingly because their gaze became captured by some small part of the scene. Mackworth and Bruner reported, in accordance with other workers, a mean fixation duration of about 350 msec for adults and only slightly longer for children.

They contrasted these values with those for fixations during reading (see Section IV,E) which were 500 msec or more for beginning readers, steadily reducing to some 240 msec for adults. Jeannerod, Gerin, and Pernier (1968) found, however, that fixation duration during the inspection of pictures varied with the complexity of the picture, with more complex pictures actually eliciting shorter fixations.

Yarbus (1967, Chapter 7) completed the description of his pioneering studies of eye movement with a discussion of the viewing of pictures, and published many eye-movement records to illustrate the discussion (see Fig. 4). His records show clearly that although patterns of scanning do show some similarities across observers, the pattern of scanning varies markedly with the picture, the observer, and, in particular, with the observer's purpose for viewing a picture on a particular occasion (see Fig. 5). Once again it appears that an observer is quite purposive and highly skilled in selecting the parts of a picture that are important or of interest for the task in hand; he examines them in turn, and thus builds up his knowledge of the picture.

2. Eye-Movement Measurement and Mental Processes

Two studies described below suggest how the imaginative use of eye-movement measurement may, in the future, contribute to the understanding of mental processes. Cooper (1974) had subjects face an array of drawings of common objects or animals while listening to a story. The story included mention of the pictured objects and animals, both directly by name and indirectly by associated word or pronoun. Even though no fixation or inspection instructions were given, Cooper found that subjects tended to look at a picture just when that object or animal was named or referred to in the story. Most interestingly, he found that when pictures were fixated at appropriate times in the story, the fixation very often commenced during the actual pronounciation of the critical story word, indicating a close, continuing relationship between understanding a speech message and spontaneous looking. The linguistic sensitivity of looking and the observed short fixation latencies suggest that with appropriate verbal tasks and visual arrays this method will be a valuable tool for the study of cognitive processes: The eye-movement record makes some aspects of a subject's comprehension processes continuously visible.

Just and Carpenter (1976) recorded eye movements of subjects performing speeded tasks presented visually; for example, deciding whether or not two adjacent outline figures represented the same three-dimensional

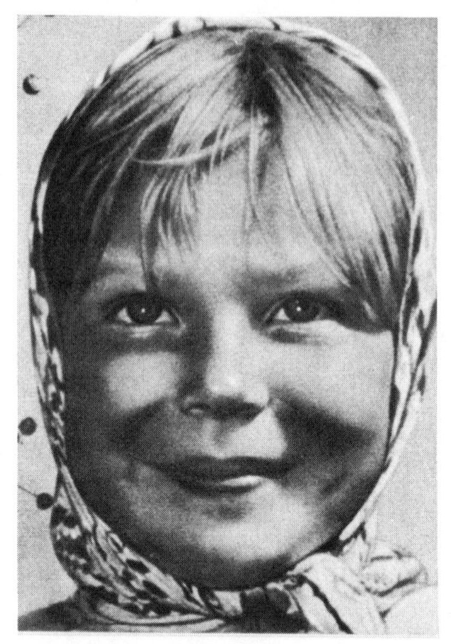

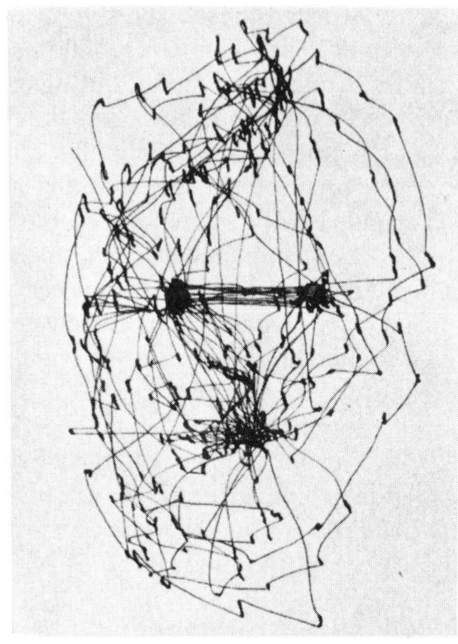

FIG. 4. A record of eye movements during a 3-min free examination of a portrait photograph. [Reproduced from A. L. Yarbus, *Eye Movements and Vision*. Translated from Russian, L. A. Riggs (Ed.).New York: Plenum Press, 1967.]

object. Shepard and Metzler (1971) proposed a mental rotation model for this task on the basis of the total time subjects took to respond. However, by recording fixation locations and durations Just and Carpenter obtained more information from their subjects and so could propose a more detailed model of how the task was carried out. They have also presented strong arguments for the correctness of the assumption on which their analysis must depend—that the symbol or element being fixated is the one currently receiving processing attention. Once again, eye movements betray something of the workings of the mind.

D. Scanning and Social Perception

Eyes have long had a special place in ritual and have posed particular problems as artists sought in their representations to stimulate the response to a living gaze (Gombrich, 1973). More prosaically, social psychologists have investigated the importance in social interaction of the perception of where another person is looking and how he moves his eyes in response to us and to our own looking behavior. Looking behavior certainly does contribute to social interaction: The adjustment of level of intimacy, synchronization of successive contributions to a conversation,

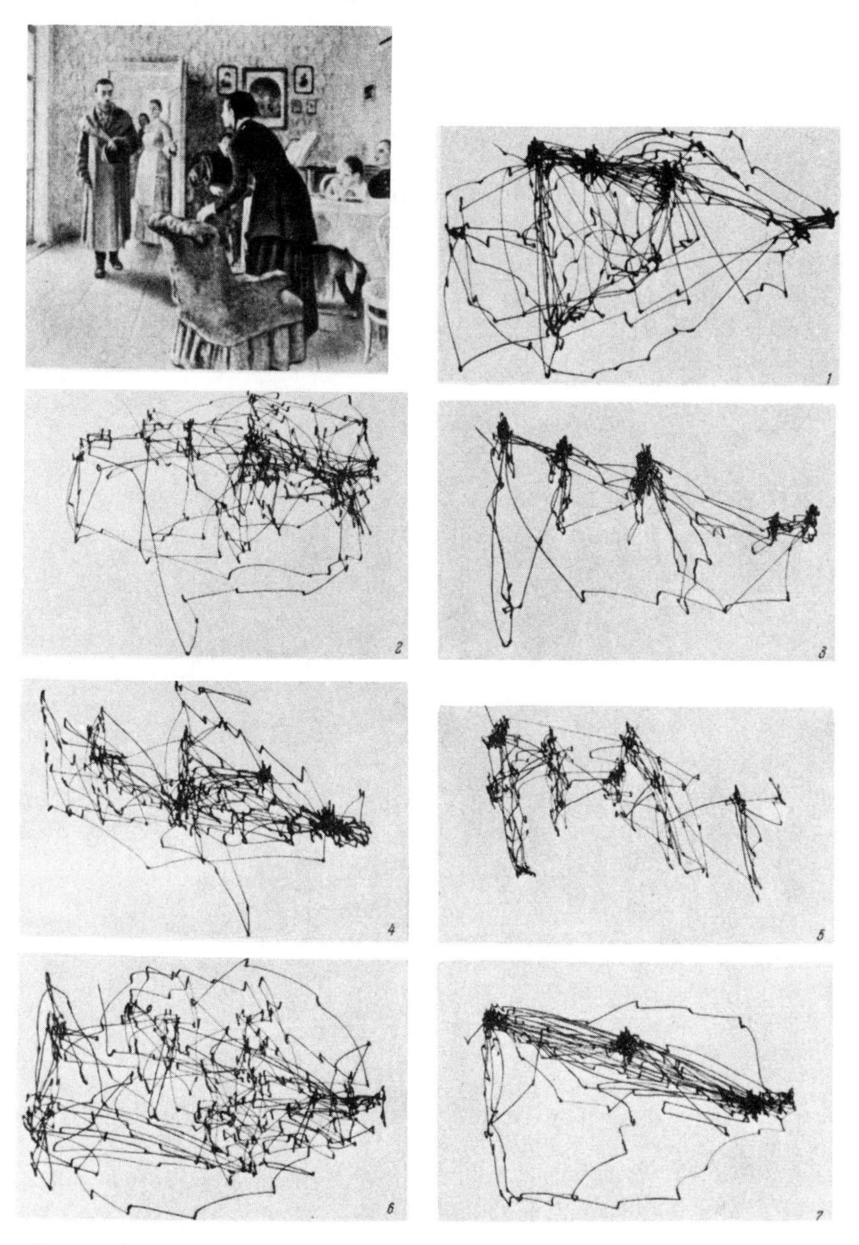

FIG. 5. Seven eye movement records, each lasting 3 min, by a single subject. In (1) the subject made a free examination of the picture. The subsequent records show the result of the subject being asked: (2) to estimate the material circumstances of the family; (3) to estimate the ages of the people; (4) to guess what activity had been interrupted by the arrival of the visitor on the left; (5) to remember the clothes worn by the people; (6) to remember the positions of people and objects in the room; and (7) to estimate how long the visitor had been away from the family. [Reproduced from A. L. Yarbus, *Eye Movements and Vision*. Translated from Russian, L. A. Riggs (Ed.).New York: Plenum Press, 1967.]

and seeking and sending of information about interpersonal attitudes are some of the aspects of interaction to which looking makes a contribution. Beyond such general statements, however, experimental work has so far given only fragmented and largely unconvincing results, of which Argyle and Cook (1975) have given a comprehensive review. One problem is that any special significance of looking directly into the eyes of another, rather than just generally toward the face, must rely on an accurate perception of our direction of gaze, yet people may not be particularly good at making this judgment (Lord & Haith, 1974).

E. Peripheral Vision

It must be peripheral vision that gives us a stable visual world around us, within which we can scan as we choose. Yet we know sadly little about peripheral vision and how it plays its important role in the two-stage model of vision. Photopic acuity certainly does drop steeply from the fovea, by about 50% over 1°, but vision can still be usefully detailed in the parafovea. Anstis (1974) has reviewed the literature and argues that from about 4 to 30° eccentricity acuity decreases linearly in all directions from the fovea. Perception of color is progressively lost and movement thresholds rise into the periphery, but the perception of movement in the periphery can still be useful and salient (Biederman-Thorson, Thorson & Lange, 1971; Breitmeyer & Ganz, 1976), an arrangement that is clearly highly adaptive for primates.

Is the whole of the peripheral field condemned never to receive attention? Helmholtz (1962, p. 455) reported that it is indeed possible to attend selectively to a point in the periphery. Trevarthen (1968) considered the possibility that some trading of sensitivity between central and peripheral vision can occur in accordance with the momentary relative importance of the two visual systems. Grindley and Townsend (1968) reported some observations and Zinchenko and Vergiles (1972) found that subjects can voluntarily move their attention over a stabilized image, but this question, as many others concerning peripheral vision, remains to be fully explored.

What little discussion there has been of the periphery concerns the first-order periphery: those points lying on the horopter, or surface through all points that are seen in single vision for some particular convergence of the eyes. Points lying nearer or farther than the horopter are seen in double vision and may be considered to be in the second-order periphery, yet we continually make combinations of saccades and vergence movements that allow us to fixate as we chose on such points. Our ignorance of vision in the second-order periphery is close to total.

To take an example, what is the role of peripheral input in reading? A reader makes on the average some four fixations each second, usually advancing along the line of text, but interspersed with regressive movements for an occasional second look and, of course, with long leftwards saccades to the start of the following line. The spacing of successive fixations and the frequency of regressions vary greatly with the difficulty of the text, the reader's skill, and his purpose in reading the text on a particular occasion. Average fixation duration, however, remains fairly constant. About 10% of time is spent in making saccades between fixations. But how is the placing of fixations determined? Does the reader make a centrally planned sequence of fixations within a broad field built up from peripheral input, or does he rely on peripheral input during one fixation in deciding where the next fixation should be? These alternatives are certainly too polarized and too simple for such a flexible activity as reading, but the question does give a basis for theoretical analysis (Haber, 1976; Hochberg, 1970) and for experimental attack. Bouma and de Voogd (1974) presented successively to subjects small overlapping segments of text at a controlled rate. In this way the subjects saw a succession of images similar to that seen in normal reading, but without making eye movements and with no control over the step size and presentation rate. Both silent and oral reading of normal speed were possible for a surprisingly wide range of step sizes and presentation rates, even when irregular step sizes and rates were used. Bouma and de Voogd concluded that the careful placing of successive fixations is not necessary for reading, and therefore no rapid peripheral analysis need be carried out on the text ahead to guide future eye movements. But their procedure relied on crude subjective estimates of when normal fluent reading was possible. Also, their conclusion ignores the occurrence of regressions, which must certainly be programmed on the basis of current analysis of the text being read (Hochberg, 1975). On the other hand, Rayner (1975), using a contingent system that allowed him to change the text being read during a saccade, found that information about word length, and even meaning, was available a little ahead of fixation, and so might be used in directing the next saccade. Even so, information available in advance need not necessarily be used to control succeeding fixations, and in any case there was evidence of information being available only about one or two words ahead of fixation. So reading, too, still remains mysterious, but the technology of contingent systems, together with a theoretical orientation recognizing the important and complex role of eye movements in all visual perception, should allow rapid progress.

V. PROBLEMS INTRODUCED BY EYE MOVEMENTS

We can move our fixation points with very great freedom and, as discussed in Section IV, we benefit greatly by being able to do so. But the sudden drastic changes in the retinal image produced by eye movements must surely, it would seem, disrupt perception. In this section we consider how stable perception can be achieved, despite the occurrence of eye movements.

Consider what does *not* happen when you make a saccade: Even though the saccade lasts for at least 30 msec—a long time in visual terms—you do not see any blurring, nor a brief dark period, nor any gross movement of the visual world. Evolution must of course provide for such perceptual stability, and at one level it is sufficient to follow Gibson (1968) and describe saccades as simply moving the retina across the visual array, which remains objectively and perceptually stationary. But questions of mechanism still remain. Three closely interrelated questions will be considered here. What may be perceived during a saccade? How is visual stability maintained from fixation to fixation? Finally, what is the role of visual persistence and masking-like interactions between successive inputs?

This section is primarily concerned with saccades and the problems they pose for perceptual stability. Tracking movements also give gross changes in the retinal image, but of course the changes occur much more slowly than with saccades. Vision during the movement and the maintenance of a stable perceptual world have not been studied so extensively for smooth tracking movements as for saccades, but it does appear that a fairly stable perception of direction is maintained during tracking movements (Matin, 1972, pp. 355–359; Ward, 1976). However, the rates of motion of a tracked target and of other stimuli in the field, and the distances they move, are frequently judged incorrectly (Stoper, 1973). In particular, if a target is tracked as it moves over a stationary background the speed of the target is underestimated and the background is seen as moving slowly in the opposite direction (Mack & Herman, 1973).

A. Perception during Saccades

That a saccade does not give a perception of blur, nor of a dark period, led to theorizing that input during a saccade is rejected by the visual system. All input during a saccade might be ignored, or only blurred input. But saccades made across a light source giving very brief flashes at

50–200 Hz give a percept of several flashes spread out in space, so vision cannot be totally suppressed during saccades (Lamanski, 1869; and see Section III,A). Many investigators have studied perception when stimuli (of very brief duration to minimize blur) were presented during saccades. But methodology has often been poor, the results have been confused, and many important experiments remain to be done. Fortunately, Matin (1974) has provided a brief but comprehensive review of the literature and an excellent discussion of saccadic suppression and related issues.

The main conclusion is that the absolute threshold for the detection of very brief flashes in the dark is raised by about .3–.5 log unit during a saccade. A similar small degree of suppression has been found for word recognition and for acuity tasks (Volkmann, 1962). It also occurs during microsaccades as well as for normal large saccades. It may even be found in the absence of all visual stimulation: Riggs, Merton, and Morton (1974) found an increased threshold during saccades for the detection of visual phosphenes induced electrically. Zuber, Stark, and Lorber (1966) reported that the normal slight constriction of the pupil to a brief flash may be almost abolished by presenting the flash during a saccade.

Considering the wide variety of experimental conditions and procedures that have been used, the .3–.5-log-unit increase in absolute threshold for brief stimuli in the dark has been found with surprising consistency. The time course of suppression does, however, vary somewhat with stimulus wavelength, saccade length, and other experimental conditions.

Suppression typically commences 50–100 msec before the start of the saccade, reaches a maximum early in the saccade and does not end until 50–100 msec after the completion of the eye movement. That suppression occurs when the eye is stationary before and after a movement and in the absence of any background field supports the inference made from experiments using very brief stimuli that this suppression must be imposed by the visual system, rather than being a consequence of rapid contour movement over the retina. Suppression is usually considered to be one central consequence of making a saccade, but Richards (1969) has conducted ingenious experiments suggesting that suppression may be caused by mechanical distortion produced peripherally in the retina by the very great forces needed to rotate the eye saccadically. He argues that shear within the retina causes suppression by increasing background neural activity, rather than by reducing the response to the stimulus itself. Riggs *et al.* (1974) have criticized some of the work on technical grounds, but the findings of Richards cannot be ignored; they remain to be replicated and reconciled with other approaches to suppression.

B. The Maintenance of Stability

The maintenance of stability from one fixation to the next might seem to pose no problem, for the size and direction of the intervening saccade is presumably known and so the appropriate adjustment can be made when interpreting the new, shifted retinal image. However, the compensation could be carried out in any of several different ways and, in any case, the speed and accuracy of the adjustment remain impressive.

There are two obvious contending models of the mechanisms involved. The first assumes the existence of a position sense in the eye: Continuous signals indicate the current orientation of the eye and so allow the retinal image to be related to absolute spatial coordinates. The second supposes that whenever a muscle command is sent to the eye a copy of this command is used to interpret the new retinal image that is received soon afterwards, assuming that the eye has indeed changed its orientation precisely as planned. Gregory (1973, Chapter 7) has given a good elementary discussion of these two possibilities, calling them the *inflow* and *outflow* theories, respectively. He has also described the classical observations that have lead to strong preference for the outflow theory. MacKay (1973) has given a more detailed discussion of the problem.

Although suitable receptors are present in human eye muscles, Brindley and Merton (1960) seemed to have demonstrated that the eye had no position sense, by showing that the lightly anaesthetised eye could be physically moved by the experimenter without the subject's having any perception of the direction or extent of the movement. But Skavenski (1972) has shown, by using a less uncomfortable and more sensitive procedure, that inflow information about eye position is available and can, for example, be used by the subject to maintain eye position in the dark with fair accuracy. Skavenski, Haddad, and Steinman (1972) also showed, however, that inflow information appears to be ignored if normal outflow information about an intended eye movement is available.

The debate over how compensation for saccades is accomplished seems in one respect quite unrealistic: Why should such precise computation be necessary to hold the world still (MacKay, 1973)? In virtually every single case in which a retinal image shift is corrected for the effects of a saccade, the result is zero change. Why not simply assume that the world does not move during a saccade, or at least why not make this assumption unless a crude compensation calculation indicates that a large movement of the world has occurred? A similar suggestion was made during the discussion in Section IV,A,2 on the invisibility of fixational eye movements. The possibility may easily be tested by measuring the threshold for detection of a stimulus movement that occurs during a saccade. This experiment has, in fact, been done: Mack (1970) and Bridgeman, Hendry, and Stark (1975) concur in finding that movements whose extent is less than 20–30% of the size of the saccade are not seen. So the world could make quite large movements and, provided they were precisely synchronized with saccades, these would not be noticed. This result also explains why one cannot see one's own eye movements in a mirror. Looking in a mirror must the the simplest way of obtaining stimulus change contingent on eye movement, and the observation that one's own eye movements are invisible has often been considered puzzling. The explanation now becomes simple: The movement of the image is, of course, perfectly synchronized with the eye movement, but for all normal viewing distances this image movement is considerably smaller than the shift in position of the fixation point, and so cannot be seen.

Another approach to the question of the maintenance of stability is to ask how perception of direction changes during the saccade itself. L. Matin and his co-workers have taken this approach by presenting brief flashes during saccades in the dark and asking subjects to localize these flashes with respect to comparison stimuli presented before the saccade commenced. Matin (1972) gives a summary of these experiments in the course of an excellent and detailed discussion of the maintenance of stability. The results obtained with this experimental procedure suggest that perception of direction can be quite inaccurate during a saccade. Perception of direction changes from being appropriate for the previous fixation to being appropriate for eye position following the saccade over a much longer time than is spent by the eye in motion. However, the visual fields used were impoverished and the response used relied necessarily on the subject's memory, so it is hard to relate the results to normal perception. Hallett and Lightstone (1976a, 1976b) used a different response. Their subjects were instructed to make saccades to the positions of successive stimuli presented briefly in darkness. The subjects could do so, and could even make an accurate eye movement to the position of a light only seen during a previous saccade. They therefore demonstrated, by using the very natural fixation response, that not only was input from a stimulus accepted during a saccade, but the absolute position of such a stimulus could also be accurately assessed.

The model of stability that postulates that compensation is based on outflow information, but is only crude, needs to be studied further with richer visual fields. The stability of peripheral parts of the field needs also to be explored: I have found in informal observations that brightly colored objects in the near periphery may sometimes be seen to move slightly with saccades. The possible contribution of continued vision of the world during the saccade itself also needs further investigation. For example, I know of no report of the consequences of using a rich visual field and simply turning the lights off for a short period during a saccade: If input during the saccade is necessary, then this procedure may disrupt the maintenance of stability.

C. The Relevance of Visual Masking

There is much more to vision than the detection of brief flashes in the dark, which was the main concern of Section V,A. The decrease in sensitivity of a mere .5 log unit that is observed with impoverished visual stimulation is probably of little relevance to normal vision and cannot explain the no-blur, no-blackness percept. Only recently have workers started to present more complex stimuli during the saccade and to study the effects of richer background fields presented before, during, or after the eye movement. Mitrani, Mateeff, and Yakimoff (1970) found that horizontal stripes are more easily seen than vertical during a horizontal saccade, suggesting that blur does contribute. Of even more interest are two other recent findings: Mitrani, Yakimoff, and Mateeff (1973) and Yakimoff (1973) showed that vision during a saccade is much less efficient in the presence of a stationary background field containing contour; and Matin, Clymer, and Matin (1972) and Mitrani *et al.* (1973) showed that interactions between stimuli visible before or after a saccade, with a stimulus presented during the saccade, cause particularly strong suppression of perception of the latter. These findings implicate immediately the whole field of visual masking, which is concerned with the disruption of perception of one stimulus by another close to the first in space and time (see Ganz, in Volume 5 of *Handbook of Perception*).

Alpern (1970) speculated that metacontrast (visual masking in which the two stimuli are adjacent rather than overlapping) may be related to saccadic eye movements. It is certainly an attractive idea that some masking interaction may help avoid any clash between the percepts obtained from two successive fixations, despite normal visual persistence of the first (Dick, 1974). E. Matin (1974, 1975) has made an initial detailed consideration of the contribution that masking concepts and results can make to the understanding of perception during saccades and the maintenance of stability. She describes the increasing importance of neurophysiological findings: For example, the identification in the monkey of a class of cortical cells selectively responsive to retinal image motion of more than 100° sec^{-1}. Such rapidly moving images would scarcely ever be encountered in normal vision except during saccades, and so these cells would appear to be specialized for contributing to perceptual processing related to saccadic eye movements.

Following some observations by MacKay (1970), Brooks and Fuchs (1975) studied the visibility of brief flashes when seen by the fixating eye, but against a background field that was moved very rapidly. A similar pattern of retinal stimulation was given by a brief flash presented against a stationary background to the saccading eye, and a similar large reduction in visibility for the flash was found in the two conditions.

Breitmeyer and Ganz (1976) have given a persuasive explanation of the masking aspect of saccadic suppression in terms of sustained and transient channels. They argue that transient channels can indirectly inhibit the response of more slowly responding sustained channels. Transient channels are strongly stimulated by the rapid image movement during a saccade (or by the rapid movement of the background field in the study of Brooks and Fuchs, 1975) and therefore inhibit the still continuing response of sustained channels to the visual field of the previous fixation. This inhibition ensures that the response of sustained channels does not persist into the next fixation.

The transient channel response, and therefore also the inhibition, would be greater for a vertically than for a horizontally striped field, and therefore this theory can also explain the results of Mitrani *et al.* (1970), originally interpreted simply as implicating retinal blur. It is, however, not so easy to explain the finding of Mitrani *et al.* (1973) that there is little suppression when a background is presented only during the saccade and that the field present at the start of the saccade is most effective in producing suppression.

In summary, although there is a small amount of centrally imposed suppression, masking-like interactions among stimuli are principally responsible for the normal smooth perceptual transition from one fixation to the next.

A final example of a type of masking experiment that should increase

S V Y

FIG. 6. An idealization of the percept consistent with the results of Davidson, Fox, and Dick (1973). A row of five equally spaced letters was presented briefly just before the subject made a saccade from the second to the fourth position in the row. Immediately after the saccade, a masking grid was presented at the absolute fourth position, now corresponding to the retinal position of the second letter. Would the second or fourth letter be masked? The results suggested that the second letter was masked, but that the mask itself was localized at the fourth position, without, however, masking the fourth letter. [Redrawn from M. L. Davidson, M. J. Fox, & A. O. Dick. Effect of eye movements on backward masking and perceived location. *Perception & Psychophysics*, 1973, *14*(1), 110–116.]

our understanding of perception in relation to saccades is that of Davidson, Fox, and Dick (1973). A standard masking task was used, but presentation was coordinated with eye movements so that a saccade occurred between the presentation of the letters that were to be reported and the masking stimulus. Would it be retinal or absolute position that determines where in the letter array masking occurs? The results suggested (see Fig. 6) that masking depended on retinal position, but, paradoxically, all the stimuli that were seen were localized correctly at their absolute positions in space. Clearly, we still have much to discover about perception in relation to eye movements.

References

Alpern, M. Movements of the eyes. In H. Davson, (Ed.), *The eye* (Vol. 3). (2nd ed.) New York: Academic Press, 1970. Pp 1–252.

Alpern, M. *Eye movements*. In D. Jameson & L. M. Hurvich (Eds.), *Handbook of sensory physiology* (Vol. VII/4). New York: Springer-Verlag, 1972. Pp. 303–330.

Anstis, S. M. A chart demonstrating variations in acuity with retinal position. *Vision Research*, 1974, **14,** 589–592.

Arend, L. E. Spatial, differential and integral operations in human vision. *Psychological Review*, 1973, **80,** 374–395.

Argyle, M., & Cook, M. *Gaze and mutual gaze*. Cambridge: Cambridge Univ. Press, 1975.

Atkinson, J. Properties of human visual orientation detectors: a new approach using patterned after-images. *Journal of Experimental Psychology*, 1973, **98,** 55–63.

Bach-y-Rita, P., & Collins, C. C. (Eds.), *The control of eye movements*. New York: Academic Press, 1971.

Barlow, H. B. Slippage of contact lenses and other artifacts in relation to fading and regeneration of supposedly stable retinal images. *Quarterly Journal of Experimental Psychology*, 1963, **15,** 36–57.

Berger, R. J. Oculomotor control: a possible function of REM sleep. *Psychological Review*, 1969, **76,** 144–164.

Berger, R. J., & Walker, J. M. Oculomotor coordination following REM and non-REM sleep period. *Journal of Experimental Psychology*, 1972, **94,** 216–224.

Biederman-Thorson, M., Thorson, J., & Lange, G. D. Apparent movement due to closely spaced sequentially flashed dots in the human peripheral field of vision. *Vision Research*, 1971, **11,** 889–903.

Bouma, H., & de Voogd, A. H. On the control of eye saccades in reading. *Vision Research*, 1974, **14,** 273.

Bridgeman, B. Hendry, D., & Stark, L. Failure to detect displacement of the visual world during saccadic eye movements. *Vision Research*, 1975, **15,** 719–722.

Breitmeyer, B. G., & Ganz, L. Implications of sustained and transient channels for theories of visual pattern masking, saccadic suppression, and information processing. *Psychological Review*, 1976, **83,** 1–36.

Brindley, G. S. Two new properties of foveal after-images and a photochemical hypothesis to explain them. *Journal of Physiology*, 1962, **64,** 168–179.

Brindley, G. S., & Merton, P. A. The absence of position sense in the human eye. *Journal of Physiology*, 1960, **153,** 127–130.

Broadbent, D. E. *Decision and stress*. London: Academic Press, 1971.

Brooks, B. A., & Fuchs, A. F. Influence of stimulus parameters on visual sensitivity during saccadic eye movement. *Vision Research*, 1975, **15,** 1389–1398.

Brown, D. R., Cosgrove, M. P., Kohl, G. A., Fulgham, D. D., & Schmidt, M. J. Stabilized images: Probe analysis of pattern and colour analytic mechanisms. *Vision Research*, 1975, **15,** 209–215.

Buswell, G. T. *How people look at pictures*. Chicago: Univ. of Chicago Press, 1935.

Clark, M. R. A two-dimensional Purkinje eye tracker. *Behavior Research Methods & Instrumentation*, 1975, **7**(2), 215–219.

Clowes, M. B. Some factors in brightness discrimination with constraint of retinal image movement. *Optica Acta*, 1961, **8,** 81–91.

Collewijn, H., Van der Mark, F., & Jansen, T. C. Precise recording of human eye movements. *Vision Research*, 1975, **15,** 447–450.

Cooper, R. M. The control of eye fixation by the meaning of spoken language. *Cognitive Psychology*, 1974, **6,** 84–107.

Cornsweet, T. N., & Crane, H. D. Accurate 2-dimensional eye-tracker using first and fourth Purkinje images. *Journal of The Optical Society of America*, 1973, **63,** 921–928.

Cowey, A. The basis of a method of perimetry with monkeys. *Quarterly Journal of Experimental Psychology*, 1963, **15,** 81–90.

Cunitz, R. J., & Steinman, R. M. Comparison of saccadic eye movements during fixation and reading. *Vision Research*, 1969, **9,** 683–693.

Davidson, M. L., Fox, M. J., & Dick, A. O. The effect of eye movement and backward masking on perceptual location. *Perception & Psychophysics*, 1973, **14,** 110–116.

Dick, A. O. Iconic memory and its relation to perceptual processing and other memory mechanisms. *Perception & Psychophysics*, 1974, **16**(3), 575–596.

Ditchburn, R. W. *Eye movements and visual perception*. London: Oxford Univ. Press, 1973.

Ditchburn, R. W., Fender, D. H., & Mayne, S. Vision with controlled movements of the retinal image. *Journal of Physiology*, 1959, **145,** 98–107.

Ditchburn, R. W., & Ginsborg, B. L. Vision with a stabilized retinal image. *Nature*, 1952, **170,** 36–37.

Evans, C. R. Further studies of pattern perception and stabilized retinal image: the use of prolonged after-images to achieve perfect stabilization. *British Journal of Psychology*, 1967, **58,** 315–327.

Fiorentini, A., & Ercoles, A. M. Involuntary eye movements during attempted monocular fixation. *Fondazione Giorgio Ronchi Atti*, 1966, **21**(2), 199–217.

Gerrits, H. J. M., & Vendrik, A. J. H. The influence of stimulus movements on perception in parafoveal stabilized vision. *Vision Research*, 1974, **14,** 175–180.

Gibson, J. J. What gives rise to the perception of motion? *Psychological Review*, 1968, **75,** 335–346.

Gombrich, E. H. Illusion and art. In R. L. Gregory & E. H. Gombrich (Eds.), *Illusion in nature and art*. London: Duckworth, 1973. Pp. 193–244.

Gopher, D. Eye movement patterns in selective listening tasks of focussed attention. *Perception & Psychophysics*, 1973, **14,** 259–264.

Gould, J. D. Eye movement during visual search and memory search. *Journal of Experimental Psychology*, 1973, **98,** 184–195.

Gregory, R. L. *Eye and brain* (2nd ed.). London: World Univ. Library, 1973.

Grindley, C. C., & Townsend, V. Voluntary attention in peripheral vision and its effects on acuity and differential thresholds. *Quarterly Journal of Experimental Psychology*, 1968, **20,** 11–19.

Haber, R. N. Control of eye movements during reading. In R. A. Monty & J. W. Senders (Eds.), *Eye movements and psychological processes*. Hillsdale, New Jersey: Erlbaum, 1976. Pp. 443–450.

Hallett, P. E., & Lightstone, A. D. Saccadic eye movements towards stimuli triggered by prior saccades. *Vision Research*, 1976, **16,** 99–106. (a)

Hallett, P. E., & Lightstone, A. D. Saccadic eye movements to flashed targets. *Vision Research*, 1976, **16,** 107–144. (b)

Hebb, D. O. Concerning imagery. *Psychological Review*, 1968, **75,** 466–477.

Heckenmueller, E. G. Stabilization of the retinal image. *Psychological Bulletin*, 1965, **63,** 157–169.

Helmholtz, H. von. *Physiological optics* (Vol. III). J. P. C. Southall (Ed.), New York: Dover, 1962.

Hendry, D. P. Saccadic velocities determined by a new perceptual method. *Vision Research*, 1975, **15,** 149–151.

Hess, E. H. Pupillometrics. In N. S. Greenfield, & R. A. Steinbach (Eds.) *Handbook of Psychophysiology*, New York: Holt, 1972, Pp. 491–531.

Heywood, S. Voluntary control of smooth eye-movements and their velocity. *Nature*, 1972, **238,** 408–410.

Hochberg, J. Components of literacy. In H. Levin & J. P. Williams, *Basic studies on reading*. New York: Basic Books, 1970. Pp. 74–89.

Hochberg, J. On the control of saccades in reading. *Vision Research*, 1975, **15,** 620.

Howard, I. P., & Evans, J. A. The instrumentation of eye torsion. *Vision Research*, 1963, **3,** 447–455.

Jeannerod, M., Gerin, P., & Pernier, J. Déplacements et fixations du regard dans l'exploration libre d'une scène visuelle. *Vision Research*, 1968, **8,** 81–97.

Jones, B., & Kabanoff, B. Eye movements in auditory space perception. *Perception & Psychophysics*, 1975, **17,** 241–245.

Judd, C. H., & Buswell, G. T. Silent reading: A study of various types. *Supplementary Education Monographs*, 1922, No. 23, 510.

Just, M. A., & Carpenter, P. Eye fixations and cognitive processes. *Cognitive Psychology*, 1976, **8,** 441–480.

Kaufman, L., & Richards, W. Spontaneous fixation tendencies for visual forms. *Perception & Psychophysics*, 1969, **5,** 85–88.

Keesey, U. T. Effects of involuntary eye movements on visual acuity. *Journal of the Optical Society of America*, 1960, **50,** 769–774.

Kennard, D. W., Hartman, R. W., Kraft, D. P., & Boshes, B. Perceptual suppression of after-images. *Vision Research*, 1970, **10,** 575–585.

Lamanski, S. Bestimmung der Winkelgeschwindigkeit der Blickbewegung, respective Augenbewegung. *Pflügers Archiv für die gesamte Physiologie des Menschen und der Tiere*, 1869, **2,** 418–422.

Lévy-Schoen, A. *Etude des mouvement ocularies*. Paris: Dunod, 1969.

Lord, C., & Haith, M. The perception of eye contact. *Perception & Psychophysics*, 1974, **16,** 413–416.

Lord, M. P., & Wright, W. D. The investigations of eye movements. *Reports on Progress in Physics*, 1950, **13,** 1–23.

Mack, A. An investigation of the relationship between eye and retinal image movement in the perception of movement. *Perception & Psychophysics*, 1970, **8,** 291–298.

Mack, A., & Herman, E. Position constancy during pursuit eye movement. *Quarterly Journal of Experimental Psychology*, 1973, **25,** 71–84.

MacKay, D. M. Evaluation of visual threshold by displacement of retinal image. *Nature*, 1970, **225,** 90–92.

MacKay, D. M. Visual stability and voluntary eye movements. In R. Jung (Ed.), *Handbook of Sensory Physiology*, (Vol. VII/3A). New York: Springer-Verlag, 1973. Pp. 307–331.

Mackworth, J. F., & Mackworth, N. H. Eye fixations recorded on changing visual scenes by the T.V. Eye Marker. *Journal of The Optical Society of America*, 1958, **48,** 439–445.

Mackworth, N. H., & Bruner, J. S. How adults and children search and recognize pictures. *Human Development*, 1970, **13**(3), 149–177.

Mackworth, N. H., & Morandi, A. J. The gaze selects informative details within pictures. *Perception & Psychophysics*, 1967, **2,** 547–551.

Marshall, W. H., & Talbot, S. A. Recent evidence for neural mechanisms in vision leading to a general theory of sensory acuity. *Biological Symposia*, 1942, **7,** 117–164.

Matin, E. Saccadic suppression. *Psychological Bulletin*, 1974, **81**(12), 899–917.

Matin, E. The two-transient (masking) paradigm. *Psychological Review*, 1975, **82**(6), 451–461.

Matin, E., Clymer, A. B., & Matin, L. Metacontrast and saccadic suppression. *Science*, 1972, **178,** 179–182.

Matin, L. Eye movements and perceived visual direction. In D. Jameson & L. M. Hurvich (Eds.), *Handbook of sensory physiology* (Vol. VII/4). New York: Springer-Verlag, 1972. Pp. 331–380.

McCall, R. B. *Early childhood: The development of self regulatory mechanism*. In D. N. Walcher & D. L. Peters (Eds.), New York: Academic Press, 1971. pp. 107–140.

Merchant, J., Morrissette, R., & Porterfield, J. L. Remote measurement of eye direction allowing subject motions over one cubic foot of space. *I.E.E.E. Transactions on Biomedical Engineering*, 1974, **BME-21, 4,** 309–317.

Mitrani, L., Mateeff, S., & Yakimoff, N. Smearing of the retinal image during voluntary saccadic eye movement. *Vision Research*, 1970, **10,** 405–409.

Mitrani, L., Yakimoff, N., & Mateeff, S. Saccadic suppression in the presence of structured background. *Vision Research*, 1973, **13,** 517–521.

Monty, R. A. An advanced eye-movement measuring and recording system. *American Psychologist*, 1975, **30,** 331–335.

Monty, R. A., & Senders, J. W. (Eds.), *Eye movements and psychological processes*. Hillsdale, New Jersey: Erlbaum, 1976.

Mourant, R. R., & Rockwell, T. H. Strategies of visual search by novice and experienced drivers, *Human Factors*, 1972, **14**(4), 325–335.

Neisser, U. *Cognitive Psychology*. New York: Appleton, 1967.

Noton, D., & Stark, L. Scanpaths in saccadic eye movement while viewing and recognizing patterns. *Vision Research*, 1971, **11,** 929–942.

Pritchard, R. M., Heron, W., & Hebb, D. O. Visual perception approached by the method of stabilized retinal images. *Canadian Journal of Psychology*, 1960, **14,** 67–77.

Rashbass, G. New method for recording eye-movements. *Journal of The Optical Society of America*, 1960, **50,** 642–644.

Rayner, K. The perceptual span and peripheral cues in reading. *Cognitive Psychology*, 1975, **7,** 65–81.

Reder, S. M. One-line monitoring of eye position signals in contingent and non-contingent paradigms. *Behavior Research Methods & Instrumentation*, 1973, **5,** 218–228.

Richards, W. Saccadic suppression. *Journal of The Optical Society of America*, 1969, **59,** 617–623.

Riggs, L. A. Visual acuity. In C. H. Graham (Ed.), *Vision and visual perception*. New York: Wiley, 1965. Pp. 321–349.

Riggs, L. A., Merton, P. A., & Morton, H. B. Suppression of visual phosphenes during saccadic eye movements. *Vision Research*, 1974, **14,** 997–1011.

Riggs, L. A., Ratliff, F., Cornsweet, C., & Cornsweet, T. N. The disappearance of steadily fixated visual test objects. *Journal of The Optical Society of America*, 1953, **43,** 495–501.

Robinson, D. A., A method of measuring eye movement using a scleral search coil in a magnetic field. *I.E.E.E. Transactions on Biomedical Electronics*. **BME-1,** 1963, 137–145.

St-Cyr, G. J. Signal noise in the human oculomotor system. *Vision Research*, 1973, **13,** 1979–1991.

Schuck, J. R. Factors affecting reports of fragmenting visual images. *Perception & Psychophysics*, 1973, **13,** 382–390.

Shepard, R., & Metzler, J. Mental rotation of three dimensional objects. *Science*, 1971, **171,** 701–703.

Simon, H. A., & Gilmartin, K. A simulation of memory for chess positions. *Cognitive Psychology*, 1973, **5,** 29–46.

Skavenski, A. A. Inflow as a source of extraretinal eye position information. *Vision Research*, 1972, **12,** 221–229.

Skavenski, A. A., Haddad, G., & Steinman, R. M. The extraretinal signal for the visual perception of direction. *Perception & Psychophysics*, 1972, **11,** 287–290.

Slater, A. M., & Findlay, J. M. The measurement of fixation position in the newborn baby. *Journal of Experimental Child Psychology*, 1972, **14,** 349–364.

Slater, A. M., & Findlay, J. M. The corneal reflection technique and the visual preference method: Sources of error. *Journal of Experimental Child Psychology*, 1975, **20,** 240–247. (a)

Slater, A. M., & Findlay, J. M. Binocular fixation in the newborn baby. *Journal of Experimental Child Psychology*, 1975, **20,** 248–273. (b)

Sokolov, E. N. Higher nervous functions: The orienting reflex. *Annual Review of Physiology*. 1963, **25,** 545–580.

Steinbach, M. J., & Pearce, D. G. Release of pursuit eye movements using after-images. *Vision Research*, 1972, **12,** 1307–1311.

Steinman, R. M. & Cunitz, R. J. Fixation of targets near the absolute foveal threshold. *Vision Research*, 1968, **8,** 277–286.

Steinman, R. M., Cunitz, R. J., Timberlake, G. T., & Herman, M. Voluntary control of microsaccades during maintained monocular fixation. *Science*, 1967, **155,** 1577–1579.

Steinman, R. M., Haddad, G. M., Skavenski, A. A., & Wyman, D. Miniature eye movement. *Science*, 1973, **181,** 810–819.

Stoper, A. E. Apparent motion of stimuli presented stroboscopically during pursuit movement of the eye. *Perception & Psychophysics*, 1973, **13,** 201–211.

Trevarthen, C. B. Two mechanisms of vision in primates. *Psychologische Forschung*, 1968, **31,** 299–337.

Tryon, W. W. Pupillometry: A survey of sources of variation. *Psychophysiology*, 1975, **12,** 90–93.

Volkmann, F. C. Vision during voluntary saccadic eye movements. *Journal of the Optical Society of America*, 1962, **52,** 571–578.

Wade, N. J. Figural effects in after-image fragmentation. *Perception & Psychophysics*, 1974, **15,** 115–122.

Wallach, H., & Lewis, C. The effect of abnormal displacement of the retinal image during eye movements. *Perception & Psychophysics*, 1966, **1,** 25–29.

Walls, G. L. The evolutionary history of eye movements. *Vision Research*, 1962, **2,** 69–80.

Ward, F. Pursuit eye movements and visual localization. In R. A. Monty & J. Senders (Eds.), *Eye Movements and Psychological Processes*. Hillsdale, New Jersey: Erlbaum, 1976. Pp. 289–297.

Weisstein, N. What the frog's eye tells the human's brain: Single cell analyzers in the human visual system. *Psychological Bulletin*, 1969, **72,** 157–176.

Westheimer, G., & McKee, S. P. Visual acuity in the presence of retinal image motion. *Journal of the Optical Society of America,* 1975, **65**(7), 847–850.

Weymouth, F. W., Anderson, E. E., & Averill, H. L. Retinal mean local sign: a new view of the relation of the retinal mosaic to visual perception. *America Journal of Physiology*, 1923, **63,** 410–411.

Wyman, D., & Steinman, R. M. Small step tracking: Implications for the oculomotor dead zone. *Vision Research*, 1973, **13,** 2165–2172.

Yakimoff, N. Saccadic suppression of a stimulus presented on a background of horizontal or vertical grating. *Comptes rendus de l'Académie bulgare des Sciences*, 1973, **26**(12), 1693–1695.

Yarbus, A. L. *Eye movements and vision*. (B. Haigh, trans.). New York: Plenum, 1967.

Young, L. R., & Sheena, D. Survey of eye movement recording methods. *Behavior Research Methods & Instrumentation*, 1975, **7**(5), 397–429.

Zinchenko, V. P., & Vergiles, N. Yu. *Formation of visual images* (B. Haigh, trans.). New York: Consultants Bureau, 1972.

Zuber, B. L., Stark, L., & Cook, G. Microsaccades and the velocity-amplitude relationship for saccadic eye movements. *Science*, 1965, **150,** 1459–1460.

Zuber, B. L., Stark, L., & Lorber, M. Saccadic suppression of the pupillary light reflex. *Experimental Neurology*, 1966, **14,** 351–370.

Chapter 7

PERCEPTUAL LEARNING*

RICHARD D. WALK

I. INTRODUCTION

Perceptual learning is a broad topic that is probably represented in every volume of this series. Perceptual learning refers to the learning of the stimulus for perception. The brain is bombarded by sights, sounds, smells, tastes, things to touch, and temperatures and movements of the body. During the course of development and continuously in our daily lives, we learn to adjust to the stimulus aspects of our environment. The stimulus may affect more than one modality. The butcher who hefts a piece of meat, cuts off a piece, and then shows that he has precisely estimated your order has learned not only the way to estimate the physical weight of the meat but also, as the size–weight illusion shows, to disre-

* The preparation of this chapter was supported, in part, by Grant MH-25,864 from the National Institutes of Health.

HANDBOOK OF PERCEPTION, VOL. IX

ISBN 0-12-161909-5

gard extraneous visual cues from the sight of the meat. Because of possible intersensory effects, one cannot describe perceptual learning as a process whereby the individual learns only about the physical properties of the stimulus. Certain aspects of the stimulus may also be more easily learned than others, and this may change with increased age or experience.

Perceptual learning is difficult to separate from cognitive learning. Cognition is part of perceptual learning, but the study of cognition goes beyond that of perceptual learning. The two topics may be partially intertwined, but each has its own area of specialization. The organism, we must remember, does not know when it is engaging in perception and when it is demonstrating cognition.

Is perceptual learning different from other kinds of learning? Hilgard and Bower (1975) define learning as follows:

> Learning refers to the change in a subject's behavior to a given situation brought about by his repeated experiences in that situation, provided that the behavior change cannot be explained on the basis of native response tendencies, maturation, or temporary states of the subject (e.g., fatigue, drugs, etc.) [p. 17].

If one reads further one finds that the authors do not believe that sharp distinctions can be made among learning and native response tendencies (e.g., imprinting, song acquisition, and song refinement in some birds), maturation (the environment is implicated in most maturation), or temporary states like fatigue and habituation. Those who study perceptual learning are very interested in phenomena like imprinting, interaction of the environment and maturation, and temporary states such as habituation. What is peripheral to the classical study of learning is of central importance to the study of perceptual learning. Thus the study of perceptual learning has developed independently of the traditional study of learning. The major theorists in the study of perceptual learning are Eleanor and James Gibson, Hebb, Held, and Piaget. Eleanor Gibson's book (Gibson, 1969) is the single most influential expository work in this field at the present time. The positions of these individuals will be mentioned where relevant in the discussion that follows.

The aim of the present chapter is to describe the breadth, the scope, and the issues involved in the study of perceptual learning. Perceptual learning theories are primarily *visual* perceptual learning theories, and this chapter will try to integrate the modalities insofar as research (which is primarily visual perceptual research) makes this possible. The plan of this chapter is first to discuss the relation of perceptual learning to the study of development and then to discuss the major issues in the field. First, however, a few examples of perceptual learning will be described, and then an experiment that provides a concrete reference point for a few of the issues will be discussed.

Books on perceptual learning which might be consulted are William Epstein, *Varieties of Perceptual Learning* (1967), P. C. Dodwell, *Perceptual Learning and Adaptation* (1970), an edited book that reprints many important source articles for the study of perceptual learning, and Eleanor Gibson, *Principles of Perceptual Learning and Development* (1969).

II. EXAMPLES OF PERCEPTUAL LEARNING

Every sense modality easily shows examples of perceptual learning. Some examples are the skilled reader who rapidly assimilates prose material, the champion tennis player whose body anticipates every movement of the ball, the conductor who picks out unerringly the violin that hits a slightly false note that is undetected by the audience, the art expert who spots a skilled forgery at a glance, the mountain climber who looks with equanimity at depths that used to scare him silly. These are general examples of perceptual learning, but perceptual learning is also involved in specific preferences. Consider the individual who prefers Norman Rockwell to Picasso, the one who likes string quartets and hates rock and roll, the person who chooses hamburgers with ketchup over Oysters Rockefeller, the bourbon drinker who cannot stand scotch, the person who prefers Sneaky Pete to Pouilly Fuissé.

Perceptual learning in the taste modality can be highly developed. Here is an example from wine tasting (Root, 1974).

> The competing sommeliers were called upon to identify 14 different vintages. It turned out that all 14 were the same wine of the same year; the contest boiled down, therefore, to naming the *clos,* the individual plot of land on which each wine has been grown, some of them only a few acres apart. *All* of the contestants realized that they were faced with 14 wines of the same category, and named correctly the wine and the year. The winner identified the 14 *clos* correctly; two others were right on 13 [p. 20].

Tea tasters also have skills that seem beyond our reach. An expert was asked to determine the elements of a blend that an American company was about to market. " 'I detect,' he answered, 'a rather good Assam, a run-of-the-mill Darjeeling, a mediocre Ceylon and, of course, the tea bag' [Root, 1974, p. 16]." Root (1974) believes that such skills are within the reach of most of us; perhaps some perceptual psychologist will try to test his hypothesis.

Most of the research on perceptual learning has concentrated on a few areas. Visual perception of the young child is one such area. The bulk of this research is in the first 6 months of age. It essentially began with Fantz's demonstration that children could discriminate patterns soon after birth and that they soon could discriminate more complex visual patterns (Fantz, 1961, Fantz, Fagan, & Miranda, 1975; Salapatek, 1975; Bond, 1972). Imprinting is a process whereby animals, such as birds, become attached, as they would to their parents in natural conditions, to any of a variety of objects, including human caretakers. The attraction is both visual and auditory. The conditions for imprinting have been extensively investigated (Hess, 1973). An individual who wears prisms that displace the visual image to one side quickly learns to adapt to the change and reach accurately for visual objects. This situation is a research model for studying the process of adaptation, with many different theoretical explanations for the change (Freedman, 1968; Kornheiser, 1976; Welch, 1974). The process of picture perception (how we come to discriminate pictures) is a last such area, having been given an impetus by the apparent demonstration by Hudson (1960) that Africans have difficulty discriminating depth in pictures. Much cross-cultural and developmental research has been conducted on this topic (Miller, 1973; Hagen, 1974). The development of *cognitive maps,* the process by which individuals learn the perceptual features of the environment, has attracted much research interest (Siegel and White, 1975).

An illustrative experiment on the perceptual learning of the identity of birdsongs was conducted (Walk & Schwartz, 1975) to illustrate a few of the issues. Ten birdsongs, chosen as representative of the Middle Atlantic states, were placed on tape. The birdsongs were those of the Baltimore oriole, cardinal, common goldfinch, mockingbird, purple finch, robin, song sparrow, towhee, tufted titmouse, and winter wren. The subjects first listened to a tape of all of the songs and were asked to guess the name of each birdsong on each trial. They were furnished in advance with the names of the birds. They next heard four repetitions of the entire repertoire, each in a different order, where the birdsong was identified a few seconds after it was played. The posttest was in still a different order and, like the pretest, gave no information on the identity of a song. The subjects were run in groups, with one group told they might make notes to help identify the songs. Dots and dashes were mentioned as of possible help to identify the high or low notes or the speed of the song. The other group was given no instructions about using notes, though they were free to do so.

Those who were told about a visual code and who also used it increased from chance performance on the pretest (10% correct) to 53% correct on the posttest, whereas those for whom no visual code was mentioned only improved to 24% correct on the posttest. The individuals who were told about a visual code but did not use it were like the control group, with only 25% correct on the posttest. Thus, perceptual learning was helped by coding the auditory stimulus into a visual aid.

The visual code helps to highlight how subjects might code the distinctive characteristics of the birdsong. Figure 1 shows the visual codes used by some of the subjects for two birdsongs, the cardinal's (an easy song to learn) and the mockingbird's (one that benefitted markedly from the use of the visual code). The distinctive characteristics or features of the cardi-

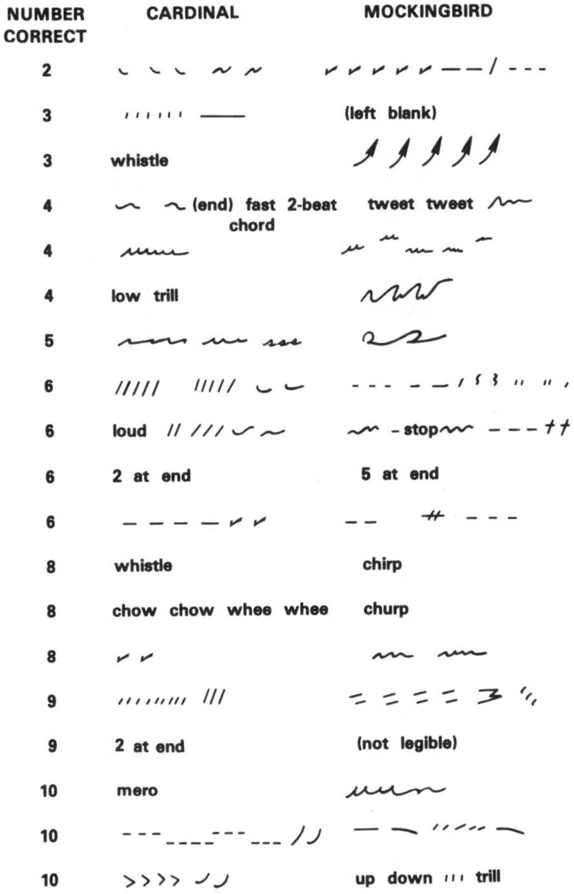

NUMBER CORRECT	CARDINAL	MOCKINGBIRD
2	⌣ ⌣ ⌣ ~ ~	✓ ✓ ✓ ✓ ✓ — — / - - -
3	,,,,,,, ——	(left blank)
3	whistle	↗ ↗ ↗ ↗ ↗
4	~ ~ (end) fast 2-beat chord	tweet tweet /~~
4	~~~~	~ ~ ~~ ~~ –
4	low trill	~~~~
5	~~~ ~~ ~~	~~
6	///// ///// ⌣ ⌣	- - - – — , , , ,, ,, ,
6	loud // /// ~ ~	~~ - stop ~~ - - - ++
6	2 at end	5 at end
6	— — — — ✓ ✓	- - # - - -
8	whistle	chirp
8	chow chow whee whee	churp
8	✓ ✓	~~ ~~~
9	,,,,,,,,, ///	= = = = 3 ,,,
9	2 at end	(not legible)
10	mero	~~~~
10	- - - _ _ _ _ - - - _ _ _ //	— — ,,,,,, —
10	>>>> //	up down ,,, trill

FIG. 1. Visual codes used by subjects to help learn two representative birdsongs, those of the cardinal and the mockingbird. The number of correct identifications (out of 10) for the subject is also shown.

nal's song were coded in somewhat similar ways in a number of protocols with a common accent on the two beats at the end. Codes used for the mockingbird, on the other hand, had few obvious similarities.

The control group was inferior in performance to the group using visual codes on all birdsongs, but performance was still reasonably good on three songs: those of the Baltimore oriole, the cardinal, and the winter wren. Some of the other songs were identified fairly accurately through the use of a visual code: those of the mockingbird, the robin, the towhee, and the tufted titmouse. One birdsong, that of the common goldfinch, did not benefit much from the use of the visual code.

This experiment shows that some auditory characteristics are salient and fairly easy to learn, others become identifiable when a visual "crutch" is allowed, and still others may be too subtle for easy learning. Perhaps a more extended training period would better help differentiate the characteristics of such songs.

Many years ago Gagne and Gibson (1947) showed that pointing out the distinctive features of airplane silhouettes helped to identify the airplanes. The birdsong study demonstrates how subject-generated visual distinctive features helps perceptual learning in the auditory modality. Visual codes may be particularly appropriate for auditory perceptual learning, and Pick (1970) contends that the auditory stimulus is translated into a visual code, citing evidence that the localization of sounds is helped by visual reference points even though the subject cannot see the location of the sound or where he points. From the beginning, then, the notion of distinctive features (which may differ from subject to subject), of the differentiation of the stimulus, and of the influence of one modality on another are mentioned as possible issues in perceptual learning.

III. DEVELOPMENT AND PERCEPTUAL LEARNING

The nature–nurture issue, the extent to which the child or animal must learn to perceive, has been the focus of much research in perception. Infants less than 1-month old demonstrate discrimination of visual patterns (Fantz, 1961), colors (Bornstein, Kessen, & Weiskopf, 1976), properties of sound (Eimas, Siqueland, Jusczyk, & Vigorito, 1971), and tastes (Nowlis & Kessen, 1976). Perceptual learning refers to the change in perception over time, based on experience. Mere maturation of the nervous system, which might bring with it increased discrimination, would not be sufficient to demonstrate perceptual learning. For example, visual acuity improves from birth, when the infant discriminates $\frac{1}{8}$ in. stripes at a distance of 10 in. (40 min or arc), to the discrimination of $\frac{1}{64}$ in. stripes at this distance (5 min of arc) for the 6-month-old child (Fantz, Ordy, & Udelf, 1962). This is usually considered to be maturation, not perceptual learning. But we know that maturation is influenced by the visual stimulation from a normal visual environment; an organism kept in the dark would discriminate poorly, perhaps less well than the newborn. This phenomenon has been well documented with animals (Riesen, 1966). We cannot separate maturation from experience here—the two are intertwined.

The sensory modalities mature at different rates. Gottlieb (1971) has pointed out the sequence of sensory development in the embryo. The tactual system is first to develop, followed by the vestibular system, the auditory system and, finally, the visual system. The auditory system is functionally well developed at birth, whereas the visual system is much slower to develop—the fovea is not fully developed until 4 months of age, and accommodation and convergence are not completely developed until several months after birth. Theories of perceptual learning are visual theories that have developed from data that show fairly slow visual development. Would the same theories have been formed from auditory data?

The earlier view of perceptual development, well-expressed and often quoted, was that the world of the young infant was a "buzzing confusion" that slowly became the articulated structured world of the adult through the accretions of experience (James, 1890). For many years, psychological methods permitted no more accurate a view. The demonstration of fairly good visual acuity in the neonate and a wealth of other research on pattern perception, space perception and hearing has markedly changed our view of the young child. The book, *Infant Perception: From Sensation to Cognition*, edited by Cohen and Salapatek (1975), describes this research in detail. Thus, the world appears to be reasonably well articulated to the neonate and becomes more differentiated with time. Also, the infant becomes easier to test, and therefore to answer research questions, as it matures. We may occasionally overestimate the amount of perceptual learning because of the sheer difficulty of testing young organisms, inferring learning from any improvement. The problem with perceptual development and learning is not just the interaction of maturation and learning, but also the difficulty in finding a response that accurately assesses the perceptual processes of the infant. The trend of the last 15–20 years is clear, however: the better the methodology the more the infant seems to discriminate. Perhaps the next trend will be a better understanding of the infant's limitations; then the role of perceptual learning in perceptual development will be better understood.

At the moment, auditory research seems to show advanced development more than does visual research, though future research developments may change this. The infant recognizes its own mother's voice from a few days to 1 week of age (André-Thomas & Autgaerden, 1963, 1966; Hammond, 1970) and may recognize her by smell at 6 days of age (MacFarlane, 1975), though Russell (1976) did not find the mother recognized by smell until 6 weeks of age. The best known research on the infant's reaction to faces was summarized by Gibson (1969), who found that a real face was not preferred to a colored photo of a face until 3 months of age. However, Carpenter, Tecce, Stechler, and Friedman (1970) may have shown visual recognition of the mother by 2 weeks of age. The Carpenter *et al.* (1970) study contrasted a live mother with an inert manikin for the significant 2-week difference, but this cannot be considered in the same way as a live mother compared to a live stranger. A comparison where live mother and live stranger were used had only an overall significant difference, not one at 2 weeks; further, the infants and their mothers were black and the experimenter–stranger was white. Thus, early visual recognition of the mother must be accepted with some caution and most of the research is against it. A third auditory study, Mills and Melhuish (1974), also found recognition of the mother's voice by infants, but the infants were somewhat older, 20–30 days of age, than in the earlier cited studies of infant recognition of the mother's voice. The weight of the evidence is in favor of very early auditory recognition of the mother; her recognition by vision has yet to be firmly demonstrated.

The infant appears to discriminate the stimulus properties of the world amazingly well, but this does not mean that all modalities develop at the same rate. The present evidence is that auditory development is ahead of visual development. The role of the different modalities cannot be ignored in the discussion of issues that follows.

IV. THEORETICAL AND RESEARCH ISSUES

Theoretical and research issues to be discussed include the following: distinctive features and prototypes, differentiation as a concept for development and learning, the constructionist position, the importance of reinforcement for perceptual learning, critical periods and early learning, an extended treatment of activity and perceptual learning, and a discussion of intersensory relationships.

A. Distinctive Features and Prototypes

Let us suppose that we are trying to learn common properties of objects so that we can differentiate them from other objects: flowers, for example. It is not difficult to tell a rose from a daylily. There are 5000 varieties of roses and 12,000 varieties of daylilies. Fairly distinctive characteristics are qualities like color, size of the blossom, and height of the plant. We next begin to notice more complex qualities of the shape, such as, for roses, variations in the shape of the petal and the number of petals in a blossom. Daylilies characteristically have six petals that vary in such characteristics as edges (ruffled, fluted, frilly, lacy), texture (heavy, crisp, crepelike), and width of petal. The overall shape, curvature of the petals, and the mixture of colors can be analyzed to such an extent that the neophyte is made giddy just reading some of the descriptive terms (e.g., 24-in. height, medium-late bloomer, fragrant, 4-in. blossom, pale yellow background and throat, petals rose dusted with eye formation and ruffled edges, sepals rose-edged and crepy throughout). Yet, we could learn to identify many roses or daylilies, if not thousands, certainly hundreds. How would we do it? Gibson (1969) stresses the notion of distinctive features. An experiment with letter-like forms has helped popularize the distinctive-feature notion, one first used in linguistics by Jakobson and Halle (1956).

An experiment with letter-like forms (Gibson, Gibson, Pick, & Osser, 1962) used 12 basic forms with variations. Each basic form was changed in a systematic, similar way: (*a*) rotated on its axis or reversed (as a right–left reversal); (*b*) changed from straight lines to curves; (*c*) given gaps or added lines (called a *break-and-close* transformation); or (*d*) made to slant or tilt back (called a *perspective transformation*). See Fig. 2. The children improved with age and by age 8 only the perspective transformations were difficult. The break-and-close transformations were always easy, line-to-curve transformations moderately difficult, and rotation and reversal errors were very high at age 4 and, by age 6–7, easier to distinguish than the line-to-curve transformations. As is well known, *b* and *d* are difficult letters for young children to distinguish.

The Gibson *et al.* (1962) experiment thus demonstrates that some distinctive features are more difficult to learn than others and that there is an improvement with age in discriminating forms based on known distinctive features. It also suggests a possible basic approach to the learning of forms that might help identify difficulties young children have in learning to read.

The letter-like forms were of 12 basic types and the distinctive features were similar variations on the basic forms. One might consider each basic form a prototype from which the distinctive features add variations. An experiment designed to test the role of prototypes, or images, as compared to distinctive features is that of Pick (1965). She used letter-like forms of the type previously used by Gibson *et al.* (1962). The stimuli are

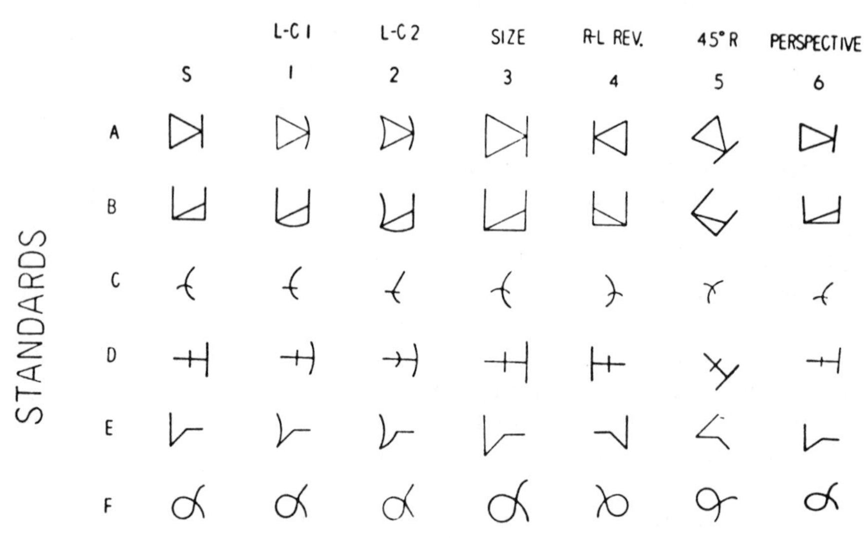

FIG. 2. Standards and transformations used by Pick (1965).

shown in Fig. 2. The subjects were first given three of the basic forms of standards, such as A,B, and C under column S in Fig. 2. They were then given a pack of cards that contained copies of each standard and three transformations of it (Line-to-Curve 1, L–C 2, and size in Fig. 2 are examples). The subject was trained to pick only exact copies of the standard. In the transfer task some subjects (prototype group) used the old standards and new transformations (rotation, reversal, and perspective transformations are examples). Another group (distinctive features) had the same transformations (e.g., L–C 1, L–C 2, size) and new standards (D,E,F). A control group had new standards and new transformations. Only one trial and no correction was given on the transfer task.

Pick (1965) found that the distinctive-feature group was superior to the prototype group, which, in turn, was superior to the control group. A similar experiment was carried out tactually with the forms as raised lines on smooth metal and the subject first felt the standard and then the transformations of it. In this experiment, prototype and distinctive-feature learning were equally advantageous compared to the control. Since the tactual task was successive (in that the subject felt the stimuli successively with the same hand), a simultaneous tactual group, with the subjects exploring the forms tactually with both hands at the same time, was added. The distinctive-feature group was again superior.

Caldwell and Hall (1970) challenged the interpretation of the Pick (1965) experiment. They felt that the distinctive feature group was given more information about the experimenter's concept of same and different than was the prototype group. A subject might learn that a line-to-curve transformation was incorrect but not that a right–left reversal was incorrect. They added an "overlay" group, in which forms were placed on top of each other to show that the form had to be exactly the same, along with a group that had a broader range of distinctive-feature variations. The Caldwell and Hall procedure reduced the superiority of the distinctive-feature group, and they suggested that prototype learning and distinctive feature learning are not different kinds of learning. Aiken and Williams (1973) used prototypes made with computer-generated patterns along with their variations and concluded that both prototype and distinctive feature learning were important for the task. A similar conclusion was reached by Aiken (1969) with an auditory task.

The notion of using distinctive features as a way to describe complex stimulus properties has struck a responsive chord among psychologists. It encourages psychologists to classify the stimulus properties of the real world, a difficult but interesting task. Distinctive features are tied to context and both perspective transformations (Schaller & Harris, 1974), and reversals (Singer & Lapin, 1976) might be easier in another context. This does not diminish the usefulness of the distinctive-feature notion, however.

To return to our illustration about daylilies, one finds that at first one's prototypical daylily has been differentiated only from other flowers. A next differentiation is based on color, perhaps, but further differentiation takes into account the shape of the flower, the features of the edges, etc. One prototype has become differentiated into several subprototypes, more prototypes of these, and so forth. Each prototype and subprototype is differentiated from the others by its own distinctive features. Both are useful notions in classification. The Pick (1965) experiment found that tactual learning favored prototype learning if the forms were felt successively. This is an interesting notion: Different conditions influence the way complex stimulus features are learned. Durant and Yussen (1976) did not find that introducing a memory, or successive factor, into the visual situation diminished the advantage of the distinctive-feature group, but this does not mean that more research on this topic might not be warranted.

The notion of distinctive features is no challenge to the notion of *images*, or *schemas*, as Gibson (1969) makes clear in her book. We will return to related aspects of this topic.

B. Differentiation

The concept of differentiation is used centrally for perceptual learning in the theories of both Gibson and Werner. Differentiation has a multiplicity of meanings. My dictionary shows that it can mean something as simple as discrimination, or it can mean a development from the simple to the complex. A last definition is that of "modification of body parts for performance of particular functions," and "the sum of processes whereby apparently indifferent cells, tissues, and structures attain their adult form and function [*Webster's Seventh New Collegiate Dictionary*, 1969, p. 232]."

If the dictionary shows complexities of meaning, psychological use of the concept shows even more. This has led to disagreements or misunderstanding, as will be made clear (if not made clear, at least pointed out). A good source for discussion of differentiation is the index of the third edition of the *Manual of Child Psychology* (Mussen, 1970). Particularly helpful are discussions by Pick and Pick on differentiation as applied to perceptual development, McNeill on differentiation and the development of language, and Kagan and Kogan on differentiation as a concept in cognitive development. Both Pick and Pick and Kagan and Kogan point out that differentiation means different things to different theorists.

Werner applies the concept of differentiation to the individual as he differentiates the environment and becomes independent of it. Along with this process comes increasing hierarchical integration. Gibson's emphasis is more on differentiation of the stimulus and of its qualities and relationships. She refers to "reduced generalization and increasing precision of response to fine differences along a stimulus dimension . . . detection of invariant relations and structure previously not responded to [Gibson, 1969, p. 94]."

A misunderstanding, or disagreement, concerning the emphasis of Gibson on the specificity of discrimination is reflected in recent comments. Zeaman and House (1974) contrast the position of Gibson and Gibson (1955), with its emphasis on differentiation as distinguishing undetected features, and that of the Tighes (Tighe & Tighe 1966, 1972), with its viewpoint that differentiation involves the discovery of more general attributes (such as form or color). Zeaman and House (1974) write: "The objective indices of differentiation are opposite for the two theories—ability to discriminate, according to the Gibsons, and ability to generalize according to the Tighes [Zeaman & House, 1974, p. 177].'' The stress that Gibson (1969) has made upon both the discrimination of undetected differences and also upon the discovery of ''invariant relations and structure'' shows that both types of differentiation are encompassed by her theory. This is, however, a point that can be confusing.

A more frontal assault on her theory was made by Eimas (1975) in his discussion of discoveries in linguistic research where the young infant can apparently distinguish more linguistic differences than can the adult. As was pointed out earlier, theories of perceptual learning are essentially theories of visual perception, and the young infant appears more competent aurally than visually. Eimas was referring to research that shows that infants can distinguish the difference between *r* and *l*, whereas Japanese adults have great difficulty in making this distinction. Eimas (1975) writes,

> the course of development of phonetic competence is one characterized by a loss of abilities over time, if specific experience is not forthcoming. This is in marked contrast to most theories of perceptual development (e.g., Gibson, 1969) that assume a gradually increasing ability to extract and utilize the relevant or critical information provided by the external world. It would be of considerable theoretical importance to determine whether this course of development is restricted to speech perception or whether there are analogous phenomena in the development of processing capacities for other forms of information [p. 346].

The Gibsonian theory does stress the increasing match of the organism with the environment. The linguistic environment of the Japanese child does not include the *r–l* distinction and gradually the child can no longer perceive it. We have then a certain paradox: With increasing specification of the linguistic environment the child loses the ability to make certain phonetic distinctions that seem to be present early in development. This is not a disconfirmation of the notion of differentiation—it is, in fact, in agreement with some aspects of the Gibson description of development—but the question remains as to the difficulty of the term *differentiation*. Perhaps, because of its multiplicity of meanings and different usage by different individuals, someone can find a better, less confusable, term.

C. The Constructivist Position and Perceptual Learning

The notion that the organism constructs the world rather than merely reacting to it is an old one, and it is very different from a view like Gibson's that the organism extracts more and more information from the invariant features of the environment. A modern constructivist position is, or rather *was*, that of Neisser (1967) who stated it as follows:

> The central assertion is that seeing, hearing, and remembering are all acts of *construction*, which may make more or less use of stimulus information depending on circumstances [p. 10].

A more recent view is that of Lewis and Brooks (1975) who state, "Structure is defined as origination within the organism, not within the stimulus [p. 104]. They criticize the Gibsons as maintaining that the organism is passive (which seems like a misrepresentation, if one has read the Gibsons) and that it absorbs information from a structured environment. The constructivist position focuses on the organism not, as does the Gibsons' position, the stimulus. Neisser, however, changed his mind, and in a later book stated that the problem in such an approach was the veridicality of perception. If this were so, the notion of construction would seem unnecessary. "Perception, like evolution," he states, "is surely a matter of discovering what the environment is really like and adapting to it [Neisser, 1976, p. 9]."

Lewis and Brooks (1975) did not offer research that would permit a choice between a constructivist position and the Gibsonian one. They did find different looking patterns in male and female infants, with males looking more at pictures of male infants and females looking more at pictures of female infants. This research is not crucial for the constructivist position and they have made no claim that it is.

One possible constructivist area of perceptual learning is related to the study of *field dependence,* which is allied to Werner's theoretical views. Individuals differ in their susceptibility to the visual field; they have difficulty in finding visual patterns hidden in larger patterns and in setting themselves to the gravitational upright in a tilted room (Witkin, 1959). Females are purported to be more field-dependent than males. The research in this area is not very impressive, however, and recent research has questioned the sex differences obtained (Fairweather, 1976, Morell, 1976).

The constructivist position has had great difficulty in obtaining hard research data to back up its claims.

The Gibsonian position can handle notions like schemata, but the preference is to avoid these somewhat mentalistic concepts. *Schema, meaning, attention, expectation, selectivity, image, template:* these are difficult concepts, and individuals disagree heatedly on the meanings of the terms. For some of them, there is even argument over whether they have any usefulness at all. The average researcher feels as though he is carefully picking his way through a hostile minefield where at any moment a booby trap set by an unfriendly philosopher may explode in his face.

Neisser (1976) tries to avoid the more mentalistic notions of perception as a constructive process, stressing the veredicality of perception, but keeping the schema notion to account for some internal system, or plan, related to types of perceptual actions. The position of Gibson (1969) does encompass these possibilities; the possibility of schemas is there, along with other cognitive concepts. But the focus is on the stimulus, and this is a valuable focus that might be lost if too much stress were allowed on the internal structure and plans related to perception. Earlier it was remarked that the organism does not know where perception leaves off and cognition begins. True, but the research issues are different. The position of the Gibsons is primarily a perceptual one and that of Neisser a cognitive one. The two interweave, but each has its separate emphasis.

D. Reinforcement

Reward has never been considered to be of importance for perceptual learning. Imprinting, for example, occurs with a particular object during the sensitive period for imprinting. The object need not be a warm mother who leads the chick to food, but may be an inert piece of wood. It is the critical period of limited span that sets imprinting apart from association learning. Association learning may occur past the sensitive period but it is qualitatively different and not as strong as imprinting (Hess, 1973).

Habituation is a simple form of perceptual learning that can be demonstrated in a wide variety of species. Eimas *et al*. (1971) presented infants with a sound like *ba* and monitored their sucking responses. Sucking responses were initially high, probably because of the attention given to the new sound, but gradually declined as the infants habituated to the sound. A control group continued to decline in sucking rate, but the infants presented with stimuli from a different adult phonetic category, a *pa* sound, recovered temporarily to the novel stimulus. The recovery to a different sound showed that they had learned something from the initial sound, enough to put the new sound in the category of being a novelty. Although one can say that paying attention to novel sounds has obvious survival value, the reinforcement is obscure.

Reinforcement has been raised as an issue in relation to some animal studies carried out by Eleanor Gibson, Herbert Pick, Thomas Tighe, and myself some years ago. Our first experiment (Gibson & Walk, 1956) placed metal triangles and circles on the walls of the cages of albino rats.

The controls had no triangles or circles on their walls. The experimental group learned to discriminate triangles from circles faster than the controls. Further research to tease out the variables in this particular effect was somewhat disappointing, as Gibson (1969) has pointed out in a brief review of some of these experiments. We interpreted the effect as an attentional one from the experience of viewing the stimuli in the cages, which then transferred to help the animal notice the forms in the discrimination situation. Kerpelman (1965), however, interpreted the results as showing *nondifferential reinforcement*. Kerpelman's experimental design was one where two time periods (2 hr or 22 hr) were paired with the presence or absence of forms in relation to food. The animals ate from food trays with forms above them, with the forms present for 2 hr or 22 hr. Another group had forms in the cage for 2 hr or 22 hr, but not while the animals ate from the food trays. A control group had no forms at any time. The experiment had two slow-learning groups: the control group and the group with forms present for 22 hr and taken out at feeding time. A nondifferential-reinforcement explanation favors the superior learning of the two groups fed in the presence of the forms, but has difficulty with the superiority of the group with forms exposed for 2 hr at a time no food was present. Kerpelman used a novelty–stimulus-change explanation, congruent with a perceptual-learning one, for this group. Any reinforcement group is also an attention group since the animals were fed near the forms and the forms were changed daily, helping assure attention. The suppressed performance of the group with forms for 22 hr that were taken out for feeding could be embarassing to interpretations of perceptual learning, but, since the forms signaled the absence of food, they may have suppressed attention to them as signaling food in the later discrimination situation. At its most conservative, an attention explanation is as good as the nondifferential reinforcement one, as each explanation had one negative group.

In any event, the curiosity and exploratory research shows that perceptual learning without reinforcement is possible for the rat. An experiment by Dember is representative (1956). Rats were placed on the gray arm of a **T** maze that had one white arm and one black arm, with a glass barrier preventing access to either arm. After 15 min in the gray arm the animal was removed while the **T**-maze arms were changed to all white or all black. The animals predominantly entered the changed arm, thus showing the attraction of novelty and also that they showed simple perceptual learning.

Is reward important for perceptual learning? Attention, curiosity, and novelty are much better in describing the basis for perceptual learning.

E. Critical Periods and Early Learning

A critical period is a time when learning must take place or not occur at all. Early learning, while not fulfilling the criteria of a critical period, may also exert a strong influence on later behavior. We all know that most superior motor-skill performers have started at an early age. Ballet dancers start between the ages of 10 and 13. Skilled tennis players tend to start early and to come from states like California or Florida, or other states with mild winters, whereas hockey players are likely to come from Canada, where ice skating and ice hockey can be practiced a large part of the year.

The skeptical psychologist might hold that in such areas a large part of the population engages in tennis or ice hockey and, with a large pool, the probability is maximized that a few superior individuals will be identified. Perhaps any dedicated person could achieve success in such sports if practice were intense enough, even if the training started at age 20. We do not know, of course, but any student of perceptual learning who reads the book about the great tennis player Arthur Ashe (McPhee, 1969) cannot help but be impressed with how such skills are perfected. Ashe was playing tennis every day, often all day, long before he was 10 years old.

Examples of early learning, some of it meeting the standards of a critical period, can probably be found in every sense modality. Motor learning, cited above, is not well documented experimentally, but examples from audition, taste, and vision are well known.

The critical period for mallard ducks and chicks is 13–16 hr after the bird is hatched. In the laboratory it may become imprinted to a colored piece of wood or the wooden model of a duck; in the field the animal is imprinted to the parent. Imprinting to natural biological objects, such as the parent, is stronger than it is to artificial lures, but an animal that has been imprinted on a human may avoid its own species and try to court the human being (Hess, 1973). Another example of imprinting is the attachment of the mother goat or sheep. If a newborn animal is kept away from its mother for 2 hr she will reject it.

Critical periods for early learning are not as well documented with human beings and the time periods are considerably longer. Absolute pitch is the ability to identify the pitch of a musical note without reference to other notes. While this rare ability has been attributed to hereditary factors, its genesis is dependent on the age when musical training begins. Sergeant (1969) investigated absolute pitch of those with various amounts of musical training. The highest proportion of individuals possessing absolute pitch was found among the specially gifted, and they began musical training at 5.6 years of age, on the average. Teachers at the Royal College of Music began training at 8 years of age, on the average, and only 33% possessed absolute pitch. A control group, even though they had some musical training starting at age 10, had no one who possessed absolute pitch. The musicians were most accurate in naming notes when these were played on the instrument they themselves played during their childhood, even when they no longer played the instrument. Madsen, Edmondson, and Madsen (1969) obtained related findings with the discrimination of F-sharp, where the subjects had to judge where the note ascended, descended, or did not change during a time interval. Those with the longest period of musical training performed the best. Musical training may also be related to hemispheric localization. Bever and Chiarello (1974) found that experienced musicians recognize simple melodies better in the left hemisphere, whereas the inexperienced recognize them best in the right hemisphere. They attributed this to the specialization of the left hemisphere for analytical functions.

An interesting area of early learning is that concerning the development of songs in birds. The white-crowned sparrow, for example, must memorize a model song between 14 and 50 days after hatching, a time when the bird does not sing; otherwise, the bird will develop an abnormal song. For most birds the birdsong becomes crystallized during the first singing season and remains fixed for the rest of its life. This is irreversible learning. Human language may have similar features. The song sparrow seems to have an innate template, in that it can produce a normal song even if raised by canaries, though it also produces abnormal songs. The white-crowned sparrow cannot produce a normal song without exposure to the normal song (Marler & Mundinger, 1971).

The normal mallard duckling prefers the maternal call of the mallard to that of the chicken. Gottlieb (1976) has shown that this preference is based on a complex interplay of maturation and experience. The mallard embryo begins to vocalize in the shell before it is hatched. If the tympaniform membranes are muted by applying collodion before vocalization begins in the shell, the isolated devocal duckling is deficient in recognizing the maternal call at 24 hr after hatching, but is not deficient 48 hr after hatching. The improvement is a maturational one that is not maintained in the absence of stimulation: By 65 hr after hatching the isolated devocal duckling is again deficient. However, if the devocal duckling is exposed to a tape recording of other ducklings' "contact–contentment" calls prior to hatching and shortly thereafter, the deficiency is eliminated. Thus, the importance of the embryo's own vocalization is demonstrated, as is the importance of conspecifics to maintain the behavior in its absence.

Human speech shows examples of perceptual learning that appear to be irreversible. The *r–l* distinction that is perceived by the infant and not by the adult is one example (Miyawaki, Strange, Verbrugge, Liberman, Jenkins, and Fujimura, 1975). French speakers do not make as sharp a distinction as English speakers do between voiced and voiceless consonants (*b* versus *p*, *d* versus *t*, *g* versus *k*), meaning that they do not perceive the distinction as categorically as do English speakers (Caramazza, Yeni-Komshian, Zurif, & Carbone, 1973). Spanish infants divide the voice onset dimension from −60 to +60 msec into three categories (a *ba–pa* distinction for us) and their parents not only divide it into two categories, but have a different boundary than do the infants (Lasky, Syrdal-Lasky, & Klein, 1975). Streeter (1976) studied Kikuyu infants in Kenya and suggested a nature–nurture interaction even at this young age (2 months). This is because the infants discriminated two voicing contrasts that we perceive as *b*—a discrimination not made by American infants, but one made in their own language. The infants also discriminated a *b–p* distinction their language did not make, suggesting innate factors, along with the learned one in the two *b* series. The issues for human speech are far from jelled: the critical periods for learning, the periods of sensitivity, the question of irreversibility, even an agreement on what constitutes the innate phonetic feature detectors, if such exist, is far from settled.

Taste is a type of perceptual learning that can exert powerful effects. Bronstein, Levine, and Marcus (1975) showed that infant rats initially prefer the diet of the mother. They call this "the nongenetic, cross-generational transfer of information [p. 295]." In the long run, these same rats preferred a sweeter diet than that of the mother, but the survival value of the first choice is obvious. Rats are extraordinarily sensitive to the taste of anything that makes them ill (Garcia, Hankins, & Rusiniak, 1974), but illness is associated with the taste modality. Note that the illness cannot occur for several hours after ingestion; this severely challenges traditional reinforcement theory's stress on temporal contiguity of stimulus and reward. They will learn to avoid saccharin for a long period if X rays subsequently make them sick. However, they do not learn visual or auditory cues associated with the sickness, though they can easily learn to avoid electric shock by these same visual or auditory cues. The notion of the amodal equivalence of information as represented by the quotation "it matters not by what sense I know a pig sty" is severely damaged by this experiment. Modalities are not equivalent for the rat. Apparently human beings also have an aversion to food consumed before illness (Garcia *et al.*, 1974) but this is not as well documented.

Some studies have shown an apparent influence of past experience on food preferences, but at what stage in the development of the organism the experience must occur in order to be effective is as yet unknown. Greene, Desor, and Maller (1975) tested monozygotic and dizygotic twins of different races and socioeconomic classes on preferences for concentrations of sucrose, lactose, and sodium chloride. Hereditary contributed nothing to preferences: Monozygotic and dizygotic twins did not differ. But males of the ages tested (9–15 years) preferred higher concentrations of sucrose and lactose than females, and both socioeconomic status (SES) and race influenced preferences. The blacks preferred more concentrated solutions of sucrose, lactose, and sodium chloride than did the whites. The low-SES blacks preferred higher concentrations of sucrose and sodium chloride than did the high-SES blacks. Caucasian, high-SES subjects preferred more concentrated lactose than did the low-SES whites. The low hereditary factor, coupled with such well-known findings as a preference for saltier foods among low-SES groups, make relatively early experiences (when or for how long is unknown) a likely reason for the findings. In India, illiterate laborers from the Karnataka region prefer high concentrations of sour and bitter substances (citric acid and quinine sulfate), whereas Indian medical students find such concentrations unpleasant. The diet of the Indian laborers has many sour foods, particularly the tamarind fruit, which is very sour (Moskowitz, Kumaraiah, Sharma, Jacobs, & Sharma, 1975).

Obviously, more research is needed on taste. Many of us remember when our bourbon tasted best with ginger ale, whereas we are now content to have it with a little water. The tonic of gin and tonic (quinine water) tastes fine even without gin; some like Campari with soda, others avoid it. Perceptual learning of taste preferences is a fascinating topic.

The influence of early visual experience has been left for last. Early visual experience is part of imprinting, and it may be extremely strong (Hess, 1973), but most early visual experience has to be fairly extreme to have definite, noticeable effects. A fairly brief period of monocular deprivation, in which the lid of the eye is sewn shut, can make a kitten virtually blind in that eye. The sensitive period, for cats, is in the second to fourth month after birth and is virtually irreversible (Hubel & Wiesel, 1970). These findings have been reviewed frequently elsewhere (Barlow, 1975; Rothblat & Schwartz, 1978). Human beings must have strabismus corrected by the age of 2 if they are to have normal binocular vision (Banks, Aslin, & Letson, 1975). The maximal sensitive period for humans is estimated to extend from ages 1–4, approximately. These are harsh insults, with profound physiological effects on the ocular dominance of visual cells of the visual cortex. This is severe perceptual learning that is irreversible. Of more interest is the apparent modification of the orientation specificity of cortical visual cells as a result of exposure to an environment of vertical or horizontal stripes (Blakemore & Cooper, 1970; Hirsch & Spinelli, 1970). These effects are not as permanent and are more controversial (Stryker & Sherk, 1975).

Research with humans does show that Cree Indians raised in tepees rather than in our environment of verticals and horizontals (the "carpentered world") are less likely to have as much loss in sensitivity to obliques as are those raised in cities (Annis & Frost, 1973). But the effect is a small one, of theoretical but not much practical interest. Similarly, citizens of the industrialized nations are more susceptible to the Müller–Lyer illusion than are natives who live outside of our carpentered environment (Segall, Campbell, & Herskovits, 1966). But the Müller–Lyer illusion diminishes with practice. Generally, except for the extreme insults cited above, human beings appear to live in much the same visual environment. Effects as dramatic as those commonly found in audition have not been documented.

To investigate early experience and perceptual learning is to raise again the question of modality differences. Perceptual learning is present in all modalities but the notion of critical periods, sensitive periods, and irreversibility appears to be most applicable to audition and some extreme insults to vision such as early strabismus. We do not know too much about taste for human beings but this is a burgeoning area of research. Anecdotally, the period from about ages 10–15 may be critical for some kinds of sensorimotor learning like ballet dancing, hockey playing, or tennis, but this is speculation.

F. Activity and Perceptual Learning

1. Importance of Activity

Theories are generally agreed that activity is important for perceptual learning. The importance of activity for perceptual learning is particularly stressed by the Gibsons, Held, and Piaget. A classic experiment on adaptation to prisms is that of Held and Hein (1958). They had subjects wear prisms that displaced the image to one side. Some subjects viewed their hand and actively moved it back and forth. Others viewed their hand while it was passively moved. Only the active movement facilitated adaptation to the prisms. The topic of adaptation to prismatic adaptation has become extremely complicated, with books or extended treatments devoted to it (Freedman, 1968; Kornheiser, 1976; Welch, 1974), but it is generally agreed that active motion does facilitate adaptation. Another experiment is that of Gibson (1962) where some subjects actively felt forms while others had the forms pressed into their hands by the experimenter. Active touch was much more efficient in discriminating the forms.

2. Components of Activity

The exact features of activity that aid perceptual learning do not yield to a simple description. The learner must examine the object, by whatever sense modality, very attentively, with probably a slightly different strategy for almost every task. Here are two descriptions of active use of taste, one for food, the other for wine:

> You catch the aroma before you put the food in your mouth, you chew it slowly, and after you swallow you wait, because the peak flavor experience, with all the overtones and so forth, happens anywhere from thirty seconds to a full minute after you swallow the bite [Anon. 1976].

> The sommelier sips a minute quantity of wine from a flat, shallow silver *taste-vin* designed to present the largest possible surface of the liquid it contains to the taste-enhancing air. He rolls it around his mouth to coat the taste buds on his tongue, sucking in more air as he does so, a good way to strangle if you are not expert at this technique. He does not swallow the wine, for fumes from his stomach might falsify his judgment of the next bottle [Root, 1974, p. 20].

Despite the importance of activity, we do not know just how its benefits are obtained. One attempt at delineating the components of activity followed from the finding by Hunter (1954) that the blind are more accurate in the discrimination of curvature than are the sighted. Davidson (1972) replicated Hunter's results with the blind and the sighted, and he also videotaped the hands of the subjects as they performed the task. The blind used more fingers than the sighted and they *gripped* the stimulus as they scanned it haptically. When the sighted were trained to use the grip procedure they improved and performed as well as the blind. The preferred blind method of exploration was one that allowed maximal haptic contact and pressure to isolate the distinctive properties of the stimulus, a method of scanning that could be taught.

3. Search and Scanning

The process of visual search behavior during development appears to progress from a piecemeal, awkward scanning of the stimulus to a more refined, selective one (Bond, 1972; Day, 1975; Salapatek, 1975). From soon after birth until about 1 month of age, the infant looks at contours in a small portion of the visual field (e.g., the apex of a triangle, the edge of a square) and tends to look more at external than at internal contours, looking at the hair and chin of the face, for example, more than at the eyes and the mouth. By 2 months of age, more internal features are scanned for simple forms (Salapatek, 1975). The progression from the scanning of external features to the study of internal features seems to require some time, since Caron,Caron, Caldwell, and Weiss (1973) found that the head and eyes were more salient for 4-month-old children than they were for those 5 months old. For the 5-month-old infants, the mouth was as salient as the eyes, and the head was no longer as salient as the internal features of the face.

The very young infant is more likely to be captured by the visual stimulus than is the older one—the older infant is soon bored. This process starts at an early age. The 1-month-old child fusses when looking at a blank field, but the newborn does not (Salapatek, 1975). The 4-month-old child looks more frequently at the stimulus pattern than the 5-month-old (Caron *et al.*, 1973); the 6-month-old looks more frequently than the 13-month-old child (Kagan & Lewis, 1965).

Day (1975), in a review of developmental trends in visual scanning, describes the course of scanning as follows: The maturing child focuses more on those aspects of the stimulus that are most informative and searches these more exhaustively, more efficiently, and with more speed. The strategy for the acquisition of this information becomes more efficient with age, and the strategy is maintained across changes in the situation. A few examples will suffice. They all use photographs of the eye movements of the subject during the process of visual search.

The search patterns of 3-, 4-, 5-, and 6-year-old children was studied by Zinchenko, Van Chzhi-tsin, and Tarakanov (1963). The infants were asked to search a simple pattern and then to identify it later. The youngest children looked at the lens of the camera (which was inside the figure) and did not look at its outline. The 4-year-old children began to improve in eye sweeps, but still had few eye movements directed to the contour of the figure. By age 5 the contour was discovered, but attention was confined to a limited portion of it, and by age 6, eye movements were almost entirely along the contour, the most informative part of the figure. With age, search time decreased, the number of eye movements per second increased, and the errors of identification decreased. It is puzzling that the 5-year-old children, with fairly restrictive search of the figure, were about as accurate as the six-year-olds, each making no errors of identification. A familiar picture of bears from a children's book yielded no age differences in search behavior. This showed to Zinchenko *et al.* (1963) that when the young child is interested he can search as adequately as an older child.

The experiment by Vurpillot (1968) presented children with a silhouette of houses in which each member of a pair had 6 windows. The child judged whether the windows of the contrasting houses were the same in internal pattern (some, for example, had curtains, others did not) or whether they differed in some way. The children, aged 3, 5, 6½, and 9, differed markedly in correct judgments and in search strategy. The youngest children only searched (looked at) 6–7 windows of 12 possible, on the average, before guessing—the strategy of search did not differ for pairs that were the same or different. In contrast, the older children looked at almost all windows when the pairs were the same, but they looked at different pairs as a function of differences—more fixations for pairs with only one different window, for example. The 6½-year-olds were very similar to the 9-year-olds, meaning that on this task, at any rate, the strategy of visual search had been well developed by school age.

A final example is one by Mackworth and Bruner (1970) on the way 6-year-old children and adults search and recognize pictures. The children were less likely than the adults to search the informative areas, especially on blurred pictures. The children searched the pictures less exhaustively, had short eye fixations on the sharply focused pictures but longer sweeps on blurred ones, whereas adults tended to be more cautious. During recognition the children were less likely than adults to look at the same areas they had seen before, and they thus demonstrated less consistency of visual search. The differences between adults and children were not terribly large; the 6-year-olds seem to be developing adult-like efficient search behavior and the two prior studies (Zinchenko *et al.*, 1963; Vurpillot, 1968) show that by age 6 a child's visual scanning is fairly well-developed.

The development of search and scanning activity for haptic perception seems to be no different than for vision. The child of 3–4 is rather passive in haptic activity, using the palms, and holding and palpating the object in a gross sense. By the age of 6–7 the child is more active with the fingers and fingertips, and the method of exploration corresponds better with the distinctive properties of the object being explored (Zaporozhets, 1965; Abravanel, 1968).

The theories most comfortable with the data on the searching of individual forms are those of Hebb (1949) and the Soviet psychologists (see Gibson, 1969, for a review, see also Zaporozhets, 1965). Hebb stressed that only some sort of figure unity (distinction of figure and ground) was present at birth and that form perception was learned through scanning of the features of forms and becomes skilled with practice as the forms and their component features are learned. Both vision and touch can fit such a theory; one must wonder again how auditory development can fit in here. However, some interesting observations of Condon and Sander (1974a,b) are related to such a view. They found that infant movements are synchronized with adult speech, even on the day of birth. This auditory–motor synchrony has some similarity to the visual search patterns.

The search and scanning research confronts us again with those old notions of schemas, plans, and expectancies. The initial gross search becomes more attuned to the stimulus features scanned, and then some details drop away, leaving the observer free to concentrate on the critical features related to the task at hand. Hochberg (1972), in a discussion of picture perception, expresses this nicely, "our perception of a static picture . . . requires that successive glances be integrated over time. . . . and consists therefore mostly of memories and expectations that reflect a . . . rapid and intimate interaction with the world [p. 92]."

4. REAFFERENCE

Reafferent activity is activity initiated by an organism in its environment. Such initiated movement gets feedback from the environment as a result of the movement. In the experiment by Held and Hein (1958), the subject wore prisms that displaced the visual image to one side. Subjects who looked at their own hands while they actively moved them were experiencing *reafference*. Another group, a passive movement group, looked at their hands while someone else moved their hands through the same motions as the active group. This was *exafference*—visual stimulation uncorrelated with movement of the subject. These subjects did not adapt, and the experiment thus demonstrates the importance of reafference for visually guided behavior.

The strong version of this position, primarily identified as that of Richard Held, holds that there is no adaptation without reafference. The theory is based on an analysis of movement by von Holst (1954), in which reafference and exafference are distinguished, but the extension to adaptation is Held's. A further extension of the theory holds that reafference is necessary for perceptual development. An experiment with kittens (Held & Hein, 1963) is illustrative. Kittens were kept in the dark for 8–12 weeks and then split into two groups—one group actively locomoted through the environment, whereas the other group was pulled through the environment by the active group, receiving (in theory, at least) the same visual stimulation but without concurrent feedback from the environment as a result of the organism's own movements. Only the active group developed visual depth discrimination on the visual cliff.

A number of experiments have shown prismatic adaptation without active movement (see Kornheiser, 1976, for a review). For example, Wallach, Kravitz, and Lindauer (1963) showed some adaptation to prisms if the subjects simply viewed their own legs.

What of the importance of reafference for normal development? Miller and Walk (1977) used younger kittens than did Held and Hein (1963), starting the visual exposure at $2\frac{1}{2}$ weeks rather than 8–12 weeks of age.

The active and the passive groups were very similar, but a replication of the original Held and Hein (1963) conditions found a marked difference between active and passive groups. The lack of improvement of the older passive group was impressive. Active movement is indeed very important for the *maintenance* of normal visually guided behavior. In another developmental experiment, Walk and Bond (1971) revealed visual reaching by monkeys not allowed to see their hands while they were raised—results somewhat contrary to Held and Bauer (1967), who had found a marked disruption of visually guided reaching for such animals.

The *strong* position of the reafference hypothesis may be disproved for both adaptation and perceptual development, but Held and his coworkers have stressed an important influence on behavior. Kornheiser (1976) points out that active movement is much more effective than passive movement for adaptation. This may seem to be no more than a reiteration of other positions that stress the importance of activity (the Gibsons, Piaget, the Soviet view, for example). This is too simple a view. The prismatic experiments have demonstrated the complex multimodal nature of perceptual learning and adaptation (see Kornheiser, 1976). Much of that research was inspired by the strong adaptation position taken by Held.

5. Sensorimotor Activity: Applications

An applied offshoot of an emphasis on activity is the possibility that experimentally increased activity can be beneficial. The retarded have been the focus of a number of such programs and the work of psychologists Held and Piaget, particularly, has been cited as conceptual support for them. Is there any experimental support for such programs? This is very difficult to determine. Neman *et al.* (1974) conducted a study with mentally retarded adolescents whose average IQ was 40 and whose age ranged from about 12 to 18, averaging 15. One group had sensorimotor patterning with mobility exercises that included creeping, crawling, rolling, etc.; visuomotor training such as visual-pursuit training and convergence training; and sensory training with geometric forms, auditory discrimination, and memory tasks. A second experimental group stressed general activity and had a number of games and projects that provided stimulation. A passive control with no extra stimulation was also included. The experiment lasted 6 months.

The sensorimotor group improved more than the other groups in measures of visual perception, mobility, and language, but not in general intelligence. The improvements were not dramatic, but any improvements would have great theoretical and possible practical interest. The interpretation, statistics, subject selection and procedure were sharply criticized by Zigler and Seitz (1975) and strongly defended by Neman (1975).

The articles cited above have given enough information to show that the results of sensorimotor training as applied to the retarded have generally been disappointing. Perhaps a psychologist who believes in the benefits of activity, as most of us do, would have expected some gains for the targeted groups. The length of the Neman (1974) experiment—6 months—should be long enough to assess whether the extra training is beneficial. The heat of the exchange between Zigler and Seitz (1975) and Neman (1975) shows that the topic is controversial. As is common in research, we must await future research to decide the issue.

G. Intersensory Relationships

The interest of this section is whether there is perceptual learning of sense modality interrelations. Learning has been hypothesized for intermodality relations whether there seems to be, experimentally, any learning or not. Much of the research in this area is complex and confusing, refusing to yield to a set of reasonable generalizations. It is, to coin a phrase, a real can of worms. The major problems in this area can be considered as relating to either the integration of modalities or the influence of one modality on another. Like most categories, they spill over a bit.

1. Integration of Modalities

One may consider the sensory modalities to be separate at first and integrated only from maturation and experience, or consider them to be integrated at first and separate later, or consider them to be integrated at all times.

The primary exponent of the learned nature of sensory integration is Birch and his co-workers (Birch & Belmost, 1964; Birch & Lefford, 1967). If a child looks at a triangle and then actively feels a number of comparison forms, making a judgment of *same* or *different* for each (such as stars, diamonds, circles, squares, and triangles), the improvement depends on age. Younger children make more errors in such a task than do older ones. The inference that cross-modality integration improves with age lacks an important control: Errors within modalities (actively feeling a standard and then the comparison ones, for example) also diminish with age. In this case, improvement in the less precise modality, touch, parallels cross-modal improvement. Since within-modality improvement parallels between-modality improvement, one cannot maintain that sensory integration is a specific property of experience. Freides (1974) has reviewed a number of studies which document the demise of the hypothesis of the learned nature of sensory integration.

The opposite point of view, that the senses are unified in infancy and are differentiated with experience, is maintained by Bower (1974). Bower notes that many senses can respond to equivalent stimulation. Localization in space can be accomplished by at least four modalities: vision, audition, touch, and olfaction. When the young blind child moves its eyes toward a sound, for example, the integration is there from the start. The sighted child reaches out toward a visible object and is upset if the visible object is not tactually felt (Bower, Broughton, & Moore, 1970). This position is supported by the data better than the earlier (Birch & Lefford, 1967) one, but it is difficult to test experimentally.

A third position, that the senses are integrated at all times and that improvement is within a modality, is particularly applicable to Gibson (1969). The studies cited by Freides (1974), for example, that show within-modality improvement but no particular cross-modality improvement support this position. An example of early cross-modal integration is an experiment by Bryant, Jones, Claxton, and Perkins (1972). Infants whose average age was 9 months were shown two objects, and then an unseen object that produced a noise was pressed into the infant's hand. When the child was shown the two objects again, it reached predominantly for the tactual object that had produced a noise. Similarly, when shown an object that made a noise, the infants later reached toward it in preference to another. Within-modality and between-modality performances were similar. Gibson (1969) referred to the increased use of higher-order invariant stimulus information that improves with age, accessible to any relevant modality. The increased ability of the child to search efficiently and localize the distinctive features of objects in vision and touch, mentioned earlier, would be an example of invariant properties that are better abstracted by the older child.

2. Intermodality Influence

The information from one modality influences judgments in another. The size–weight illusion, cited at the beginning of the chapter, shows that judgments of weight are influenced by the visual size of an object, that is, the subject expects a large object to weigh more than a small object, and, if the two are of equal weight, the larger is judged to be lighter than the smaller. This section will deal with a number of possible intermodality influences: the developmental dominance of modalities, the effect of deprivation in one modality on other modalities, the transfer of information from one modality into another during development, the differentiation of one modality from the influence of another during development, and the effect of special experience in one modality on other modalities. A few examples will be cited to illustrate each topic; no attempt will be made to be exhaustive.

The philosophical (Berkeleian) view of touch regarded it as a more proximal sense that was therefore dominant over vision, the more distal modality. Judgment of visual distance would therefore be a cross-modal integration of tactual–kinesthetic distance. In this view, touch educates vision. No evidence has been found for this position. Bower, Broughton, and Moore (1970) presented infants with a virtual visual object (as seen through goggles with polaroid filters). The infants reached for it, as mentioned before, and cried when no object was felt. The object placed in the hand was not palpated by the infants, and they did not even look at it until they were 3 months of age. Thus, vision was dominant over touch. Rock and Harris (1967) cite a number of studies showing the dominance of vision over touch with adult subjects. An object seen through a prism may look large, even though its physical size is small. The subject that feels the object matches the visual appearance more than the tactual one, but the tactual match is easy as long as the eyes are closed.

An example of the way the deprivation of one modality may affect another is that of the blind or the deaf child. The classical view was that the handicapped child would develop the other senses more acutely, and an often cited example of this view was that of *facial vision*, or the ability of the blind to locate walls before they bumped into them. In such cases, the blind reported the distal information to be localized on the face, but research showed (Supa, Cotzin, & Dallenbach, 1944) that the information used was auditory, not tactual, and that normally sighted individuals, blindfolded, were able, with training, to utilize the auditory distance cues as well as the blind. The better judgment of curvature by the blind was cited earlier (Davidson, 1972), but the influence of blindness on other modalities seems to be relatively mild. Rice (1970) has cited research that shows the blind to be better in some echodetection and echolocation tasks (though a sighted subject was just as good). Warren, Anooshian and Bollinger (1973) cited some advantages of late blindness for complex tasks, whereas the early blind were better on simpler tasks. Their hypothesis was that the visual frame of reference helped the late blind, and this would be congruent with Pick's (1974) hypothesis, to be considered shortly. But one can conlude that, generally, neither facilitative nor depressive effects are observed. The same applies to the deaf (Reynolds, 1978). However, if one considers improvement in reading to be perceptual learning, as it undoubtedly is, then the deaf are handicapped. But reading is an auditory–visual skill, dependent on language development, and language development is usually retarded in the deaf (Reynolds, 1978).

As the child matures, one modality may become more influential than another on some tasks, even though both are dealing with the same general stimulus information. McGurk and MacDonald (1976) had subjects view a film and repeat what they heard the model saying. But the sound track did not correspond to the lip movements of the model. A model whose voice on tape was saying "ga-ga", might be seen with lips saying "ba-ba." Children were much more tied to the auditory voice than were the adults. Adults were more likely to report the visual lip sounds, some fusion ("da-da"), or a combination like "gabga." Auditory development, as was mentioned, proceeds faster than visual development, and this may account for the children's greater fidelity to the auditory stimulus. The adults were more influenced by vision. Pick (1974) has taken the position that nonvisual spatial information is coded into the visual modality, no matter from which modality. For example, we code the voice as coming from the moving lips of the ventriloquist's dummy rather than from the ventriloquist. Auditory localization is also more accurate with the eyes open to see the environment when the subject cannot see the sound source (Pick, 1974). Pick also cites research that shows that visual experience may interfere with tactual discrimination. The blind discriminate line-to-curve transformations as well as rotations, but the sighted do not, and the blind with some visual experience are intermediate. Thus, visual perceptual learning may interfere with tactual performance.

A generalization of the Pick hypothesis is that sensory information is translated into the modality best fitted for the task. He suggests that texture qualities such as roughness or smoothness might be coded into the tactual modality, and food qualities might be coded into the gustatory–olfactory modality (Pick, 1974). Auditory presentation of information is superior to visual presentation for short-term memory, though not, interestingly, for long-term memory (Penney, 1975).

Although vision may be the modality for the encoding of some information, the influence of vision on some tasks may actually decrease. In a conflict between visual and gravitational vertical, the younger child is more influenced by the visual environment than is the older one. If a room is tilted to one side, with walls not at the gravitational vertical, younger children are more apt to align themselves more in congruence with the visual environment, and less so with the gravitational one, than are older children (Witkin, 1959). A related experiment was performed by Lee and his co-workers. An infant was placed on a floor inside a room that could move independently of the floor. When the room walls moved toward the standing child, 14–16 months old, the child tended to sway, stagger, or fall—a reaction which the researchers termed *visual proprioception* (Lee & Aronson, 1974). Adults, on the other hand, were less affected by the movement of the room (Lishman & Lee, 1973). Thus, proprioception became more independent of vision with age. The classic studies of Piaget with conservation of amount, weight, and volume show that the developing child becomes less susceptible to visual appearance with maturity. A child that sees a liquid poured from a bottle into another one that is taller and thinner is less likely to judge that the quantity of liquid has increased, for example. An influence of vision on body stance that changes with experience is found in studies of paratroopers. The paratrooper must learn to exit from an airplane with the proper body form; this is learned in jumping from a tower where the free fall is snubbed by canvas straps. That the body form is reasonably simple to learn without the visual conflict is shown by the marked improvement in performance when the height of the tower above the ground is lessened. Prior experience with heights (as asked on a questionnaire) or experience with sports like football or swimming also improved performance on the tower (Walk, 1959). Experience helped to overcome the influence of visual height on performance.

Individual differences of two types must be considered: those produced by the experimenter and those resulting from prior experience by the subject. Canon (1970, 1971) hypothesized that a conflict of sensory information from two modalities would lead to adaptation in the unattended modality. He studied both visual and auditory attention in a situation where the target could be identified both visually and through audition. The subjects wore prisms to displace visual stimulation and a pseudophone that systematically shifted auditory information by the same amount. When subjects attended to the visual stimulus, they had little visual adaptation, but a great deal of auditory adaptation; with auditory attention, more visual than auditory adaptation was observed. Visual and proprioceptive discrepancies with prismatic visual displacement were studied by Kelso, Cook, Olson, and Epstein (1975). Subjects that attended to the visual modality adapted to the proprioceptive–kinesthetic modality and those that attended to the proprioceptive modality adapted visually.

Ballet dancers are very body oriented, and Kahane and Auerbach (1973) tested them in a prismatic adaptation task compared to controls. The ballet dancers adapted very little, whereas the nondancers adapted selectively to the adapting hand. This is congruent with Canon's attention hypothesis, since the dancers did not adapt to bodily parts, as nondancers did. Another study of ballet dancers was that of Gruen (1955), in which he studied adaptation to the Witkin (1959) tilting room. The dancers were more accurate than nondancers when they used body cues by moving the chair to the gravitational upright, but were not better than controls when they sat passively in the tilted chair and tried to adjust the room surrounding them to the upright. Sex differences, often hypothesized in field-dependent situations, were found for both controls and dancers in the external, passive, tilted-room condition, but were not present for the dancers (though still found for the controls) in the active situation in which the chair was set upright to gravity. Experience, then, affected both visual dependency and sex differences, and one may hypothesize that sex differences in perceptual experiments simply reflect differential perceptual learning.

Samuel (1976) found that individuals of high athletic ability adapted less than those of low athletic ability in a prismatic adaptation task. This confirms Kahane and Auerbach (1973), and both results are congruent with the analysis of Harris (1965) that the prismatic task is one of proprioceptive adaptation, since those with hypothesized attention to body cues (ballet dancers, athletes) adapted less. Relatedly, Warren and Platt (1975) found that proprioceptive adaptation was related to measures of visual abilities. Samuel's subjects of high artistic experience had more prismatic adaptation than those of low artistic experience, which is also congruent with the attention hypothesis. But those with artistic experience were much better on embedded figure tasks, whereas those of high athletic experience did poorly (Samuel, 1976). The attention hypothesis thus ties together both prior experience and experimentally induced attention.

Different modalities are of differential efficiency in the pickup or processing of information common to more than one modality. One presumes that the complex dominance of modalities would be evidence that the more efficient modality normally dominates equivalent information, even though attention or experience can change this. Goodnow (1969) systematically investigated an indirect finding of Pick and Pick (1966) on the difference in the way haptic and visual information was processed. She found that curvature was highly discriminable by vision, whereas the orientation of forms was not. Conversely, haptic exploration easily discriminated form orientation, but had difficulty with curvature.

While the separate modalities may process different types of common information somewhat differently, we must be cautious about downgrading intermodal unity. It can be very impressive. An example is the remarkable experiments that convert an optical image into a tactile display on the back of the subject (White, Saunders, Scadden, Bach-y-Rita and Collins, 1970). The experience from the tactual display on the back is much like seeing: The subjects learn monocular cues to depth, they duck to an expansion pattern, and they can observe the kinetic depth effect (White *et al.* 1970). A blind subject described it as nothing at all like his experience with the piecemeal world of touch (Guarniero, 1974). Yet this "vision" is based on touch, and blind subjects learn with minimal difficulty to respond to some of the higher-order invariances (Gibson, 1966) that are common and taken for granted in vision, and accessible to touch with the proper instrumentation.

V. CONCLUSION

This introduction to the study of perceptual learning has shown that perceptual learning is a broad area and an active one, seething with research. The future will bring more and more research that moves beyond the visual modality and into audition, taste, odor, and motor processes. The new surge beyond traditional boundaries should not neglect sensory integration and the unity of the senses. Sense modalities may not be quite as unified as some theoretical positions claim, but they do interact and have unexpected unities; a separate perceptual learning for each modality would be a loss. Many of the questions across modalities are common ones; the brain cannot consider the modalities as separate, as does the researcher.

To redress the focus on visual research, this review has tried, particularly, to bring in new research on the auditory modality, and it has tried to stress modality integration and interaction. Perhaps more should have been written about motor learning. Every indication is that motor processes are moving into the mainstream of perceptual research (see Stelmach, 1976). Consider this: I have not ridden a horse in over 20 years, but the muscles still "know"; they are ready (and they would get just as sore, or more sore, than the last time). Bicycle riding, automobile driving, ice skating, and swimming-pool diving are all "muscle-memory" types of perceptual learning that last a long time. As a book on ballet states "once the habit has been ingrained, it remains in the muscles for a long time [Mazo, 1974, p. 49]." And later: "Someone asked to do the opposite of what her kinesthetic memory tells her is correct can become very confused [Mazo, 1974, p. 52]." These examples are a challenge for our research. Is it not of some theoretical importance that many motor skills apparently must start so early? Richard Stockton won $40,000 in a tennis match in Philadelphia by beating Jimmy Connors. The two tennis players, now in their twenties, first met in championship play at the Orange Bowl Juniors when Stockton was 10 and Connors was 9.

Early learning seems to be more applicable to audition, however, than to any other sense modality. Here feature detectors seem to decay at an early age, when the first language is learned. Motor learning is not so dramatic, but early learning seems very important for ballet and for sports. There is no evidence for any decay of feature detectors in motor learning, and in vision it is controversial. Are there tastes we can no longer taste? Is the present generation done in by pablum and processed baby food, too many visits to McDonald's, doomed to miss the subtle flavors the French chef spends years developing? Are there smells that can no longer be smelled? Perhaps Shakespeare's rose (which would smell as sweet by any other name) no longer smells the same to us because of our exposure to modern plastic packaging and deodorants. Shakespeare experienced a broader range of smells than we do, fine for keeping the feature detectors in tune. Some of us can still remember outhouses and manure in the streets. To open one door may be to close another; only the feature detectors know for sure.

The field of perceptual learning has been shaped more by issues than by theories, and the research issues have forced consideration of the multimodal nature of perceptual learning. What may have seemed like a simple research issue, the basis for adaptation to visual stimulation displaced by prisms, has found kinesthetic–proprioceptive, visual, and auditory components (Kornheiser, 1976). The newly discovered complexity helps to underscore the integration and the interdependency of all of the senses.

Insofar as any one theory has tried to encompass this diverse and tangled field, one must cite the Gibsons', particularly as represented by their books, *The senses considered as perceptual systems* (J. J. Gibson, 1966), *Principles of perceptual learning and development* (E. J. Gibson, 1969), and *The psychology of reading* (E. J. Gibson and H. Levin, 1975). The new theories will have a difficult time meeting such high standards, and let us hope they will be encompassing theories, not small ones that attempt to integrate only a small subset of the field.

Individual differences in perceptual learning and adaptation based on prior experience were cited as part of the section on sensory integration. These individual differences are a natural outgrowth of the influence of experience on perception, but they seem fairly minor. They will not, let us hope, bring back the ghost of each person in his or her own perceptual world. Such individual differences make one think of A. Conan Doyle's Sherlock Holmes or of William James. Sherlock Holmes was fond of deducing an individual's occupation and past history from small, almost indiscernible traces. William James warned that we all soon become "walking bundles of habits. . . . We are spinning our own fates, good or evil, never to be undone [James, 1890, p. 197]." If mothers only knew what a decision it truly was to send a child to learn to play the piano or to take ballet lessons, they might, James warns, pause and think. Recent research backs James up, but not to the extent that A. Conan Doyle described.

Perceptual learning research is becoming more "ecologically relevant" without losing scientific rigor. Those of us in the field find an ever-expanding series of challenging questions.

Acknowledgments

I want to thank my son, David, for his incisive editorial comments (but do not blame him if I sometimes disregarded good advice) and for stimulating discussions, particularly on perceptual learning as related to sports.

References

Abravanel, E. The development of intersensory patterning with regards to selected spatial dimensions. *Monographs of the Society for Research in Child Development*, 1968, **33** (Serial No. 118).

Aiken, E. G. Auditory discrimination learning: Prototype storage and distinctive feature mechanisms. *Perception & Psychophysics*, 1969, **6,** 95–96.

Aiken, L. S., & Williams, T. M. A developmental study of schematic concept formation. *Developmental Psychology*, 1973, **8,** 162–167.

André-Thomas, & Autgaerden, S. Audibilité spontanée de la voix maternelle, audibilité conditionée de toute autre voix. *La Presse Médicale*, 1963, **71,** 1761–1764.

André-Thomas & Autgaerden, S. *Locomotion from pre- to postnatal life*. London: Spastics Society/ Heinnemann, 1966.

Annis, R. C., & Frost, B. Human visual ecology and orientation anistotropies in acuity. *Science*, 1973, **182,** 729–731.

Anon. Talk of the town. *The New Yorker*, March 8, 1976, p. 30.

Banks, M. S., Aslin, R. N., & Letson, R. D. Sensitive period for the development of human binocular vision. *Science*, 1975, **190,** 675–677.

Barlow, H. B. Visual experience and cortical development. *Nature*, 1975, **258,** 199–204.

Bever, T. G., & Chiarello, R. J. Cerebral dominance in musicians and non-musicians. *Science*, 1974, **185,** 537–539.

Birch, H. G., & Belmont, L. Audio-visual integration in normal and retarded readers. *American Journal of Orthopsychiatry*, 1964, **34,** 852–861.

Birch, H. G., and Lefford, A. Visual differentiation, intersensory integration and voluntary motor control. *Monographs of the Society for Research in Child Development*, 1967, **32** (Serial No. 110).

Blakemore, C., & Cooper, G. F. Development of the brain depends on the visual environment. *Nature*, 1970, **228,** 477–478.

Bond, E. K. Perception of form by the human infant. *Psychological Bulletin*, 1972, **77,** 225–245.

Bornstein, M. H., Kessen, W., & Weiskopf, S. The categories of hue in infancy. *Science*, 1976, **191,** 201–202.

Bower, T. G. R. The evolution of sensory systems. In R. B. MacLeod & H. L. Pick, Jr. (Eds.) *Perception: Essays in Honor of James J. Gibson*. Ithaca, New York: Cornell Univ. Press, 1974. Pp. 141–152.

Bower, T. G. R., Broughton, J. M., & Moore, M. K. The coordination of vision and tactual input in infancy. *Perception & Psychophysics*, 1970, **8,** 51–53.

Bronstein, P. M., Levine, M. J., & Marcus, M. A rat's first bite: The nongenetic cross-generational transfer of information. *Journal of Comparative and Physiological Psychology*, 1975, **89,** 295–298.

Bryant, P. E., Jones, P., Claxton, V., and Perkins, G. M. Recognition of shapes across modalities by infants. *Nature*, 1972, **240,** 303–304.

Caldwell, E. C., & Hall, V. C. Distinctive-features versus prototype learning re-examined. *Journal of Experimental Psychology*, 1970, **83,** 7–12.

Canon, L. K. Intermodal inconsistency of input and directed attention as determinants of the nature of adaptation. *Journal of Experimental Psychology*, 1970, **84,** 141–147.

Canon, L. K. Directed attention and maladaptive "adaptation" to displacement of the visual field. *Journal of Experimental Psychology*, 1971, **88,** 403–408.

Caramazza, A., Yeni-Komshian, G. H., Zurif, E. B., & Carbone, E. The acquisition of a new phonological contrast: The case of stop consonants in French–English bilinguals. *Journal of the Acoustical Society of America*, 1973, **54,** 421–428.

Caron, A. J., Caron, R. F., Caldwell, R. C., & Weiss, S. J. Infant perception of the structural properties of the face. *Developmental Psychology*, 1973, **9,** 385–399.

Carpenter, G. C., Tecce, J. J., Stechler, G., & Friedman, S. Differential visual behavior to human and humanoid faces in early infancy. *Merrill-Palmer Quarterly*, 1970, **16,** 91–108.

Cohen, L. B., & Salapatek, P. (Eds.), *Infant perception: From sensation to cognition* (2 vols.). New York: Academic Press, 1975.

Condon, W. S., & Sander, L. W. Synchrony demonstrated between movements of the neonate and adult speech. *Child Development*, 1974, **45,** 456–462. (a)

Condon, W. S., & Sander, L. W. Neonate movement is synchronized with adult speech: Interactional participation and language acquisition. *Science*, 1974, **183,** 99–101. (b)

Davidson, P. W. Haptic judgments of curvature by blind and sighted humans. *Journal of Experimental Psychology*, 1972, **93,** 43–55.

Day, M. C. Developmental trends in visual scanning. In H. W. Reese (Ed.), *Advances in child development and behavior* (Vol. 10). New York: Academic, 1975. Pp. 153–193.

Dember, W. N. Response by the rat to environmental change. *Journal of Comparative and Physiological Psychology*, 1956, **49,** 93–95.

Dodwell, P. C. (Ed.), *Perceptual learning and adaptation*. Baltimore, Maryland: Penguin, 1970.

Durant, M. J., & Yussen, S. R. Effect of memory on distinctive features and schema learning. *Journal of Experimental Psychology: Human Learning and Memory*, 1976, **2,** 315–321.

Eimas, P. D. Auditory and phonetic coding of the cues for speech: Discrimination of the (r–l) distinction by young infants. *Perception & Psychophysics*, 1975, **18,** 341–347.

Eimas, P., Siqueland, E. R., Jusczyk, P., & Vigorito, J. Speech perception in infants. *Science*, 1971, **171,** 303–306.

Epstein, W. *Varieties of perceptual learning*. New York: McGraw-Hill, 1967.

Fairweather, H. Sex differences in cognition. *Cognition*, 1976, **4,** 231–280.

Fantz, R. L. The origin of form perception. *Scientific American*, 1961, **204**(5), 66–72.

Fantz, R. L., Fagan, J. F., III, & Miranda, S. B. Early visual selectivity. In L. B. Cohen & P. Salapatek (Eds.), *Infant perception: From sensation to cognition* I. *Basic visual processes*. New York: Academic Press, 1975. Pp. 249–345.

Fantz, R. L., Ordy, J. M., & Udelf, M. S. Maturation of pattern vision in infants during the first six months. *Journal of Comparative and Physiological Psychology*, 1962, **55,** 907–917.

Freedman, S. J. (Ed.), *The neuropsychology of spatially oriented behavior*. Homewood, Illinois: Dorsey, 1968.

Freides, D. Human information processing and sensory modality: Cross-modal functions, information complexity, memory and deficit. *Psychological Bulletin*. 1974, **81,** 284–310.

Gagne, R., & Gibson, J. J. Research on the recognition of aircraft. Chapter 7 in J. J. Gibson (Ed.), *Motion picture training and research*. Report No. 7, Army Air Force Aviation Psychology Program, Research Reports, U.S. Government Printing Office, Washington, D.C., 1947.

Garcia, J., Hankins, W. G., & Rusiniak, K. W. Behavioral regulation of the milieu interne in man and rat. *Science*, 1974, **185,** 824–831.

Gibson, E. J. *Principles of perceptual learning and development*. New York: Appleton, 1969.

Gibson, E. J., Gibson, J. J., Pick, A. D., & Osser, H. A. A developmental study of the discrimination of letter-like forms. *Journal of Comparative and Physiological Psychology*, 1962, **55,** 897–906.

Gibson, E. J., & Levin, H. *The psychology of reading*. Cambridge, Massachusetts: MIT Press, 1975.

Gibson, E. J., & Walk, R. D. The effect of prolonged exposure to visually presented patterns on learning to discriminate them. *Journal of Comparative and Physiological Psychology*, 1956, **49,** 239–242.

Gibson, J. J. Observations on active touch. *Psychological Review*, 1962, **69,** 477–491.

Gibson, J. J. *The senses considered as perceptual systems*. Boston, Massachusetts: Houghton-Mifflin, 1966.

Gibson, J. J., & Gibson, E. J. Perceptual learning: Differentiation or enrichment? *Psychological Review*, 1955, **62,** 32–41.

Goodnow, J. J. Eye and hand: Differential sampling of form and orientation properties. *Neuropsychologia*, 1969, **7,** 365–373.

Gottlieb, G. Ontogenesis of sensory function in bird and mammals. In E. Tobach, L. R. Aronson, & E. Shaw (Eds.), *The biopsychology of development*. New York: Academic Press, 1971. Pp. 67–128.

Gottlieb, G. Conceptions of prenatal development: Behavioral embryology. *Psychological Review*, 1976, **83,** 215–234.

Greene, L. S., Desor, J. A., & Maller, O. Heredity and experience: Their relative importance in the development of taste preference in man. *Journal of Comparative and Physiological Psychology*, 1975, **89,** 279–284.

Gruen, A. The relation of dancing experience and personality to perception. *Psychological Monographs*, 1955, **69** (14) (Whole No. 399).

Guarniero, G. Experience of tactile vision. *Perception*, 1974, **3,** 101–104.

Hagen, M. A. Picture perception: Toward a theoretical model. *Psychological Bulletin*, 1974, **81,** 471–497.

Hammond, J. Hearing and response in the newborn. *Developmental Medicine and Child Neurology*, 1970, **12,** 3–5.

Harris, C. S. Perceptual adaptation to inverted, reversed, and displaced vision. *Psychological Review*, 1965, **72,** 419–444.

Hebb, D. O. *The organization of behavior*. New York: Wiley, 1949.

Held, R., & Bauer, J. A., Jr. Visually guided reaching in infant monkeys after restricted rearing. *Science*, 1967, **155,** 718–720.

Held, R., & Hein, A. Adaptation of disarranged hand-eye coordination contingent upon re-afferent stimulation. *Perceptual and Motor Skills*, 1958, **8,** 87–90.

Held, R., & Hein, A. Movement produced stimulation in the development of visually guided behavior. *Journal of Comparative and Physiological Psychology*, 1963, **56,** 872–876.

Hess, E. H. *Imprinting*. Princeton, New Jersey: Van Nostrand-Reinhold, 1973.

Hilgard, E. R., & Bower, G. H. *Theories of learning*. Englewood Cliffs, New Jersey: Prentice-Hall, 1975. 4th ed.

Hirsch, H. V. B., & Spinelli, D. N. Visual experience modifies distribution of horizontally and vertically oriented receptive fields in cats. *Science*, 1970, **168,** 869–871.

Hochberg, J. The representation of things and people. In E. H. Gombrich, J. Hochberg, & M. Black, *Art, perception and reality*. Baltimore, Maryland: Johns Hopkins Press, 1972. Pp. 47–94.

Holst, E. von. Relations between the central nervous system and the peripheral organs. *British Journal of Animal Behavior*, 1954, **2,** 89–94.

Hubel, D. H., & Wiesel, T. N. The period of susceptibility to physiological effects of unilateral eye closure in kittens. *Journal of Physiology*, 1970, **206,** 419–436.

Hudson, W. Pictorial depth perception in sub-cultural groups in Africa. *Journal of Social Psychology*, 1960, **52,** 183–208.

Hunter, I. Tactile–kinesthetic perception of straightness in blind and sighted humans. *Quarterly Journal of Experimental Psychology*, 1954, **6,** 149–154.

Jakobson, R., & Halle, M. *Fundamentals of language*. The Hague: Mouton, 1956.

James, W. *Principles of psychology*. New York: Holt, 1890.

Kagan, J., & Lewis, M. Studies of attention in the human infant. *Merrill-Palmer Quarterly*, 1965, **11,** 95–127.

Kahane, J., & Auerbach, C. Effect of prior body experience on adaptation to visual displacement. *Perception & Psychophysics*, 1973, **13,** 461–466.

Kelso, J. A. S., Cook, E., Olson, M. E., & Epstein, W. Allocation of attention and the locus of adaptation to displaced vision. *Journal of Experimental Psychology: Human Perception and Performance*, 1975, **1,** 237–245.

Kerpelman, L. C. Pre-exposure to visually presented forms and nondifferential reinforcement in perceptual learning. *Journal of Experimental Psychology*, 1965, **69,** 257–262.

Kornheiser, A. S. Adaptation to laterally displaced vision: A review. *Psychological Bulletin*, 1976, **83,** 783–816.

Lasky, R. E., Syrdal-Lasky, A., & Klein, R. E. VOT discrimination by four to six and a half month old infants from Spanish environments. *Journal of Experimental Child Psychology*, 1975, **20,** 215–225.

Lee, D. N., & Aronson, E. Visual proprioceptive control of standing in human infants. *Perception & Psychophysics*, 1974, **15,** 529–532.

Lewis, M., & Brooks, J. Infants' social perception: A constructivist view. In L. B. Cohen & P. Salapatek (Eds.), *Infant perception: From sensation to cognition*. Vol. II. *Perception of space, speech and sound*. New York: Academic Press, 1975. Pp. 101–148.

Lishman, J. R., & Lee, D. N. The autonomy of visual kinaesthesis. *Perception*, 1973, **2,** 287–294.

MacFarlane, J. A. Olfaction in the development of social preferences in the human infant. In Ciba Foundation Symposium 33 (new series), *Parent–infant interaction*. Amsterdam: Elsevier, 1975. Pp. 103–113.

Mackworth, N. H., & Bruner, J. S. How adults and children search and recognize pictures. *Human Development*, 1970, **13,** 149–177.

Madsen, C. K., Edmondson, F. A., III, & Madsen, C. H., Jr. Modulated frequency discrimination in relationship to age and musical training. *Journal of the Acoustical Society of America*, 1969, **46,** 1468–1472.

Marler, P., & Mundinger, P. Vocal learning in birds. In H. Moltz (Ed.), *The ontogeny of vertebrate behavior*. New York: Academic, 1971.

Mazo, J. H. *Dance is a contact sport*. New York: Dutton, 1974.

McGurk, H., & MacDonald, J. Hearing lips and seeing voices. *Nature*, 1976, **264,** 746–748.

McPhee, J. *Levels of the game*. New York: Bantam, 1970.

Miller, D. R., & Walk, R. D. *Visual motor experience and visual depth perception in kittens*. Unpublished paper, 1977.

Miller, R. J. Cross-cultural research in the perception of pictorial materials. *Psychological Bulletin*, 1973, **80,** 135–150.

Mills, M., & Melhuish, E. Recognition of mother's voice in early infancy. *Nature*, 1974, **252,** 123–124.

Miyawaki, K., Strange, W., Verbrugge, R., Liberman, A. M., Jenkins, J. J., & Fujimura, O. An effect of linguistic experience: The discrimination of (r) and (l) by native speakers of Japanese and English. *Perception & Psychophysics*, 1975, **18,** 331–340.

Morell, J. A. Age, sex, training, and the measurement of field dependence. *Journal of Experimental Child Psychology*, 1976, **22,** 100–112.

Moskowitz, H. W., Kumaraiah, V., Sharma, K. N., Jacobs, H. L., & Sharma, S. D. Cross-cultural differences in simple taste preferences. *Science*, 1975, **190,** 1217–1218.

Mussen, P. H. (Ed.) *Carmichael's Manual of Child Psychology* (3rd ed.) (2 vols.). New York: Wiley, 1970.

Neisser, U. *Cognitive psychology*. New York: Appleton, 1967.

Neisser, U. *Cognition and reality*. San Francisco: Freeman, 1976.

Neman, R. A reply to Zigler and Seitz. *American Journal of Mental Deficiency*, 1975, **79,** 493–505.

Neman, R., Roos, P., McCann, B. M., Menolascino, F. J., & Heal, L. W. Experimental evaluation of sensorimotor patterning with mentally retarded children. *American Journal of Mental Deficiency*, 1974, **79,** 372–384.

Nowlis, G. H., & Kessen, W. Human newborns differentiate differing concentrations of sucrose and glucose. *Science*, 1976, **191,** 865–866.

Penney, C. G. Modality effects in short-term memory. *Psychological Bulletin*, 1975, **82,** 68–84.

Pick, A. D. Improvement of visual and tactual form discrimination. *Journal of Experimental Psychology*, 1965, **69,** 331–339.

Pick, A., & Pick, H. L., Jr. A developmental study of tactual discrimination in blind and sighted children and adults. *Psychonomic Science*, 1966, **6,** 367–368.

Pick, H. L., Jr. Systems of perceptual and perceptual–motor development. In J. P. Hill (Ed.), *Minnesota symposia on child psychology* (Vol. 4). Minneapolis: Univ. of Minnesota Press, 1970. Pp. 199–219.

Pick, H. L., Jr. Visual coding of nonspatial information. In R. B. MacLeod & H. L. Pick, Jr. (Eds.), *Perception: Essays in honor of James J. Gibson*. Ithaca, New York: Cornell Univ. Press, 1974.

Reynolds, H. N. Perceptual effects of deafness. In R. D. Walk & H. L. Pick, Jr. (Eds.), *Perception and experience*. New York: Plenum, 1978. Pp. 241–259.

Rice, C. E. Early blindness, early experience and perceptual enhancement. American Foundation for the Blind, *Research Bulletin* No. 22, 1970, 1–22.

Riesen, A. H. Sensory deprivation. In E. Stellar & J. M. Sprague (Eds.), *Progress in physiological psychology* (Vol. 1). New York: Academic Press, 1966. Pp. 117–147.

Rock, I., & Harris, C. S. Vision and touch. *Scientific American*, 1967, **216**(5), 96–104.

Root, W. Of wine and noses. *New York Times Magazine*, December 22, 1974, pp. 14 *et seq*.

Rothblat, L. A., & Schwartz, M. L. Altered early environment: Effects on the brain and visual behavior. In R. D. Walk & H. L. Pick, Jr. (Eds.), *Perception and experience*. New York: Plenum, 1978. Pp. 7–36.

Russell, M. J. Human olfactory communication. *Nature*, 1976, **260,** 520–522.

Salapatek, P. Pattern perception in early infancy. In L. B. Cohen & P. Salapatek (Eds.), *Infant perception: From sensation to cognition*. (1). I. *Basic visual processes*. New York: Academic Press, 1975. Pp. 133–248.

Samuel, J. M. F. Unpublished dissertation proposal, George Washington Univ., 1976.

Schaller, M. J., & Harris, L. J. Children judge "perspective" transformations of letter-like forms as different from prototypes. *Journal of Experimental Child Psychology*, 1974, **18,** 226–241.

Segall, M. H., Campbell, D. T., & Herskovits, M. J. *The influence of culture on visual perception*. Indianapolis, Indiana: Bobbs-Merrill, 1966.

Sergeant, D. Experimental investigation of absolute pitch. *Journal of Research in Music Education*, 1969, **17,** 135–143.

Siegel, A. W., & White, S. H. The development of spatial representations of large-scale environments. In H. W. Reese (Eds.), *Advances in child development and behavior* (Vol. 10). New York: Academic Press, 1975. Pp. 9–55.

Singer, M. H., & Lappin, J. S. Similarity: Its definition and effect on the visual analysis of complex displays. *Perception & Psychophysics*, 1976, **19,** 405–411.

Stelmach, G. E. (Ed.) *Motor control: Issues and trends*. New York: Academic Press, 1976.

Streeter, L. A. Language perception of 2-month old infants shows effects of both innate mechanisms and experience. *Nature*, 1976, **259,** 39–41.

Stryker, M. P., & Sherk, H. Modification of cortical orientation selectivity in the cat by restricted visual experience: A re-examination. *Science*, 1975, **190,** 904–906.

Supa, M., Cotzin, M., & Dallenbach, K. 'Facial vision,' The perception of obstacles by the blind. *American Journal of Psychology*, 1944, **57,** 133–183.

Tighe, L. S., & Tighe, T. J. Discrimination learning: Two views in historical perspective. *Psychological Bulletin*, 1966, **66,** 353–370.

Tighe, T. J., & Tighe, L. S. Stimulus control in children's learning. In A. D. Pick (Ed.), *Minnesota symposia on child psychology* (Vol. 6). Minneapolis: Univ. of Minnesota Press, 1972.

Vurpillot, E. The development of scanning strategies and their relation to visual differentiation. *Journal of Experimental Child Psychology*, 1968, **6,** 632–650.

Walk, R. D. *Fear and courage: A psychological study*. On deposit at George Washington Univ. Library, 1959.

Walk, R. D., & Bond, E. K. The development of visually guided reaching in monkeys reared without sight of the hands. *Psychonomic Science*, 1971, **23,** 115–116.

Walk, R. D., & Schwartz, M. *Perceptual learning of birdsongs with and without a visual code*. Unpublished paper, 1975.

Wallach, H., Kravitz, J. H., & Lindauer, J. A passive condition for rapid adaptation to displaced visual direction. *American Journal of Psychology*, 1963, **76,** 568–578.

Warren, D. H., Annoshian, L. J., & Bollinger, J. G. Early vs. late blindness: The role of early vision in spatial behavior. American Foundation for the Blind, *Research Bulletin* No. 26, 1973, 151–170.

Warren, D. H., & Platt, B. B. Understanding prism adaptation: An individual differences approach. *Perception and Psychophysics*, 1975, **17,** 337–345.

Webster's Seventh New Collegiate Dictionary. Springfield, Massachusetts: G.&S. Merriam Co., 1969.

Welch, R. B. Research on adaptation to rearranged vision: 1966–1974. *Perception*, 1974, **3,** 367–392.

White, B. W., Saunders, F. A., Scadden, L., Bach-y-Rita, P., & Collins, C. C. Seeing with the skin. *Perception & Psychophysics*, 1970, **7,** 23–27.

Witkin, H. A. The perception of the upright. *Scientific American*, 1959, **200**(2), 50–56.

Zaporozhets, A. V. The development of perception in the preschool child. *Monographs of the Society for Research in Child Development*, 1965, **30** (Serial No. 100), 81–101.

Zeaman, D., & House, B. J. Interpretation of developmental trends in discriminative transfer effects. In A. D. Pick (Ed.), *Minnesota symposia on child psychology* (Vol. 8). Minneapolis: Univ. of Minnesota Press, 1974. Pp. 144–186.

Zigler, E., & Seitz, V. On "An experimental evaluation of sensorimotor patterning": A critique. *American Journal of Mental Deficiency*, 1975, **79,** 483–492.

Zinchenko, V. P., Van Chzhi-tsin, & Tarakanov, V. V. The formation and development of perceptual activity. *Soviet Psychology and Psychiatry*, 1963, **2,** 3–12.

Chapter 8

SIZE, DISTANCE, AND DEPTH PERCEPTION*

WALTER C. GOGEL

I. INTRODUCTION

A distribution of energy in the world outside the observer, the distal stimulus, cannot be consequential for perception until it becomes a proximal stimulus by activating sensory receptors such as are found in the retina of the eye. Clearly the proximal, not the distal, stimulus is the effective stimulus for perception. A basic problem in the history of research concerning the visual perception of the three-dimensional world is whether factors in addition to the proximal stimulus must be postulated to explain the range of perceptions that occur. A need to postulate factors in addition to the proximal stimulus would indicate that the proximal stimulus, to some degree, is ambiguous in specifying a perception. An inability

* The preparation of this chapter was supported by PHS Research Grant No. MH-15651 from the National Institute of Mental Health.

ISBN 0-12-161909-5

of the proximal stimulus to account for the perception completely will be called *stimulus ambiguity*. The purpose of this chapter is to indicate some of the conditions under which stimulus ambiguity is present and to consider some of the possible explanations of the perceptions that occur under these conditions.

The relation between distal and proximal stimuli is often complex, since a three-dimensional distal world must be registered on a two-dimensional retina. As will be discussed, in this transfer of three-dimensional information to a two-dimensional surface, the depth or distance dimension is coded on the retina, rather than represented on the retina directly. Other aspects of the stimulus, however, can have characteristics that are similar in the proximal and distal world. For example, size and shape are characteristics both of objects in the distal world and of stimuli on the retina. Indeed, if two-dimensional objects of different physical sizes or shapes are all in a frontoparallel plane at a constant distance from the observer, the sizes and shapes produced by the objects on the retina will be proportional to the distal sizes and shapes. It is usually found under these circumstances that the perceptions of size and shape are proportional to the sizes and shapes of the distal (and the proximal) stimulus. From these perceptions obtained with frontoparallel presentations at a constant distance, it is tempting to conclude that the size and shape of the retinal image (the core stimuli) are the prime stimuli for the perception of size and shape, and that failures of core stimuli to determine the perceptions that occur when the objects are physically at different distances or at different orientations in depth (slant) are instances of stimulus ambiguity. But, in many of these latter instances, if cues of distance or slant are available, such a conclusion would be premature. The reason for this is that the definition of the proximal stimulus must include coded proximal information as to distance or slant, whenever these are present, as well as including the core stimuli of size or shape on the eye.

The need to include proximal correlates of distance in defining the proximal stimulus for perceived size or perceived shape, whenever these are present, can be discussed in relation to Figs. 1(d) and 1(e). Figure 1(d) illustrates the situation in which two rectangles of the same shape but different physical sizes, and at different distances from the observer, produce the same size on the retina. Each rectangle is physically oriented to be in a frontoparallel plane of the observer. The size and shape of the proximal stimulus is defined by the visual angles subtended by these objects or the parts of these objects at the nodal point (N) of the eye. Figure 1(e) illustrates a situation like that of Fig. 1(d) except that the nearer of the two objects is a tilted trapezoid that produces the same shape on the retina as a rectangle in the frontoparallel plane. Thus, a

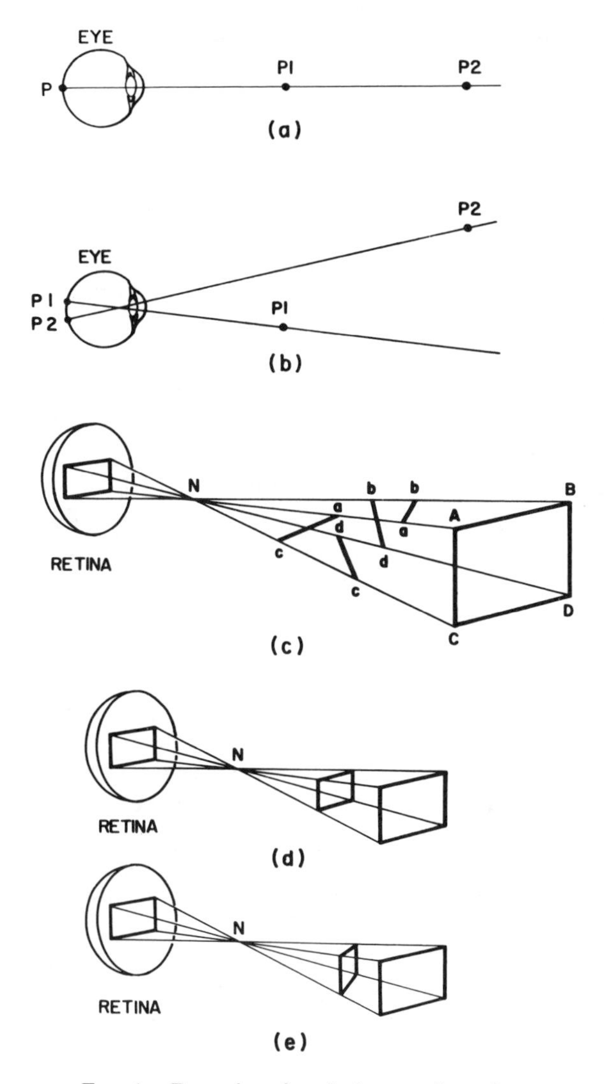

FIG. 1. Examples of equivalent configurations.

particular rectangle on the eye can result from a variety of objects of different sizes and shapes at different distances and orientations. If cues to the distance and tilt of the distal object are available in the proximal stimulus, the perceived size and shape of the object often will resemble the particular distal object producing the core stimulus, even though the core stimulus by itself is ambiguous with respect to the size and shape of the distal object. The presence of distance or slant information in the proximal stimulus reduces the stimulus ambiguity that would be present if the core constituted the entire proximal stimulus.

II. INSTANCES OF STIMULUS AMBIGUITY

A. Equivalent Configurations

All of the different distal stimuli that could produce the particular proximal stimulus, including the proximal information as to distance or slant (if available), are called equivalent configurations. A particular proximal stimulus will define a particular set or class of equivalent configurations. As the distance information in the proximal stimulus is reduced, the range of configurations included in the class of equivalent configurations increases. It is likely, however, even in the case in which the reduction of distance information is complete, that the perception will be of an object of some size and shape located at some distance from the observer. It is also likely that this perception will resemble one distal stimulus from the class of distal stimuli that could have produced the proximal stimulus. In other words, as cues of distance are increasingly reduced, the class of distal configurations that could have produced the proximal stimulus is increased. But, despite this, the perception, rather than becoming increasingly indeterminant, is likely to resemble only one of the possible equivalent configurations. The problem is to identify the factors that determine the particular perception that occurs or, more generally, to identify the factors that result in the class of perceptions being a subset of the class of equivalent configurations. If a particular perception occurs despite the fact that the proximal stimulus specifies a class of equivalent configurations, rather than specifying only a single configuration, it is clear that factors other than the proximal stimulus must be invoked to explain this perception. Equivalent configurations provide likely instances of stimulus ambiguity.

Two examples of equivalent configurations were discussed with the aid of Figs. 1(d) and 1(e). The remaining diagrams of Fig. 1 provide other examples. Figure 1(a) shows that a point of light (*P*) on the retina could have been produced by a point source located anywhere along the constant visual direction indicated by the line extending from *P* on the retina through the pupil of the eye. In the absence of proximal information regarding distance, all point sources (e.g., P_1 or P_2) along this constant visual direction are equivalent configurations. The experimental question is, Which distal stimulus of the set of possible distal stimuli will be perceived under these reduced conditions?

Figure 1(b) shows an instance of equivalent configurations slightly more complex than that illustrated in Fig. 1(a). Two point sources, P_1 and P_2, located along different visual directions produce two points of light on the retina. In Fig. 1(b) each of the points on the retina could have been produced by a distal source located anywhere along a constant visual direction. Thus, in the absence of cues of distance, the retinal stimulus is indeterminant with respect to the distance of each point source from the observer, the depth between the sources, and the distal lateral separation of the sources. The problem is to explain the spatial perception that occurs under these circumstances. Which of the many possible distal configurations is actually perceived, and why is that particular configuration perceived, rather than some other configuration from the class of equivalent configurations as defined by the proximal stimulus?

Figures 1(c), 1(d), and 1(e) indicate something of the range of equivalent configurations that, in the absence of distance cues, could produce a rectangular stimulus on the retina. Figure 1(c) illustrates that the distal stimulus could be either a rectangle (*ABCD*) or a series of four lines labeled *ab*, *bd*, *cd*, and *ac*. Obviously, any other series of four lines distributed in distance could have produced the same proximal stimulus if these lines were arranged to extend between the lines of sight (visual directions) represented by the thin lines of the drawing.

Figure 2 combines several aspects of the previous figures and illustrates a still more complex example of equivalent configurations. The portion of the figure labeled *ABCD* represents a top-view drawing of an Ames distorted room (Ames, 1961), with *ab* and *cd* representing windows on the back wall of the room. Two objects (*ef* and *gh*) of equal physical size are presented near the distance of each of the windows. Sizes and shapes on the retina, as in the previous drawings, are represented by the visual angles subtended by the stimuli at the nodal point *N* of the eye. A distal stimulus equivalent to the distorted room with the windows and objects is the rectangular room labeled *A'B'C'D'*, with the objects labeled *e'f'* and *g'h'*. Again, it is likely that only one of the large number of rooms, windows, and objects of different size and shape (all producing the same distribution of sizes and shapes on the eye) will be perceived by the particular observer at a particular moment, even though no distance cues are available to mediate between the alternative possibilities.

Figure 3 represents another example of equivalent configurations in which the class of equivalent configurations of Fig. 1(a) is enlarged to include concomitant motion. In the situation represented by Fig. 3, the observer moves his head laterally from Position 1 to Position 2 while gazing at a stationary point of light located physically at *b*. A distal stimulus producing a stimulus equivalent to that of *b* on the eye, as the

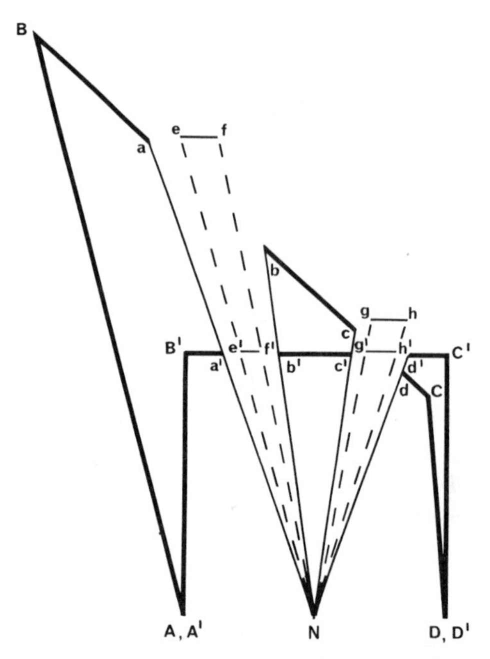

FIG. 2. A distorted room (ABCD) and a rectangular room (A′B′C′D′) that produce the same proximal stimulus. [From W. C. Gogel & D. H. Mershon. The perception of size in a distorted room. *Perception & Psychophysics*, 1968, *4*, 26–28.]

head is moved, is a point of light located at a_1 that physically moves to a_2 as the head moves from Position 1 to Position 2. Since the motion of point a is correlated with the head motion, it is called concomitant motion. The concomitant motion of a is in the same direction as that of the head. Point c represents a point of light that moves from c_1 to c_2 as the head is moved

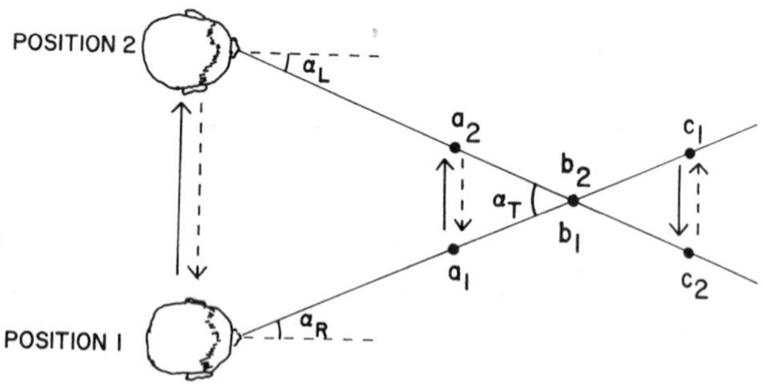

FIG. 3. Several instances of lateral motion of a stimulus point concomitant with a lateral motion of the head.

from Position 1 to Position 2. The concomitant motion of point c is in a direction opposite to the direction of motion of the head. The stationary point b, with no concomitant motion ($b_1 = b_2$), and the two concomitant motions of points a and c are three members of a class of equivalent configurations that would produce the same amounts of eye turning (if the gaze were fixed on a point) or the same motion of the point on the retina (if the direction of the gaze were always straight ahead). It would be expected that some configuration would be perceived in a case in which there were no distance cues to specify a particular distal stimulus. The problem is to determine the perception that occurs under this condition and to explain why this perception, rather than another of the possible perceptions consistent with the proximal stimulus, is obtained.

Other clear instances of stimulus ambiguity are found with reversible figures such as the Necker cube. Reversible figures are figures that can alternate perceptually between several possibilities (equivalent configurations) even though the proximal stimulus is invariant (see Hochberg, 1974a). The stimulus ambiguity found in reversible figures is particularly impressive, since, unlike the previous instances of equivalent configurations, the perception can change spontaneously from similarity with one equivalent configuration to similarity with another.

B. Cues of Distance

Cues of distance can be divided into egocentric and exocentric cues. Egocentric cues of distance are cues that determine the perceived distance of a point or an object from the observer. Exocentric cues of distance are cues that indicate the perceived depth or distance between points or objects. As will be discussed, egocentric cues of distance, theoretically at least, can remove stimulus ambiguity. Exocentric cues of distance, however, although capable of reducing the range of equivalent configurations consistent with the proximal stimulus, specify a class of equivalent configurations, rather than a unique distal stimulus.

A geometrical description of the major quantitative cues of egocentric and exocentric distance is shown in Fig. 4. The three diagrams in Fig. 4, although representing different cues of distance, are geometrically the same. In all three diagrams α_e and α_f are the angles subtended by a horizontal extent K at points e and f. The physical distance of e and f from K is D_e and D_f, and the distance (depth) between e and f is labeled d_{ef}. If the simplifying assumption is made, that $\tan \alpha = \alpha$ in radians and that D is large relative to d, it follows that

$$\alpha = K/D, \tag{1}$$

and

$$\gamma = Kd/D^2, \tag{2}$$

where $\gamma = \alpha_e - \alpha_f$, and α and γ are in radians.

Figure 4(a) represents the situation in which points e and f are at D_e and D_f and are viewed binocularly with K_1 representing the horizontal distance between the nodal points of the left and right eyes. In this diagram α_e or α_f represents the convergence of the eyes if the observer fixates a point, and 1/D represents the accommodation of the eyes (with D expressed in meters) if the observer accommodates to the distance of the point. Thus, α can represent the oculomotor cues of egocentric distance. Also, in this case, it can be shown that γ is proportional to the difference in the horizontal separation of the two points in the two eyes (i.e., $\alpha_e - \alpha_f$ is proportional to the difference between the extent of *ef* on the left and right eye—in the diagram these distances must be summed since they are opposite in direction). Thus, γ is the exocentric cue of binocular disparity. Figure 4(b) represents the situation in which points e and f are viewed from two lateral positions of the head with K_2 representing the physical distance through which the head is moved from one viewing position to the other. In this case α is the egocentric distance cue of absolute motion

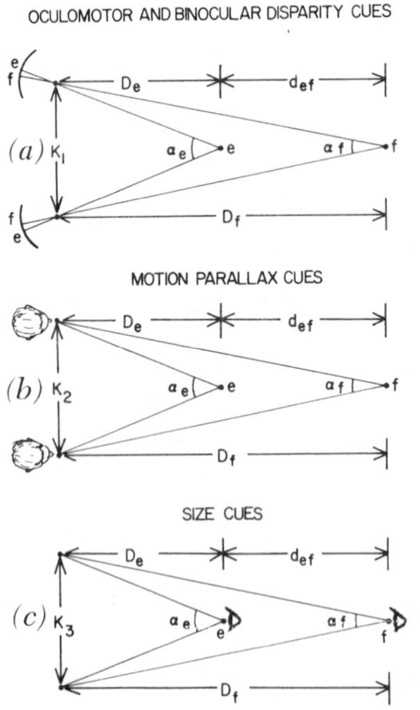

FIG. 4. Diagrams illustrating different kinds of egocentric and exocentric cues of distance. Part (a) is an illustration of oculomotor and binocular disparity cues. Part (b) illustrates motion parallax cues. Part (c) illustrates size cues.

parallax, which is defined either by the change in the viewing direction to a point or, if the gaze remains unchanged as the head is moved, by the motion of the point on the retina. Also, in this diagram, γ represents the exocentric cue of relative motion parallax resulting from the relative motion of the two points on the retina as the head is moved. Figure 4(c) represents the case in which an object of height K_3 is viewed by the observer from *e* or *f*. In this case, α is proportional to the size of K_3 on the retina. If the object is a familiar object, K_3 is its familiar size (familiar height) and K_3 and α together specify the familiar size cue to egocentric distance. Also, in this diagram, γ represents the relative size cue to exocentric distance (i.e., the change in the size on the eye as the observer moves away from the stationary object or as the object is moved away from a stationary observer).

According to Fig. 4, in Eq. (1), α represents the egocentric distance cues of accommodation, convergence, absolute motion parallax, and familiar size; in Eq. (2), γ represents the exocentric distance cues of binocular disparity, relative motion parallax, and relative size. The egocentric and exocentric cues of distance expressed by Eqs. (1) and (2) differ in a fundamental way. In Eq. (1), each value of α uniquely specifies a particular egocentric distance D. But, in Eq. 2, the relation between γ and d is different for different values of D. In other words, a given γ, rather than specifying a particular exocentric distance, defines a class of equivalent configurations of exocentric distance (d) with the magnitude of d increasing as the distance of the points defining d increases from the observer. This, of course, applies to each of the exocentric cues (binocular disparity, relative motion parallax, and relative size). A given value of any of these, defined proximally by γ, specifies a class of equivalent configurations, not a particular depth interval. The experimental problem is to determine, given a particular γ on the eye, which of the equivalent configurations will actually be perceived. As in the situations illustrated by Figs. 1–3, the theoretical problem of predicting perceptions in situations involving equivalent configurations is eliminated if information for the perception of egocentric distance (D') is available in the proximal stimulus. The resulting D' will specify a unique value of perceived exocentric distance (d') for a given value of γ. Consistent with this conclusion, it has been found that the perceived exocentric depth associated with a particular binocular disparity is modified by the perception of (or cues to) egocentric distance, (see Foley, in press; Gogel, 1972). Although the other cues of exocentric distance (relative motion parallax and relative size) have not been investigated in terms of their relation to perceived egocentric distance, it seems reasonable to suppose that the results obtained from binocular disparity will apply to these cues also. It should not be assumed, however, that if the egocentric perception of the distance to a configuration is veridical the perceived depth within the configuration necessarily will be veridical (Foley, 1968; Gogel, 1960). Nevertheless it is clear that the calibration of exocentric cues by perceived egocentric distance not only permits a particular d' to occur despite the distal uncertainty implicit in γ, but also avoids the magnitude of errors that would occur if the perception were directly proportional to γ.

It might seem redundant to distinguish between egocentric and exocentric cues of distance. Certainly if the egocentric distances to a group of objects are specified, the perceived distances between the objects are also specified by a process of subtraction. But it is easy to demonstrate that both kinds of cues are necessary to describe the perceptions of distance that occur in most situations. In general, the cues described by γ have a much lower threshold than those described by α. For example, the threshold of binocular disparity as a cue of distance is much lower than that of convergence (Gogel, 1972). The greater precision of exocentric as compared with egocentric cues also applies to motion parallax and size cues of distance and is important in extending the perception of egocentric distance to distant portions of the visual field. Although the egocentric cues of convergence and accommodation are effective only within a distance of several meters from the observer, these egocentric cues can, by determining the apparent distance of near objects, calibrate the exocentric cues between a near object and other objects more distant from the observer. It follows that the perception of the egocentric distance of a far object is the sum of the perceived egocentric distance to the near object and the calibrated perception of depth between the near and far object. This process of the addition of the egocentric and the more sensitive exocentric cues to produce a summed perception of egocentric distance permits the perception of egocentric distance to extend to distances far beyond those that would be possible from only egocentric cues.

Although egocentric cues can uniquely specify one configuration in the class of equivalent configurations defined by an exocentric cue, it should not be concluded that stimulus ambiguity will not occur in situations in which both types of cues are available. As shown by Eqs. (1) and (2), the magnitude of proximal cues of distance resulting from a constant d and K is inversely related to physical distance. Since both egocentric and exocentric distance cues are a negatively accelerated function of physical distance (D), the far portions of the visual field are less adequately represented in terms of proximal distance cues than the near portions. This inverse relation between the magnitude of distance cues and physical distance also means that at far distances many of the cues of distance will fall below threshold, resulting in a reduction of the variety of distance cues available in far portions of the visual field. In terms of both the number and magnitude of distance cues, distant portions of the visual field are severely reduced in proximal information as compared with nearer portions. It follows that stimulus ambiguity is more likely to occur in distant portions of the visual field.

In the discussion so far, stimulus ambiguity can be said to occur if the class of perceptions resulting from a proximal stimulus is more restricted than the class of equivalent configurations, with the class of equivalent configurations defined in terms of the total information available in the proximal stimulus. Under these conditions, some factor or factors other than the proximal stimulus must be postulated to explain the particular perceptions that occur. Conversely, if such additional factors are available to contribute to the perception, they should become increasingly evident as the number and effectiveness of distance information is reduced. Thus, even under circumstances in which the proximal stimulus can uniquely specify a distal stimulus geometrically, the perception might deviate from that expected from this geometry because of the contribution of these additional factors. The failure of the proximal stimulus to completely determine the perception under these conditions would be attributed to a stimulus ambiguity resulting from a reduction in cue effectiveness, rather than to the stimulus ambiguity resulting from a lack of cue specificity as found in the concept of equivalent configurations.

Stimulus ambiguity associated with either a reduction in cue effectiveness or a reduction in cue specificity is expected to occur more frequently in distant than in near portions of the visual field. Both types of stimulus ambiguity, however, can be present to different degrees for different objects in any portion of the visual field. The stimulus ambiguity that is limited to a restricted portion of the visual field or that occurs for only a particular object can be called a local stimulus ambiguity. An example of an extreme case of local stimulus ambiguity is an afterimage resulting from monocular observation of an unfamiliar object. Such an afterimage contains no proximal cues of distance. It will not be ambiguous in its *apparent* spatial location, however, but will appear to be located on the surface or wall against which it is viewed. Also, an object suspended above the floor of a visual alley is likely to have fewer cues for its perceived localization than an object resting on the alley floor. If factors in addition to the proximal stimulus contribute to the distance perception, they are expected to have a greater effect on the perceived localization of some objects or parts of the visual field as compared with others. This would occur because cue effectiveness (as well as cue specificity) often varies within the visual field.

C. Vector Analysis of Motion

The cases of stimulus ambiguity discussed so far have concerned the kind of proximal–distal uncertainty involved in equivalent configurations or in a lack of cue effectiveness. There are phenomena, however, which demonstrate stimulus ambiguity in the absence of either of these types of uncertainty. One of these is represented in Fig. 5. Figure 5 illustrates a case in which two points of light are moving physically in a frontal plane of the observer. As shown by the diagram on the left, point 1 moves right and left as point 2 moves down and up. Like the physical motion, the motion on the eye is horizontal for point 1 and vertical for point 2. The points are viewed binocularly and are perceived to be moving in a frontoparallel plane. In this case, there is no lack of specificity between the proximal and distal motion, and all the cues of motion are far above threshold. The middle diagram indicates the perception of the motion of the points. According to Johansson (1958) the two points will appear to move toward and away from each other and to move as a group diagonally between the upper left and lower right. The diagram on the right shows the analysis of the proximal motion provided by the visual system in achieving this perception of motion. The vectors P_1 and P_2 are the motion vectors corresponding to the physical (or retinal) motion shown in the left diagram. The motion specified by the vectors P_1 and P_2 can also be represented by many other combinations of vectors. For example, the motion specified by vector P_1 can be specified by the pair of orthogonal vectors R_1 and C_1 and the motion specified by vector P_2 can be specified by the pair of orthogonal vectors R_2 and C_2. The sum of vectors R_1 and C_1 is physically equal to P_1, and the sum of vectors R_2 and C_2 is physically equal to P_2. The vectors R_1 and R_2 define the physical (or retinal) motion of the points relative to each other and are called *relative motion vectors*. The vectors C_1 and C_2 define the motions common to the points and are called *common motion vectors*. The vectors P_1 and P_2 define the motions of the points independently of each other and are called *absolute motion*

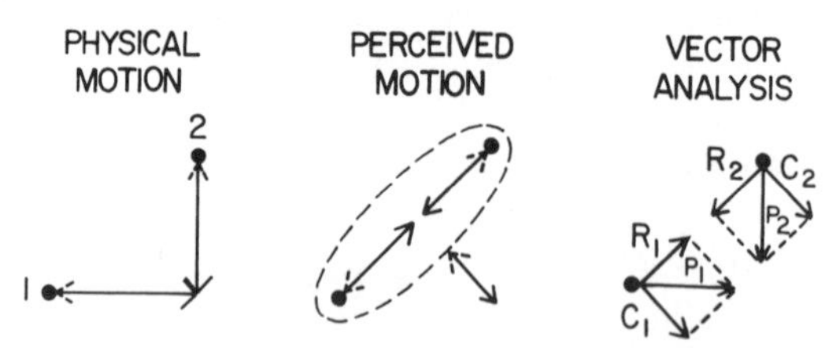

FIG. 5. The perceived motion associated with the physical motion of two points of light as determined by the vector analysis of the motion in terms of relative (R) and common (C) motion.

vectors. As is indicated by the middle diagram of Fig. 5, the perceived motion often resembles the relative and common motions, although as shown by Gogel (1974), the absolute motions will also influence the perception. According to the vector analysis, a variety of perceptions are possible, all of which are equally correct in terms of the distal motion, but all of which are different. Perhaps these should be called equivalent perceptual configurations, since the alternative perceptions are equally consistent with the proximal stimulus. A case of stimulus ambiguity is involved, however, since the set of equivalent vectors used by the visual system is not specified by the proximal stimulus. The lack of perceptual specificity in the proximal motion might be called a failure of vector specificity. It should be noted that since this type of motion analysis by the visual system is necessary whenever two or more moving points vary in separation, this kind of stimulus ambiguity occurs very frequently.

D. Cue Conflicts

If three points in an apparent frontoparallel plane are varying in separation, the relative motion cues between one point and each of the other points will be in conflict. Rather than discarding the analysis in terms of relative motion vectors under these conditions, the perceptual system gives each relative motion vector a weight that is inversely related to the separation of the points. This process is indicated in Fig. 6, which illustrates a situation in which three points of light are moving repetitively (right and left or up and down) in the directions shown by the arrows. The dashed-line square permits easy comparison between the several situations, and does not represent any stimulus present during the experiment. The physical and perceived motions in three situations are shown in the upper and lower diagrams of Fig. 6, respectively. Only the simultaneously presented points of light were visible with the remainder of the field of view dark. Although in this experiment (Gogel, 1974) a measure of apparent motion was obtained only for point 2, the expected apparent motion, consistent with the conclusions from the study, are shown for all three points.

As point 1 moves down and up, point 3 moves up and down (opposite in phase to point 1), and point 2 moves left and right. It will be noted that with these physical motions the relative motion vector of point 2 with respect to point 1 is at right angles to that of point 2 with respect to point 3. These different relative motions constitute a cue conflict. According to the relative motion vector of point 2 with respect to point 1, point 2 should appear to move in the direction from lower right to upper left whereas, from the relative motion vector of point 2 with respect to point 3, point 2

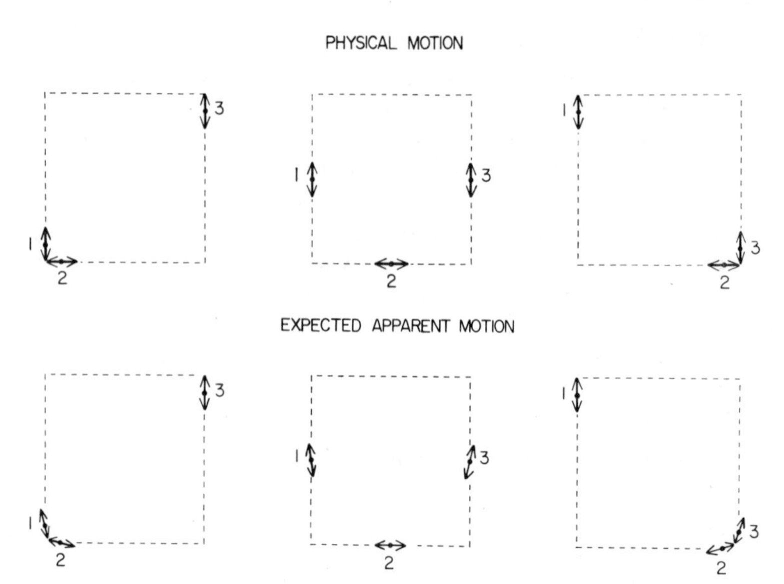

FIG. 6. Physical and expected apparent motion of three points of light from the response to relative motion modified by an adjacency effect.

should appear to move in the direction from lower left to upper right. It was found, as indicated in the diagrams, that the apparent path of motion of point 2 was determined more by the relative motion with respect to the point adjacent to, rather than displaced from, point 2. Thus, the apparent motion of point 2 was between lower right and upper left when it was near point 1, was horizontal when it was midway between points 1 and 3, and was between lower left and upper right when it was near point 3. Of importance for the concept of stimulus ambiguity, the weights perceptually given by the observer to the conflicting relative vectors differed according to a rule of visual analysis not specified by the proximal stimulus. This rule is that the effectiveness of a particular relative motion cue in a conflict of relative motion cues is inversely related to the apparent frontoparallel separation of the points or objects between which the relative motion occurs. As has been indicated, such conflicts of relative motion cues occur whenever three or more points in the same apparent frontoparallel plane move in different directions. Also, it is likely that cue conflicts occur frequently between absolute and relative cues of motion (Gogel, 1974). The perceptual resolution of conflicts involving cues of motion and of cue conflicts in general (Gogel, 1977) are ubiquitous examples of failures of the proximal stimulus to specify perceptions completely, and, therefore, provide additional examples of stimulus ambiguity.

E. Summary

It is clear that the stimulus for perception is the proximal stimulus defined in its totality (including cues of depth or distance). Situations in which perceptions occur that cannot be specified completely by the proximal stimulus are instances of stimulus ambiguity. Two kinds of situations in which stimulus ambiguity is likely to occur have been identified. One of these is the situation in which cues of distance (particularly egocentric cues of distance) are reduced. This reduction can be local or general. It can result in a lack of cue specificity, as defined by the concept of equivalent configurations or in a lack of cue effectiveness, as is often found in distant portions of the visual field. It is asserted that under these conditions the perception is often more specific or definite than would be expected from the proximal stimulus. The second kind of situation in which stimulus ambiguity is evident is one in which the visual system must provide rules for the vector analysis of proximal motion or for the resolution of informational conflicts in the proximal stimulus. These decision rules are necessary despite specific and effective cues in the proximal stimulus.

III. THEORIES OF THE RESOLUTION OF STIMULUS AMBIGUITY

A. The Core–Context Hypothesis

Theories of perception differ in their explanation of the perceptions that occur under conditions in which stimulus ambiguity is present. Of the different situations in which stimulus ambiguity is likely to occur, that of equivalent configurations resulting from some amount of cue reduction has received the most theoretical attention. One theory that offers an explanation for the perceptions occurring as a function of cue reduction is the core–context hypothesis, a product of an early form of empiricism (Boring, 1942, 1946). The core–context hypothesis differentiates between core stimuli and context stimuli. The core stimuli for perceived size and shape are size and shape on the retina. The context stimuli for these perceptions are the distance cues that specify the perceived distance of the object or its perceived tilt in depth. In reduced conditions in which only the core stimulus is present, the perception will resemble the core stimulus. For example, a rectangle slanted in distance will appear trapezoidal in shape under conditions in which no cues to perceived slant are available. In this example, the perception resembles the distal stimulus in the class of equivalent configurations that is shaped like the core stimulus. On the other hand, if the depth information is only somewhat reduced, the perception will be intermediate between that indicated by the core and by the distal stimulus. With good cues of slant, the class of equivalent configurations is increasingly restricted and the perception increasingly resembles the slanted rectangle.

A clear example of the kind of experiment that is interpreted as supporting the core–context hypothesis is the study by Holway and Boring (1941) in which the perceived sizes of objects subtending a constant visual angle at different distances were measured as a function of the number of cues of distance available. When a variety of cues of distance were available, the perceived size of the object closely followed the physical (distal) size (law of size constancy). As the distance cues were increasingly reduced, the perception of size increasingly followed the constant retinal size (law of visual angle). The core–context hypothesis also seems to be supported by studies of size and shape constancy which show that the amount of constancy is less if the observer is instructed to respond to visual angle (analytic instructions) rather than to the distal stimulus (objective instructions). On the other hand, even in the total absence of distance cues, the perception is usually that of an object larger than the proximal stimulus (which often is only several millimeters in size), and the object is perceived to be at some distance in front of the observer (rather than on the retina). The core–context hypothesis denies that the relation between proximal stimulus and perception under cue reduction need be ambiguous. According to this hypothesis, the perception can be determined by whatever proximal stimulus is present. If the proximal stimulus includes sufficient cues of distance, the perception will resemble the particular distal stimulus responsible for the proximal stimulus. If the proximal stimulus is limited to the core stimulus, the perception will resemble the core stimulus.

B. The Theory of Direct Perception

A point of view that would resolve the theoretical problem represented by stimulus ambiguity by denying that it is a phenomenon of importance is Gibson's theory of direct perception (Gibson, 1950, 1959, 1966). The position of this theory with regard to the problem of stimulus ambiguity is expressed by Gibson's psychophysical hypothesis that states: "*for every aspect or property of the phenomenal world of an individual in contact with his environment, however subtle, there is a variable of the energy flux at his receptors, however complex, with which the phenomenal property would correspond if a psychophysical experiment could be performed* [Gibson, 1959, p. 465]." This hypothesis denies that for an individual in contact with his environment there is any ambiguity between the proximal stimulus and perception. Contact with the environment implies that the observation is occurring without optical restrictions and is in an environment that is rich in information, rather than in a more reduced environment or with more restricted conditions of observation conducive to the occurrence of equivalent configurations. The core–context hypothesis would deny that there need be any discrepancy between proximal stimulation and perception under either reduced- or full-cue conditions of observation. The theory of direct perception would deny the generality of perceptual processes deduced from experiments performed under reduced conditions. Of course, some restriction in the number or presentation time of stimuli is required if stimulus factors are to be isolated experimentally. The spatial information usually encountered, however, is not only richly varied but also is redundant. Perhaps information can be restricted to cues of a particular kind without significantly reducing the ability of the distribution of proximal stimuli to specify a veridical perception. If this can be assumed, perceptual processes can be studied meaningfully in situations in which cues of only one kind, such as gradients of size or gradients of binocular disparity, extend over considerable portions of the visual field or in which continuous spatial transformations of stimuli are presented for a limited but appreciable interval of time. It is characteristic of the research generated by the theory of direct perception that the spatial and temporal units of stimulation are more extensive than those generally used by researchers with other theoretical points of view (Hochberg, 1974a). But under such circumstances errors in spatial responses can occur, particularly with respect to distant portions of the visual field. Since the theory of direct perception is essentially a theory of correct perception, such errors require explanation. In discussing the causes of deficient perception Gibson (1966, pp. 306–307) suggests that sensations can sometimes obtrude on perceptions, with this intrusion interfering with the observer's ability to attend to veridical information. Instead of the perception being determined by a compromise between the core and context stimuli, a lack of perfect constancy is explained as the result of a compromise between attending to sensory and perceptual information.

The stimulus ambiguity that is implied in a lack of cue specificity (equivalent configurations) or in a lack of cue effectiveness poses two problems for the theory of direct perception. The first problem is to predict the perceptions that will occur under the reduction conditions in which the concept of equivalent configurations is apt to apply. It seems consistent with the theory of direct perception to expect that a reduction in distance information would increase the variability of the perceptions but would not result in systematic errors. As will be discussed, there is, on the contrary, clear evidence to support the conclusion that systematic errors in perception occur with cue reduction.

The second problem that the stimulus ambiguity resulting from a lack of cue specificity or cue effectiveness raises for the theory of direct perception is that of defining the conditions under which all stimulus ambiguity is removed—if indeed stimulus specificity is ever complete. This problem has received attention in an article by Eriksson (1973) and a reply by Rosinski (1974). The mathematical analysis of the proximal information available in visually rich environments indicates that, with a few assumptions, the proximal information, if effective, is sufficient to restrict drastically the range of possible perceptions. It follows that, under many conditions, veridical perception could occur to within a scale factor or within a ratio of two perceptual components. For example, as indicated by a geometrical analysis by Purdy (1960), perceived distance along a surface can be specified within the scale factor of height above the ground if the observer assumes that the texture is at least stochastically regular and is distributed along a flat surface. As another example, the response to a flow pattern on the eye produced either by a moving observer or by a surface moving in the environment can be veridical within a scale factor if it is assumed that the motion is regular (a constant velocity). It seems, in general, that there are a number of simplifying assumptions of regularity required with respect to both the environment and behavior of the viewer if the problem of stimulus ambiguity is to be avoided, even in full-cue conditions of observation. The simplifying assumptions required for the elimination of stimulus ambiguity seem to be best suited to environments in which surfaces are flat and sizes are orderly, and to observing organisms that move in a smooth and regular manner. Observer locomotion, however, often is highly irregular, involving frequent changes in height, orientation, and velocity, and it occurs most often over irregular surfaces covered by texture, with at least local variability. It does not seem, however, that this irregularity increases the difficulty of the organism in perceptually extracting information from the environment. Also, it should be noted from the discussion with respect to Fig. 4 that the perception of egocentric distance is ubiquitously involved in the perception of exocentric distance. To assume that a perceived exocentric scale is available independently of the specification of a metric by egocentric cues is an oversimplification of the complexity of the interaction between these two types of cues. A metric perception of extent from exocentric cues is not possible in the absence of an egocentric perception. On the other hand, it is possible that exocentric distance information from several different cues could provide information that would add a metric to the exocentric information. Consider, for example, the case in which γ and K are available to the observer to be substituted into Eq. (2) for the binocular disparity between two points and also for relative motion parallax with respect to the same two points. With this information, and by the process of simultaneous solutions, the egocentric distance of the configuration from the observer theoretically can be determined (see also Purdy, 1960). Whether the perceptual process is capable of using information in this manner and whether this capability, if it occurs, should be considered as involving a direct response to the distance information is a different problem.

The stimulus ambiguity indicated by the vector analysis of motion and by the occurrence of cue conflicts also poses problems for the theory of direct perception. The vector analysis of motion and the resolution of cue conflicts by the visual system can be considered to be useful in perceptual organization. For example, as indicated in the center drawing of Fig. 5, the vector analysis of the visual system results in the two moving points being perceived as a group, with the details of this perceptual organization modified by the resolution of cue conflicts in the manner illustrated in Fig. 6. More impressive examples of this perceptual grouping are shown by Johansson (1971) in experiments in which a clear impression of form occurs from lights attached to the joints of otherwise invisible persons in motion. To describe this perceptual organization as a direct response to structure in the stimulus ignores the particular visual rules necessary in order for these perceptions to occur.

C. The Cognitive Theory

A third theoretical approach to the problem of stimulus ambiguity is the cognitive approach expressed by Helmholtz (Helmholtz, 1924; also see Hochberg, 1971). This has been amplified in the transactional point of view (Kilpatrick, 1961) and represented more recently by Gregory (1974) and Rock (1970). Two parts of this point of view might be characterized as the inductive and deductive aspects. The inductive aspect is expressed by Helmholtz's (1924) rule:

> The general rule determining the ideas of vision that are formed whenever an impression is made on the eye, with or without the aid of optical instruments, is that *such objects are always imagined as being present in the field of vision as would have to be there in order to produce the same impression on the nervous mechanism, the eyes being used under ordinary normal conditions* [p. 2].

According to this rule, perception will resemble that distal stimulus which in the past most frequently has produced the proximal stimulus. For example, since proximal stimuli usually are produced by distal stimuli, the perception will always be of an object outside the observer. This inductive aspect of the theory can be applied to resolve the stimulus ambiguity indicated in some of the previous figures. In Fig. 1(c) a rectangle, rather than a group of lines distributed in depth, is perceived, since a rectangular proximal stimulus is more frequently caused by the former distal configuration. In Fig. 2 a rectangular, rather than a trapezoidal, room is perceived, since the latter is an unlikely distal event. In Fig. 3 a point of light at *b*,rather than at *a* or *c*, is most likely to be perceived, since situations do not often occur in which an object will move concomitantly with the head. The application of Helmholtz's rule to the situations represented by Figs. 1(a), 1(b), 1(d), 1(e) and Figs. 5 and 6 is less clear. Figures 1(c) and 2 are in principle similar or identical to two of the Ames demonstrations (the chair illusion and the distorted room illusion) that have been designed to illustrate the transactional approach to perception (Kilpatrick, 1961). As will be discussed, it is possible to give interpretations to the perceptions occurring in these situations different from those indicated by the transactional approach.

The deductive aspect of the cognitive approach to perceptions is concerned with the logical structure of perceptions. This aspect also can be illustrated in terms of equivalent configurations. Under conditions of few or no cues to distance, the proximal stimulus defines a class of equivalent configurations rather than a single configuration and, presumably, the possible perceptions are limited to this same class. For example, in Fig. 1(d) the angular size or width (θ) of the rectangular shape on the eye specifies the size (S) to distance (D) ratio of the distal stimulus. If the perceptions are consistent with this proximal stimulus, the ratio of perceived size or width (S') to perceived egocentric distance (D') will be proportional to θ. In other words, if the observer is given a value of θ and either an S' or a D', he will perceive a D' or an S' consistent with that of θ. It is as though the perceptual system operates logically with the available information so that a particular θ (a given visual angle) will establish an invariant ratio of perceived size to perceived distance. This relation is called the *size–distance invariance hypothesis*,

$$S'/D' = C\ \theta, \tag{3}$$

where C is an observer constant. The size–distance invariance hypothesis expresses the range of hypotheses regarding S' or D' that logically are possible given the particular proximal stimulus and given either D' or S', respectively. More generally, any proximal stimulus that defines a class of equivalent configurations will also specify an invariance hypothesis and, therefore, a class of acceptable perceptual hypotheses. It was noted in the discussion with respect to Fig. 4 that a value of γ in Eq. (2) is a proximal stimulus which defines a class of equivalent configurations and presumably an invariance hypothesis involving d' and D'. Another invariance hypothesis is illustrated in Fig. 3. Suppose that the point of light is physically at b but appears at a. In this case, as the observer moves his head from Position 1 to Position 2 while gazing at the point of light, the point will appear to move with his head. The class of equivalent configurations (the invariance hypothesis) consistent with the change in visual direction as the head is moved from Position 1 to Position 2 is defined by Eq. (4) (Gogel & Tietz, 1974),

$$m' = A' - \alpha'_T D', \tag{4}$$

where m' is the perceived motion of the point associated with the sensed head motion A', α'_T is the sensed change in the direction of the point expressed in radians, and D' is the perceived distance of the point. As another example of an invariance hypothesis, in Fig. 1(e) the shape of the stimulus on the eye defines a relation between the perceived shape and perceived slant of the object. This is called the *slant–shape invariance hypothesis*. The perception of the size of each segment of the object determines the perceived shape of the object and, since a perceived slant can be specified as differences in the perceived distances of the parts of the object from the observer, the slant–shape invariance hypothesis is a special case of the size–distance invariance hypothesis. Theoretically, there are as many invariance hypotheses as there are different classes of equivalent configurations.

A description of the manner in which the perceptual system seems to use hypotheses is expressed by the minimum principle which states that the perception will be as simple as the proximal stimulus conditions permit (see Hochberg, 1957). This statement not only specifies the interrelation between perceptions expressed by many hypotheses of invariance, but it also suggests a criterion by which the observer selects from the equivalent configurations the particular configuration that the perception will resemble. The minimum principle asserts that the configuration perceived will resemble the simplest member of the class of equivalent configurations consistent with the proximal stimulus. For example, in Fig. 1(c) the perception of a rectangle in a frontoparallel plane, rather than of lines distributed in distance, is not only consistent with Helmholtz's rule, but is also a selection of the simpler perception in that less information is required to specify the former than the latter perception. The structuring of the perceptual world in terms of a criterion of simplicity, although not always as clear as in the example cited, can result in a visual world with objects that maintain their identity despite changes in the proximal stimuli. This, of course, is the world as it is ordinarily perceived.

The inductive aspect of the cognitive theory (Helmholtz's rule) obviously emphasizes the importance of past experience for perception, whereas the logical structure involved in the deductive aspect is not so obviously empiricistic. But, regardless of its origin, this deductive aspect is consistent with a variety of perceptions. Although it is not obvious why a constant visual angle should result in perceived size being directly related to perceived distance—the special case of Eq. (3) known as *Emmert's law*—there is, nevertheless, considerable evidence that this relation occurs (Epstein, Park, & Casey, 1963; Price, 1961). Similarly, although the exact relation between D' and d' for a constant binocular disparity is somewhat in doubt (see Foley, in press) the fact that in this case d' increases with D' is secure.

The interrelations between perceptions expressed, for example, by Eq. (3) or by any of the invariance hypotheses are illustrations of the lawfulness of perception. It does not necessarily follow, however, that such interrelations occur because perception is logical. The problem is whether these interrelations will remain even in instances where they produce perceptions that are illogical. There is evidence for this possibility. Instances of illogical perceptions are found in displays producing impossible objects (see Gregory, 1974; Hochberg, 1974b). Consider the Penrose triangle (Gregory, 1974, pp. 264–265), illustrated in Fig. 7. As is shown in Fig. 7(a), perspective cues indicated that the three sides of the triangle point in different directions in three dimensional space, so that sides *a* and *b* can never meet. Although a Penrose triangle is impossible physically, it is possible perceptually if the inconsistencies between the perspective orientation of the parts is ignored, as in Fig. 7(b). Those perceptual inconsistencies would be resolved, however, if instead of seeing the ends of sides *a* and *b* as being joined at the same distance, the observer would instead see the three dimensional extensions indicated in Fig. 7(a). The fact that Fig. 7(b) is a drawing on a two-dimensional surface cannot account for the observer perceiving the ends of *a* and *b* at the same distance, since drawings of other objects such as a Necker cube produce apparent depth. In the Penrose triangle a perception of an unreasonable object—Fig. 7(b)—occurs despite a logical alternative—Fig. 7(a). Another instance of the perception of unreasonable events is the perception of unusual sizes of familiar objects in the distorted room illustrated in Fig. 2. In this figure suppose that *ef* and *gh* are two familiar objects of the same physical size (for example, two faces), with each of these physically located near the distance of a window. Using monocular observation and

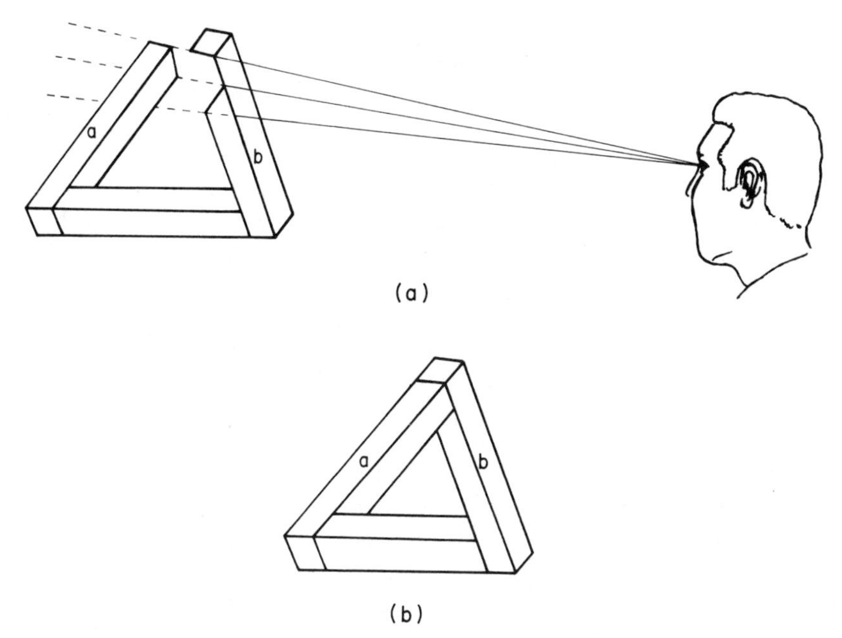

FIG. 7. Construction (a) and perception (b) associated with the Penrose impossible figure.

a restrictive aperture (to eliminate convergence and accommodation as cues to distance) the observer at *N* will *perceive* a rectangular room (A′B′C′D′) and two faces of different size (*e′f′* and *g′h′*), each perceptually located at the distance of its enclosing window. The perception of the room as rectangular is consistent with past experience, the perception of the faces as differing markedly in size is not. Logically, the resolution of this inconsistency would be to perceive a rectangular room with faces of the *same* size, but with the face in the left window more distant than that in the right window. Indeed, it might seem surprising that this perception does not occur, since there are no distance cues to indicate that either face is at the distance of its enclosing window and since the relative size cue between the two faces would facilitate the perception that the faces are at different distances. These examples of unlikely or impossible events are exceptions to the hypothesis that perceptions must be logically consistent. They present particular difficulties for this hypothesis since, as was indicated, there are options available to the perceptual system that are both reasonable and consistent with the proximal information. There are other examples also critical in their implications for the cognitive theory of perception. A physically three-dimensional outline cube is an ambiguous figure when viewed monocularly. It can be perceived correctly in its physical depth or perceived incorrectly with its far side closer than its near side. If a stem is attached to the cube so that the cube can be rotated tactually by the observer, a striking perceptual inconsistency will occur when the cube is perceived at its illusory orientation. In this case, the cube is perceived visually as rotating in a direction opposite to the felt direction in which the observer is actively rotating the stem (Gregory, 1970). The direction of apparent rotation of the cube is consistent with the apparent orientation in depth of the cube, given the change in retinal stimulation produced by the physical rotation. The perceptual inconsistency in this example is between the visual and tactile modalities. This inconsistency remains despite the alternative possibility of visually perceiving the correct orientation and, therefore, the correct rotation of the cube. A similar inconsistency in apparent rotation, in this case occurring entirely within vision, is illustrated by the display in which a tube, rigidly attached to the Ames' rotating trapezoidal window, appears during one part of the rotation to cut through the plane of the window (Ames, 1961). A necessary condition for this unlikely event is that the window (because of misleading perspective cues) will appear at an illusory orientation during part of its physical rotation. Logical consistency in the perception could be achieved in this case if the apparent orientation of the rod always followed the apparent orientation of the window. But this usually does not occur. It would seem, if logical consistency were an important criterion of the perceptual system as required by the cognitive theory, that inconsistent perceptions would be abandoned whenever a more consistent alternative was available.

In principle, the cognitive approach could explain the perceptions provided by the process of vector analysis and by the resolution of cue conflicts. Wallach (1968) has suggested that the observer might learn to prefer relative over absolute motion cues. If this occurred, the learning would have to be very detailed and effective to explain the rapid and complex perceptions that can be determined by the process of vector analysis. Still more subtle learning would be required to explain the function relating cue effectiveness and spatial separation.

There is some evidence, possibly in opposition to the cognitive theory of perception, that indicates that perceptual and cognitive laws differ. This is found in studies of size constancy in which the observer is instructed to physically equate the sizes of objects at different distances. Under such "objective" instructions, it is often found that the nearer object will be adjusted to be physically larger than the farther object (Carlson, 1960; Epstein, 1963). An explanation for this over-constancy of size is that although the observer perceives the sizes accurately, he expects that the same object will appear smaller if it is more distant from himself. Consistent with this expectation, and in an effort to respond veridically, the observer adjusts the near object to be perceptually larger than the far object. It seems that the over-constancy of size is the result of a cognitive correction based upon the expectation that near objects will *appear* larger than far objects of the same physical size. Conversely, the perception that objects appear larger or smaller than normal can be used cognitively to conclude that they are at a smaller or greater distance respectively than the distance at which they appear (Gogel, 1969a). The interrelation of the response to size and distance implied in the perceptual rule of Eq. (3) and in the above cognitive rule involve sometimes similar and sometimes different results but seemingly different processes (Gogel & Sturm, 1971). From this evidence it seems unlikely that perceptual rules have their origins in cognitive rules. If cognitive rules became perceptual rules through continued experience, a continuity of process between the two kinds of response would be expected. Metzger (1974) has pointed out another difference between the processes involved in conscious cognition and in perception. Cognition becomes more difficult as the information to be evaluated becomes more complex. On the other hand, perception becomes less equivocal as the situation being viewed becomes more complex. Nevertheless, although cognitive and perceptual processes can differ, this does not mean that there is no relation between perception and cognition. It is entirely possible that logic has its origins in perceptual consistencies rather than the converse (see Gibson, 1969). Just as Michotte (1946) would argue that the perception of mechanical causality is innate rather than determined by the laws of association, so it might be that logic at the cognitive level finds its precursors in the consistencies expressed, for example, in perceptual invariances.

D. Autochthonic Processes

There is another point of view that can be applied to the explanation of the perceptions that occur under conditions of stimulus ambiguity. This point of view as applied to the problem of equivalent configurations would suggest that perceptions increasingly approach certain limiting conditions as the cues to distance are increasingly reduced. These limiting conditions are called perceptual tendencies. One, called the *equidistance tendency*, is a tendency for objects to appear equidistant from the observer, with the strength of this tendency inversely related to directional separation (Gogel, 1965; Lodge & Wist, 1968). According to the equidistance tendency, two or more objects or parts of objects in the absence of exocentric cues of distance will usually appear to be about the same distance from the observer. The equidistance tendency can be illustrated by the

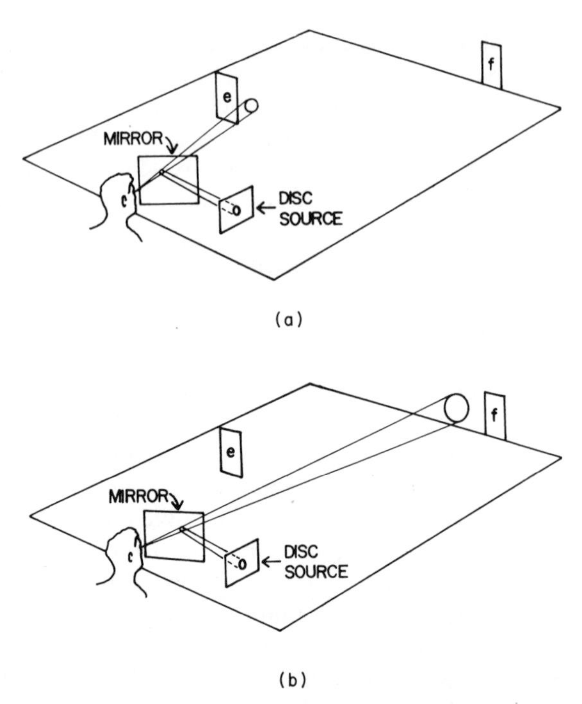

FIG. 8. Apparatus to demonstrate the equidistance tendency as a function of the directional separation of a disk from other objects (e or f).

situation represented in Fig. 8, in which a luminous disk located to the right of the observer is viewed by means of a mirror. The mirror is partially transmitting and partially reflecting so the rectangles *e* and *f* and the disk can be seen simultaneously. The observation is monocular and the viewing is through a restrictive aperture, so as to maximize the equidistance effect by eliminating the difference in the accommodative cue between the disk and the rectangles. The apparent direction of the disk can be adjusted by rotating the mirror. It will be found that if the disk is adjusted to be directionally adjacent to rectangle *e* it will appear near this rectangle in depth, as is indicated in Fig. 8(a). If the mirror is adjusted to position the disk directionally adjacent to rectangle *f*, the disk will appear near the apparent distance of rectangle *f*, as is indicated in Fig. 8(b), and, in agreement with Eq. (3), it will increase in apparent size (Emmerts' law). Figure 8 illustrates both the equidistance tendency and the variation in the strength (effectiveness) of the equidistance tendency with the relative directional separation of the objects (in this case the disk and a rectangle). It has been concluded that the equidistance tendency can modify the perceived depth between objects (e.g., the disk and a rectangle) even in situations in which some exocentric cues of distance are available between these objects (Gogel, 1965).

A second observer tendency, called the *specific distance tendency*, is the tendency in the absence of distance information for an object to appear at a relatively near distance (about 2 or 3 m) from the observer (Gogel, 1969b). One method of demonstrating this can be described in relation to Fig. 3. In the discussion of this figure, it was noted that if the point being viewed were physically located at b, moving the head from Position 1 to Position 2 would result in *no* apparent motion of the point or object only if the *perceived* location of the object were also at b. If the object, although physically at b, appeared at a or c, apparent concomitant motion in the direction of or opposite to the motion of the head would occur. Only under the condition that the physical and perceived distances are the same will the stationary object appear stationary with a lateral movement of the head. According to the specific distance tendency, an object presented without distance cues will appear to be at a distance of about 2 or 3 m. Under a partial or total reduction in cues of distance, objects physically more distant or less distant than 2 or 3 m will appear at a less or greater distance, respectively, than their physical distance. It follows that if a point of light is placed at different physical distances from the observer and is viewed with a moving head, at some particular distance the apparent concomitant motion will disappear and the point will be seen as stationary. The physical distance at which this occurs is also the distance of the specific distance tendency. The specific distance tendency measured in this manner (Gogel & Tietz, 1973), in general, is consistent with other data supporting the existence of such a tendency (Gogel, 1973). Also, as in the case of the equidistance tendency, there is evidence that the specific distance tendency can modify perceived distance despite the presence of some distance cues (Gogel & Tietz, 1973).

The equidistance and specific distance tendencies can be applied to the resolution of many examples of equivalent configurations occurring with a reduction of distance cues. Consider the application of these tendencies to the equivalent configurations of Figs. 1, 2, and 7. According to the specific distance tendency, the point of light should appear to be several meters from the observer in the situation of Fig. 1(a), and, in addition, the two points of light should appear to be equidistant in the situation of Fig. 1(b). In Fig. 1(c), the equidistance effect, with its inverse relation to directional separation, should perceptually combine the end points of the lines that are located along common directions, so that ends with the same letter (for example, the two ends labeled a or the two ends labeled b) will appear to be joined (at A or at B), and the entire object will appear with all of its parts equidistant (in the frontoparallel plane $ABCD$). In Fig. 1(d), in agreement with the specific distance tendency, the rectangle will appear at about 2 or 3 m, and because of the equidistance tendency all parts of the tilted trapezoid in Fig. 1(e) will appear to be at the same distance (in an apparent frontoparallel plane). In Fig. 2, the rear wall of the room, according to the equidistance tendency, will appear frontoparallel and will appear at about 2 or 3 m, according to the specific distance tendency. Each face in a window (*ef* or *gh*) will appear at the distance of its enclosing window, according to the equidistance tendency, and, consistent with this perceived distance and the size–distance invariance hypothesis, the physically more distant face will appear smaller than the physically nearer face. This expectation is consistent with the data available (Gogel & Mershon, 1968; Mershon & Gogel, 1975). In Fig. 7, the upper end of sides *a* and *b* of the Penrose figure appear joined because, as in Fig. 1(c), the strength of the equidistance tendency is greatest between parts of objects identical in visual direction. Also, the equidistance tendency can be applied to explain the path of apparent motion of objects attached to the monocularly observed, rotating trapezoidal window—including the apparent motion of the rigidly attached rod (Gogel, 1956).

The equidistance and specific distance tendencies resolve the stimulus ambiguity implicit in reduced conditions of observation. The existence of these tendencies is incompatible with both the theory of direct perception and the core–context hypothesis, although superficially the core–context hypothesis and the observer tendencies seem to have similar implications for perception under reduced conditions. According to the equidistance tendency, in the absence of exocentric cues of distance the object will appear frontoparallel (all parts of the object will appear equidistant) regardless of its distal orientation and, therefore, the perceived shape will be proportional to retinal shape. Also under reduction conditions, since the object appears to be at the distance defined by the specific distance tendency, the perceived size of the object according to the size–distance invariance hypothesis will vary directly with its retinal size. A difference between the core–context hypothesis and the observer tendencies is that the latter, not the former, specifies that perceived egocentric and perceived exocentric distances, rather than disappearing or becoming indefinite with the elimination of distance cues, instead acquire specific values. The core–context hypothesis states that under reduction conditions θ and S' are directly related. The size–distance invariance hypothesis states that under all conditions θ and S'/D' are directly related. By providing a D' in the absence of distance cues, the specific distance tendency permits the size–distance invariance hypothesis to apply to all conditions of observation. The observer tendencies are also opposed to the theory of direct perception in that they define perceptual states supplied by the observer (not by the proximal stimulus) toward which the perception moves as the distance cues are reduced. Because of these observer tendencies, the perceptions occurring under reduced conditions exhibit the same processes (such as the interrelation of perceived size and perceived distance) found under conditions in which many distance cues are available. Thus, the perceptions occurring under reduced conditions of observation can be considered, in disagreement with the theory of direct perception, to be as representative of perceptual processes as those occurring under any other conditions of observation. Also, in opposition to the direct theory of perception, the perceptual errors specified by these tendencies are not simply an increase in the variability of the perceptions. Finally, as will be discussed, the effects of the observer tendencies on spatial perceptions are not limited to conditions of total reduction of distance cues. It is as though distance information, when available, competes with these tendencies for the control of the perception.

The equidistance and specific distance tendencies are not expected to dominate spatial perceptions under conditions in which strong cues of distance are available. Nevertheless, these tendencies contribute to the perception of distance in a variety of visual environments. It was noted previously that distant portions of the visual field approximate conditions of cue reduction even though many objects and surfaces are present throughout the visual field. It follows that the perceived depth between distant objects can be expected to be an underestimation of physical depth because of the equidistance tendency. The underestimation of the depth between distant mountains is a possible example of this (Gogel, 1965). Also, in agreement with the specific distance tendency and the size–distance invariance hypothesis, distant houses viewed from an aircraft appear miniature in size. It was also noted in the previous discussion that cue effectiveness can be different for different objects in the visual field. The contribution of the equidistance tendency to depth perception under local reductions in cue effectiveness is illustrated by a variety of phenomena, including the following:

1. The horizon moon appears at the distance of objects on the horizon.
2. The image from a movie projector appears on the visible screen (rather than somewhere between the screen and the observer) even though observed monocularly.
3. A shadow appears extended horizontally along the floor or vertically on a wall.
4. A monocularly observed object appears to be on a table top at the distance at which the base of the object intersects the table top.
5. An afterimage appears at the distance of the wall against which it is viewed.

It seems from these and other examples that the equidistance tendency provides an important function in specifying the perceived location of many details of the visual field.

The equidistance and specific distance tendencies can be regarded as processes contributed by the observer to the perception in the interest of achieving a coherent, organized percept despite a reduction in distance information. Other examples of observer tendencies in perceptual organization are provided by the Gestalt laws of perceptual grouping such as symmetry, good continuation, common fate, and proximity (Rock, 1975). A principle that applies to phenomena more extensive than that of perceptual grouping is called the *adjacency principle* (Gogel, 1970). This principle states that the effectiveness of cues between objects in determining perceived object characteristics is inversely related to the perceived separation of the objects either frontally or in depth. Essentially this is a statement expressing a tendency toward local spatial autonomy in cue interaction or induction effects. This principle has been applied to a number of different kinds of perceptions, with an application to induced motion illustrated in Fig. 9. In the situation represented by Fig. 9, two luminous frames are presented at different stereoscopic distances from the observer. Both frames move from right to left repetitively, as is indicated by the solid and dashed arrows. Also as indicated by the arrows, the motion of the two frames is in opposite phase (e.g., as the near frame moves to the left, the far frame moves to the right), with the magnitude of the frame motions such that the far frame is not occluded by the near frame during any portion of the motion. The three filled circles in Fig. 9 represent three alternate distance positions (near, middle, and far) of a physically stationary disk of light. The two, simultaneously presented, moving frames and the single stationary disk of light are viewed in an otherwise totally dark visual field. As shown by Duncker (1939) and by others (Wallach, 1959; Brosgole, 1966), a physically stationary point or disk presented in a moving frame will appear to move opposite in phase to

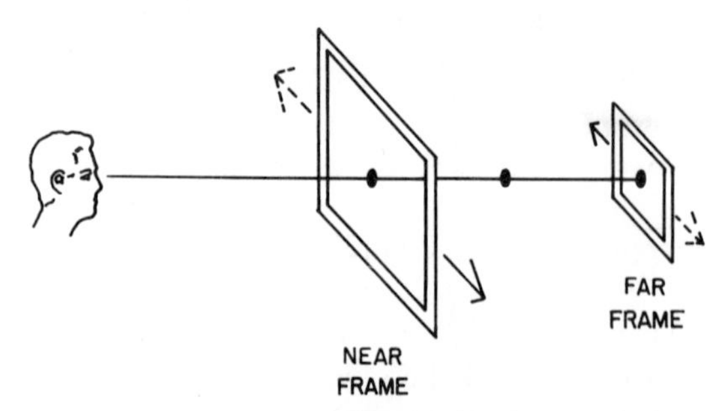

FIG. 9. Apparatus for demonstrating the effect of the depth separation between a disk and a frame upon the magnitude of the induced motion.

the physical motion of the frame. This apparent motion of the disk is called *induced motion*. In Fig. 9 the apparent motion of the disk could be produced by either the near or the far frame, with the frame responsible for the induced motion identified by the phase of the induced motion relative to the physical motion of the frame. It has been found (Gogel & Koslow, 1971) that the disk will have an induced motion determined mainly by the frame at the same apparent distance as the disk. Thus, with the disk at the near distance, the motion induced in the disk will be determined mainly by the near frame. With the disk at the far distance, the motion induced in the disk will be determined mainly by the far frame. With the disk midway in depth between the near and far frames, little or no induced motion will occur in the disk. Consistent with the adjacency principle, induced motion and other induction effects (Gogel, 1970) are most effective between objects located in the same apparent portion of the visual field. The results discussed in relation to Fig. 6 also provide an example of the importance of adjacency in determining cue effectiveness. The visual system resolves the conflict between relative cues of motion by giving a weight to a relative motion cue that is inversely related to the separation between the moving objects. Basically, the adjacency principle concerns the resolution of the stimulus ambiguity implicit in cue conflicts. Considering the range of adjacency phenomena that have been demonstrated (Gogel, 1970), it seems that the presence of cue conflicts, and the stimulus ambiguity implicit in cue conflicts, are the usual conditions under which perceptions are formed.

Neither the core–context theory nor the theory of direct perception seem to be suited to explain the occurrence and resolution of cue conflicts implied by the phenomena used to support the adjacency principle. Also, it is relevent to the cognitive theory that the weighting of relative cues in terms of adjacency implies a kind of local autonomy in perceptual consistency. The perceptual system seems to be able to accept inconsistencies in information more readily when they occur between displaced, rather than between adjacent, portions of the visual field. Thus, with the Penrose impossible figure (see Fig. 7(b)) the information at the corners of the triangle is consistent. The discrepant information occurs between the different corners. Presumably it is because the inconsistency is between spatially separated parts that it is not obvious at first glance.

IV. RÉSUMÉ

Stimulus ambiguity is present whenever the information contained in the proximal stimulus is insufficient to explain the perception. Several kinds of stimulus ambiguity have been identified. One kind is present because the proximal information (particularly under conditions of a reduction in distance cues) is insufficient geometrically to uniquely specify a distal stimulus. Despite this insufficiency, a perception of some specific distal stimulus always seems to occur, suggesting that some factor or factors in addition to the proximal stimulus must be contributing to the perception. This kind of stimulus ambiguity is implied in the concept of equivalent configurations, with the class of equivalent configurations and the amount of stimulus ambiguity becoming larger as the distance information in the proximal stimulus is decreased. Not all distance information, however, is equally effective in limiting or eliminating this kind of stimulus ambiguity. Exocentric cues of distance are less effective in this regard than are egocentric cues.

A proximal stimulus including the proximal representation of distance information can be geometrically adequate in specifying a particular distal stimulus without necessarily being able to uniquely specify the perception. This can occur in situations in which the distance cues are not present in sufficient magnitude or number to be effective. Although the total proximal stimulus may not lack in specificity, factors in addition to proximal stimulation may contribute to the percept because the proximal cues are not sufficiently effective to remove the influence of these factors.

Another kind of stimulus ambiguity results from a tendency of the visual system to respond to relative cues (cues between objects) rather than either exclusively or predominantly to absolute cues (cues between the object and the observer). This tendency for the visual system to respond to relative cues has advantages. First, the threshold for relative cues under many conditions is lower than that for absolute cues. Thus, for example, relative (exocentric) cues of distance permit the perception of distance to be extended to much farther distances from the observer than would be possible from absolute (egocentric) cues alone. A second advantage is found in the perceptual organization that results from relative cues. This is illustrated particularly by the research of Johansson (1971, 1973), in which it is shown that motion perception is often organized in terms of relative and common motion, rather than in terms of the motion of each stimulus independently. However, the tendency to respond to relative cues also has a disadvantage. The disadvantage is that different relative cues, particularly in the case of moving stimuli, can be in conflict with each other and also in conflict with absolute cues (Gogel, 1977). To resolve these conflicts, the perceptual system must weight the conflicting cues in some consistent manner. It seems that the weight given a relative cue in a perception is greater the more adjacent are the objects between which the cue occurs. Both the vector analysis of motion demonstrated by Johansson and the weighting of cues as indicated by the adjacency principle involve factors other than the proximal stimulus and reflect a subtle, though ubiquitous, form of stimulus ambiguity.

Several theories of the resolution of stimulus ambiguity were examined, particularly with respect to the kind of ambiguity implied in equivalent configurations. The core–context hypothesis denies that stimulus ambiguity must occur with a reduction in distance cues, asserting instead that with increasing cue reduction the perception is increasingly determined by the core stimulus. The theory of direct perception denies the significance for perceptual theory of responses that occur under conditions of cue reduction. The cognitive theory of perception, on the other hand, frequently uses situations involving equivalent configurations to demonstrate the effect of past experience on perception in agreement with the inductive rule of Helmholtz. Another aspect of the cognitive theory that is less obviously empiricistic concerns the logical interrelation of perceptions. Examples of this deductive aspect are found in a variety of hypotheses of invariance. Evidence in opposition to the cognitive theory is found in the logical inconsistencies in perception that persist despite reasonable alternatives and in the differences between cognitive and perceptual laws. A fourth point of view, characterized as autochthonic, involves the perceptual effects resulting from observer tendencies. Two such tendencies that define perceptual states that are increasingly approximated as distance information is increasingly reduced are the equidistance and specific distance tendencies. These can explain a variety of the perceptions that occur under conditions in which either the specificity or effectiveness of cues are reduced. A third tendency expresses the manner in which the perceptual system weights conflicting information from relative cues.

Stimulus ambiguity of one type or another, and in varying degrees, is a very prevalent condition. It remains one of the central problems of perception and the answer to the question of how this ambiguity is resolved serves to distinguish sharply between the different theoretical approaches to perception.

References

Ames, A. The rotating trapezoid. Description of phenomena. In F. P. Kilpatrick (Ed.), *Explorations in transactional psychology*. New York: New York Univ. Press, 1961.

Boring, E. G. *Sensation and perception in the history of experimental psychology*. New York: Appleton, 1942.

Boring, E. G. The perception of objects. *American Journal of Physics*, 1946, **14**, 99–107.

Brosgole, L. An analysis of induced motion. NAUTRADEVCEN, LH-48, U.S. Naval Training Device Center, Port Washington, New York, 1966.

Carlson, V. R. Overestimation in size-constancy judgments. *American Journal of Psychology*, 1960, **73,** 199–213.

Duncker, K. The influence of past experience on perceptual properties. *American Journal of Psychology*, 1939, **52,** 255–265.

Epstein, W. Attitudes of judgment and the size-distance invariance hypothesis. *Journal of Experimental Psychology*, 1963, **66,** 78–83.

Epstein, W., Park, J., & Casey, A. The current status of the size–distance invariance hypothesis. *Psychological Bulletin*, 1963, **60,** 265–288.

Eriksson, W. S. Distance perception and the ambiguity of visual stimulation: A theoretical note. *Perception & Psychophysics*, 1973, **13,** 379–381.

Foley, J. M. Depth, size, and distance in stereoscopic vision. *Perception & Psychophysics*, 1968, **3,** 265–274.

Foley, J. M. Primary distance perception. In R. Held, H. Leibowitz, & H. L. Teuber (Eds.), *Handbook of Sensory Physiology* (Vol. 8). New York: Springer-Verlag, in press.

Gibson, E. J. *Principles of perceptual learning and development*. New York: Appleton, 1969.

Gibson, J. J. *The perception of the visual world*. Boston: Houghton, 1950.

Gibson, J. J. Perception as a function of stimulation. In S. Koch (Ed.), *Psychology, a study of a science* (Vol. 1). New York: McGraw-Hill, 1959.

Gibson, J. J. *The senses considered as perceptual systems*. Boston: Houghton, 1966.

Gogel, W. C. Relative visual direction as a factor in relative distance perception. *Psychological Monographs*, 1956, **70,** 1–19 (Whole No. 418).

Gogel, W. C. The perception of depth interval with binocular disparity cues. *Journal of Psychology*, 1960, **50,** 257–269.

Gogel, W. C. Equidistance tendency and its consequences. *Psychological Bulletin*, 1965, **64,** 153–163.

Gogel, W. C. The effect of object familiarity on the perception of size and distance. *Quarterly Journal of Experimental Psychology*, 1969, **21,** 239–247. (a)

Gogel, W. C. The sensing of retinal size. *Vision Research*, 1969, **9,** 1079–1094. (b)

Gogel, W. C. The adjacency principle and three dimensional visual illusions. In J. C. Baird (Ed.), Human space perception: Proceedings of the Dartmouth conference. *Psychonomic Monograph Supplement* 3, (Whole No. 45), 1970.

Gogel, W. C. Scalar perceptions with binocular cues of distance. *American Journal of Psychology*, 1972, **85,** 477–498.

Gogel, W. C. The organization of perceived space: I. Perceptual interactions. *Psychologische Forschung*, 1973, **36,** 195–221.

Gogel, W. C. Relative motion and the adjacency principle. *Quarterly Journal of Experimental Psychology*, 1974, **14,** 425–437.

Gogel, W. C. The metric of visual space. In W. Epstein (Ed.), *Stability and constancy in visual perception: Mechanisms and processes*. New York: Wiley, 1977.

Gogel, W. C., & Koslow, M. The effect of perceived distance on induced movement. *Perception & Psychophysics*, 1971, **10,** 142–146.

Gogel, W. C., & Mershon, D. H. The perception of size in a distorted room. *Perception & Psychophysics*, 1968, **4,** 26–28.

Gogel, W. C., & Sturm, R. D. Directional separation and the size cue to distance. *Psychologische Forschung*, 1971, **35,** 57–80.

Gogel, W. C., & Tietz, J. D. Absolute motion parallax and the specific distance tendency. *Perception & Psychophysics*, 1973, **13,** 284–292.

Gogel, W. C., & Tietz, J. D. The effect of perceived distance on perceived movement. *Perception & Psychophysics*, 1974, **16,** 70–78.

Gregory, R. L. *The intelligent eye*. London: Weedenfeld, 1970.

Gregory, R. L. Choosing a paradigm for perception. In E. C. Carterette & M. P. Friedman (Eds.), *Handbook of perception* (Vol. 1). New York: Academic Press, 1974.

Helmholtz, H. von. *Physiological optics*. (Trans. from the 3rd German Ed., Vol. 3. J. P. C. Southall, Ed.). Optical Society of America, 1924.

Hochberg, J. E. Effects of the Gestalt revolution: The Cornell symposium on perception. *Psychological Review*, 1957, **64,** 73–84.

Hochberg, J. E. Perception. In J. W. Kling & L. A. Riggs (Eds.), *Woodworth and Schlosberg's Experimental Psychology* (3rd ed.). New York: Holt, 1971. Pp. 395–550.

Hochberg, J. E. High-order stimuli and inter-response coupling in the perception of the visual world. In R. B. MacLeod & H. L. Pick (Eds.), *Perception: Essays in honor of James J. Gibson*. Ithaca, New York: Cornell Univ. Press, 1974. (a)

Hochberg, J. E. Organization and the Gestalt tradition. In E. C. Carterette & M. P. Friedman (Eds.), *Handbook of perception* (Vol. 1). New York: Academic Press, 1974. (b)

Holway, A. H., & Boring, E. G. Determinants of apparent visual size with distance variant. *American Journal of Psychology,* 1941, **54,** 21–37.

Johansson, G. Rigidity, stability and motion in perceptual space. *Acta Psychologica*, 1958, **14,** 359–370.

Johansson, G. Visual perception of biological motion and a model for its analysis. Report 100, Department of Psychology, University of Uppsala, Uppsala, Sweden, 1971.

Johansson, G. Vector analysis in visual perception of rolling motion. Report 130, Department of Psychology, Univ. of Uppsala, Uppsala, Sweden, 1973.

Kilpatrick, F. P. (Ed.) *Explorations in transactional psychology*. New York: New York Univ. Press, 1961.

Lodge, H., & Wist, E. R. The growth of the equidistance tendency over time. *Perception & Psychophysics*, 1968, **3,** 97–103.

Mershon, D. H., & Gogel, W. C. Failure of familiar size to determine a metric for visually perceived distance. *Perception and Psychophysics*, 1975, **17,** 101–106.

Metzger, W. Consciousness, perception and action. In E. C. Carterette & M. P. Friedman (Eds.), *Handbook of perception* (Vol. 1). New York: Academic Press, 1974.

Michotte, A. *La perception de la causalité*. Louvain: Institut Supérieur de Philosophie, 1946.

Price, G. R. On Emmert's law of apparent sizes. *Psychological Record*, 1961, **11,** 145–151.

Purdy, W. C. The hypothesis of psychophysical correspondence in space perception. *General Electrical Technical Information Series*, 1960, No. R60ELC56.

Rock, I. Perception from the standpoint of psychology. In D. A. Hamburg, K. H. Pribram, & A. J. Stunkard (Eds.), *Perception and its disorders*. Baltimore: Williams & Wilkins, 1970.

Rock, I. An introduction to perception. New York: Macmillan, 1975.

Rosinski, R. R. On the ambiguity of visual stimulation: A reply to Eriksson. *Perception & Psychophysics*, 1974, **16,** 259–263.

Wallach, H. The perception of motion. *Scientific American*, 1959, **201,** 56–60.

Wallach, H. Informational discrepancy as a basis of perceptual adaptation. In S. J. Freedman (Ed.), *The neuropsychology of spatially oriented behavior.* Homewood, Illinois: Dorsey, 1968.

Part III

Illusions and Disorders

Chapter 9

ILLUSIONS AND HALLUCINATIONS

RICHARD L. GREGORY

I. INTRODUCTION

Illusions may be defined as perceptual departures from physical reality. There are many kinds of illusions, including distortions, paradoxes, ambiguities, and perceptions quite unrelated to present physical state or the current stimulus pattern presented to the senses (hallucinations). Such perceptual creations may be partial rather than complete, as in illusory contours (see Fig. 1a,b). There may also be rejection of what would normally be perceived—perhaps appropriately described as *negative* hallucinations.

The kinds of illusion commonly discussed are distortions and ambiguities. Distortion illusions include such very well known figures as the Müller–Lyer, the Ponzo, and the Orbison illusions. The best-known ambiguous figure is the Necker cube, which reverses spontaneously in apparent depth. This, and most, if not all, of the other illusions, can occur for objects as well as for flat figures, though there are significant differences between perception of objects and perception of figures. In a sense, any figure or picture seen as an object is illusory, for it is seen not merely as the lines and shading of which it is composed, but as an object or shape lying in a generally three-dimensional space, though the figure is physically flat. In his deservedly well-known book, *Art and illusion*, Sir Ernst Gombrich (1960) is not referring to this sense of illusion, though his discussions are highly relevant to the general problems of perception

HANDBOOK OF PERCEPTION, VOL. IX

ISBN 0-12-161909-5

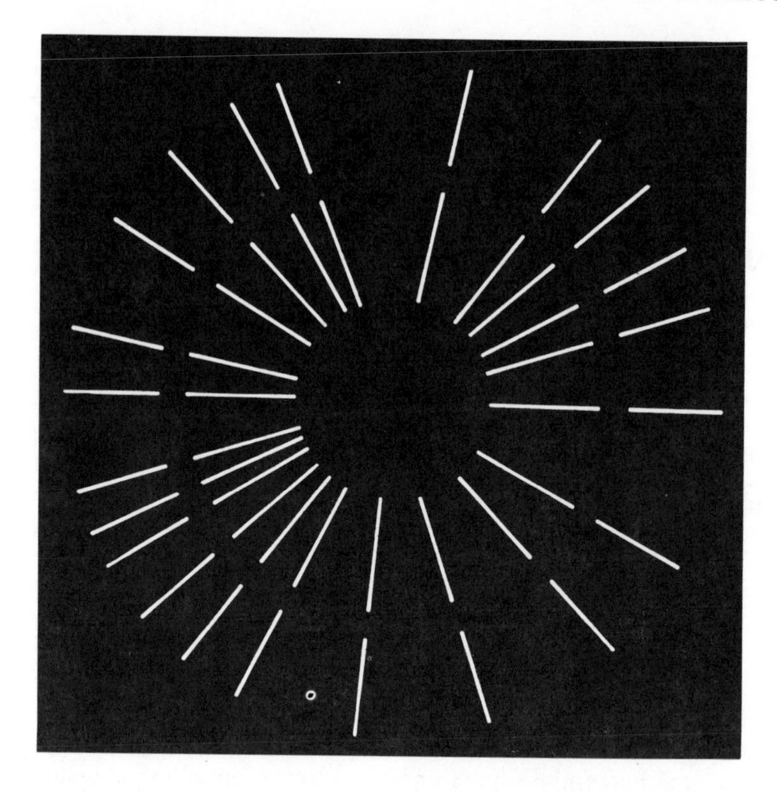

and illusion. He is referring rather to the artist's ability to create alternative perceptual realities with the techniques of painting. When these correspond closely to the usual or intended state of affairs, the artist's illusion, in Gombrich's sense, is most complete and most successful. We are interested here, however, in why some perceptions depart from the objective world of physics, rather than in the creation of alternative perceptual realities created by artists or the camera that may serve as simulations of physical reality.

This takes us beyond a narrow discussion of perception and its failings to epistemology and the philosophy of science. It takes us also to the basic issues underlying the major theories of perception. It is for this reason that illusions are not conceptually trivial, though they may *appear* trivial. Appearance of triviality has proved to be misleading throughout the history of science. Who could have thought a few hundred years ago that pith balls attracted to rubbed amber, or the behavior of lodestones or spinning tops, would turn out to be not merely the actions of toys, but significant tools for understanding the physical universe? Similarly, illusions, though apparently childishly trivial, can be regarded as significant

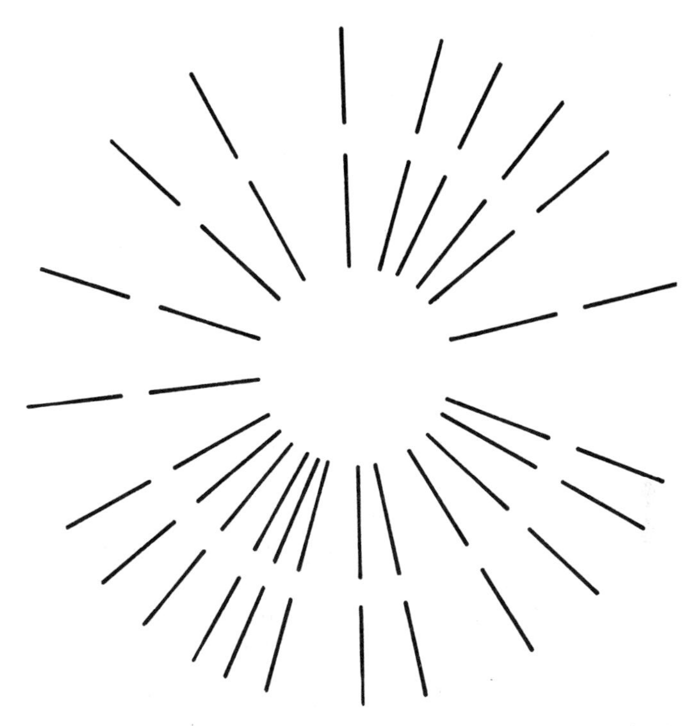

FIG. 1. (a) As the number of broken lines is increased, the illusion "ring" and "sun" becomes stronger—presumably because there is increasing evidence for nearer masking objects to "explain" the gaps. (b) With reversed contrast, the illusory contrast reverses.

tools for exploring and understanding the mechanisms and processes by which perception works.

The general position that I shall hold is that both empirical statements about the physical world and perceptions are hypotheses: In both cases they are hypotheses based on limited data and strictly inadequate assumptions, and either can be incorrect. We tend to regard scientific departures from physical fact as errors; but perceptual departures as illusions. However, in my view errors and illusions are similar, and may often be generated in similar ways. This is so when the processes and procedures of science are similar to the processes and procedures of perception. How similar they are we must try to discover.

If we take the epistemological view that statements of physical fact are *hypotheses*, then we may say that illusions are perceptual departures from hypotheses that we hold to be true of the physical world. This will include the most general hypothesis of all—that there *is* a physical world. So, it may be noted, this notion of illusion disappears with solipsism.

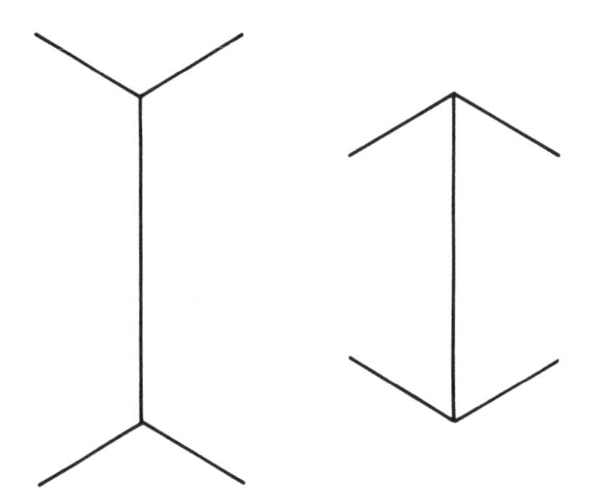

FIG. 2. The Müller–Lyer illusion. [The "shaft" of (a) appears longer than (b), though they are the same length.] These are perspective figures of corners: The distortion occurs though this is not realized.

Given that we cannot know physical states of affairs with certainty, it follows that we can never be entirely sure that we have an illusion. All we can be sure of is that perceptions sometimes depart from what we *believe* to be true of the physical world. Consider for example the Müller–Lyer figure (Fig. 2). The left-hand vertical line looks longer than its fellow to the right. We may measure to assure ourselves that the two lines are in physical fact the same length. But it is perhaps always just possible that we are mistaken in believing that the left line is not, in physical fact, longer than its fellow. To assert that this perception is a departure from physical fact, we must be as sure as possible that the left-hand line is not indeed longer; but since this, in principle, is open to doubt, whether or not we have an illusion is also open to some doubt. This gives us a link between physics and perception that is deep and interesting. It provides one of the reasons for thinking of perceptions as hypotheses. It is their interaction, I believe, that is the central problem of epistemology and the deepest issue for theories of perception. There are, however, still divided opinions on how to think about the relation between perception and physics, and therefore no agreement on how we should think about the phenomena of perception, such as illusions (Gregory, 1974).

On a Realist account, perceptions are *part of* the physical world, so Realists cannot even say that illusions are specifically perceptual phenomena. Illusions tend to be ignored by realists, for it is embarrassing to have to say that physical reality is distorted, or paradoxical, or ambiguous (and hallucination is explicitly ruled out). This very embarrassment and its reasons provide strong arguments against Realism. Realists should be completely opposed to our notion of illusion as departure from physical reality, for if perceptions are selections of reality they cannot depart from it. For them, illusions could at most be unfortunate selections that are in some way misleading. But how could they be misleading if they are not departures, or errors, of some kind? Representative theories have no such difficulty, for representations such as pictures, prose, and mathematics can be (and often are) wholly or in part distorted, paradoxical, ambiguous,or fictional. What Realism promises—and it is a uniquely attractive promise—is certain undeniable knowledge from perception. But the fact that illusions occur (and we cannot always recognize an illusion as such) shows that this is a false promise. Representational accounts make no such claim of certain knowledge, either from perception or from science. Representational epistemology holds all knowledge to be hypothetical; that is, dependent upon assumptions that cannot be tested by observation or, indeed, in any other way. This may be uncomfortable, and it is still resisted by many philosophers, including those of the logical positivist and operationalist schools; but it looks as though we have to accept it and live with it. It is only uncomfortable if we seek certainty for empirical statements and issues; but this would be dogmatism, which has a sad history both in science and society.

II. PROCESSES OF PERCEPTION

It seems clear that perception works by accepting neural activity from the sense organs. For these signals to be accepted and useful they must be decoded, or "read." This is clearly analogous to how signals from scientific instruments, such as radio telescopes or space probes,are used. In both cases, for the signals to be useful there must be assumptions of calibration, of how the sense organ, probe, or other transducer is applied, and of how the code is to be read. More general assumptions are also (perhaps always) required, such as what is to be accepted as moving or stationary, and which way is right side up or upside down? Some of these assumptions may ultimately be arbitrary or conventional (as relativity theory suggests), but they are nevertheless highly important for reading signals from sense organs or from instruments. With adequate assumptions, *signals* can be accepted as *data*.

We may then go on to say that the data (which are what are read from signals) can be used for suggesting and testing hypotheses. This, also, applies both to science and to perception. In both cases, hypotheses are the end result, to be used for control, prediction, and understanding. When they are adequate and appropriate, there is no obvious mismatch, no obvious departure from physical reality: They may be accepted as true. Often, though, they are accepted only as true in a limited domain, or as true only in restricted ways. Are there levels of error and levels of illusion? For some purposes perceptions and scientific hypotheses may be acceptable that at a deeper analysis are inadequate or wrong. For deep levels of understanding one might, indeed, be tempted to say that all perceptions are illusory. If, for example, desks or tables are bunches of electrons in violent motion, though they appear to the senses as hard, solid, static objects, then all our perceptions of tables depart from the deeper scientific account, and so might be described as illusions. But we generally restrict the notion of illusion to surface appearances: sensations, measurements, and perceptual hypotheses. This does no harm, providing we realize that this is a convention. But it has generated astonishing philosophical confusion, as for example, L. Susan Stebbing's worries about the "two worlds" of perception and physics (Stebbing, 1960). But on a representational view there are not two worlds; there are two accounts, and this is very different. When they are not mutually incompatible, and both are useful, we have no need to say that either departs from reality or is false or illusory; both can be accepted, for limited realms of discourse, as appropriate accounts of the physical world.

However this may be, we can see that if perception generates hypotheses from signals processed to give data (much as in experimental science), there are many possible sources of error capable of producing illusions. So, given this kind of account of perception, it is absurd to expect to find one simple explanation. We should, however, be able to give theoretically significant classifications to illusions, and to relate them experimentally to physiological processes. It is here that we come to an important issue that may be raised as a question: What exactly do we mean by *physiological*?

No one would doubt that the study and description of neural signals is part of physiology. So if an illusion were due to the upset of neural signals (such as adaptation or lateral inhibition) then we would have a physiological account of this illusion. But if we take our next step, to say that neural signals are (and must, to be useful) *read as data,* then it is no longer clear that we are within the realm of physiology. The doubt arises because procedures must be carried out, from accepted assumptions, to derive data from signals. But *processing procedures* are no more part of the physics of the nervous system than, for example, the statistical procedures of science are part of the electronics of the computers carrying them out. A statistical error in an experiment might be due to instrument failure or overloading—or it might be due to an inappropriate procedure being applied, with no electronic or mechanical failure. Incorrect execution of a procedure is very different from an inappropriate procedure, though resulting errors may be similar. This leads us to distinguish between illusions due to *signal transmission errors* and illusions due to *data processing errors*. It may be best to call illusions produced by the first *physiological illusions* and those by the second *cognitive illusions*. The word *cognitive* applies when an error is due to incorrect knowledge or to inappropriate processing procedures; but as these are not failures of the physiology of the perceptual system, they should not be called physiological illusions. This does not of course imply that physiological mechanisms or processes are not involved—we cannot imagine perception without them. But if they are working normally we cannot attribute errors to them. Somewhat similarly, we may need chess pieces and a board to play chess; but an error is not to be attributed to shortcomings in these, but rather to a mistake of strategy or assumptions. This distinction continues for fully automated computer chess. The distinction we are making is not special to the nervous system; it applies also to computer systems and their errors, and to all decision taking.

A mechanism is quite different from what it does, though mechanisms are necessary for anything to be done. In the same way, the cognitive process cannot be reduced to a physiological mechanism and, in the same way, cognitive illusions are in a class apart from physiological illusions (cf. Gregory, 1973).

III. PHYSIOLOGICAL AND COGNITIVE ILLUSIONS

Which illusions are physiological and which are cognitive? What criteria should we adopt for recognizing which is which and classifying them? It is here that we come to the details of interpreting and designing experiments on illusory perception. Because of our lack of detailed knowledge of the *procedures* of perception and its *physiology* (though understanding of both is growing rapidly), many illusions are difficult to classify.

It is virtually certain that such illusions as aftereffects of movement are physiological adaptations. In general, short-term adaptations (such as afterimages) seem to be of this kind, while long-term generalized adaptations (such as to wearing inverting spectacles) may be cognitive. Such effects as inability to see highly unlikely objects correctly (such as hollow molds of faces) are clearly cognitive, for they depend on our knowledge (e.g., of faces). This we can assert with confidence, because there is nothing in stimulus terms that is physiologically odd about an object such as the hollow mold of a face. This does not, however, apply to the bizarre effects of a Bridget Riley painting, or to high-spatial-frequency gratings, or to contours given by isoluminant color contrast (Gregory 1977). Here one suspects that the signal-channel physiology is upset.

A class of illusion that is virtually certainly cognitive is that of *ambiguous* illusions, such as depth-reversing Necker cubes and Rubin's object–ground reversals. These are spontaneous changes without stimulus change, and the perceptions are generally limited to a few alternatives that are likely from the observer's past experience of objects.

Consider an ambiguous figure such as the Necker cube, which is physically a flat drawing, though it appears alternately as one of two likely three-dimensional (cube) shapes. Both of these shapes are illusory, in that neither corresponds to the truly flat drawing. If, on the other hand, we look at a wire skeleton cube, this also will reverse in perceived depth, but

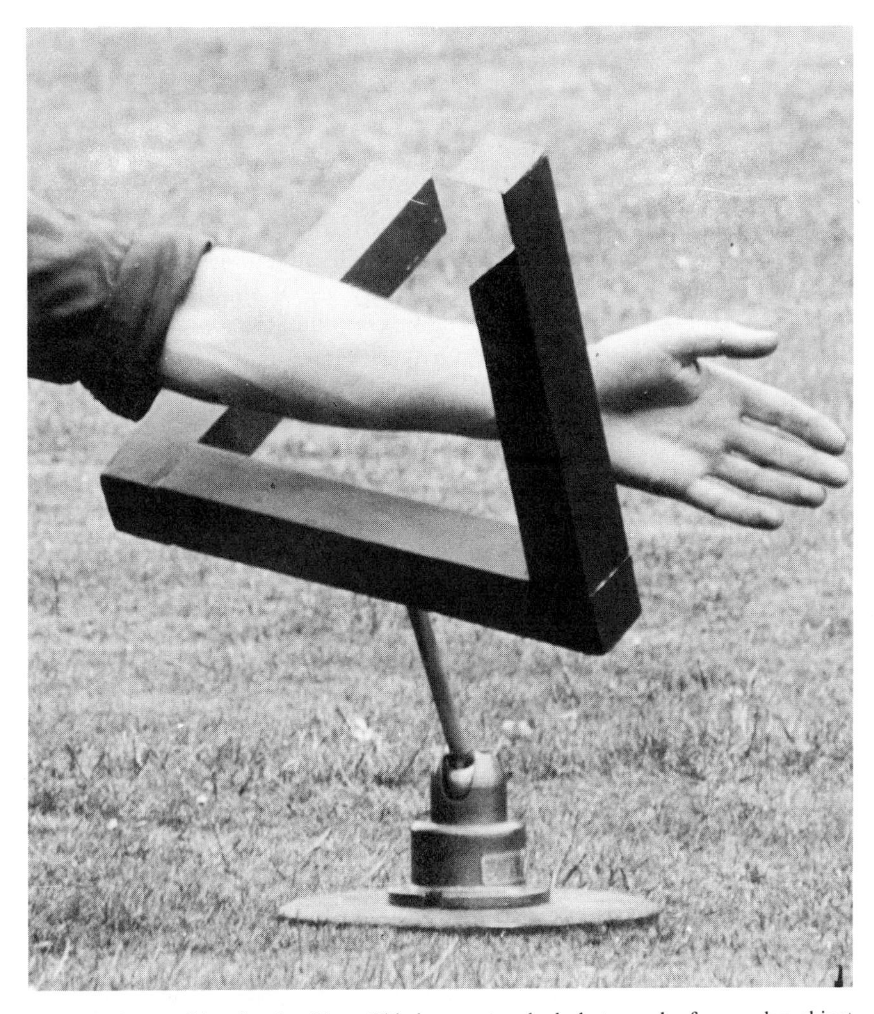

FIG. 3. Impossible triangle object. This is an untouched photograph of a wooden object which appears impossible—because the visual system accepts a false assumption, from which the "perceptual hypothesis" is developed. The perception is paradoxical. Perceptual hypotheses are not always corrected by intellectual knowledge.

now one of the perceptions is correct. So here the ambiguity is from an illusion to nonillusory perception.

What we have called *paradoxical* illusions, such as the Penrose impossible figure (Penrose & Penrose, 1958) or our impossible triangle object—see Fig. 3 (Gregory 1970), are also almost certainly cognitive illusions, for the contours and the texture and so on are normal, yet perception goes dramatically wrong. In these cases, it seems clear that inappropriate processing strategies are generating paradoxical perceptual hypotheses of three-dimensional space.

Illusory contours—see Fig. 1 (a) and (b)—are examples of a tricky case. This is so because we can put up reasonably plausible explanatory hypotheses for both physiological and for cognitive accounts. Although these explanations are very different, it is quite difficult to decide between them. There are other cases of this difficulty, as we shall see, for as in all "crucial" experiments, we have to make certain background assumptions upon which there is lack of agreement.

If it is supposed that contours are seen by the activity of what are sometimes called the *feature detectors* of the striate cortex, discovered by Hubel and Weisel (1962), then it might also be supposed that straight-line feature detectors are triggered and signal through the gaps (see Fig. 1). This is to say that line detectors are tolerant of gaps, much as retinal movement detection is tolerant of spatial gaps, yielding the phi phenomenon (Gregory 1972). But why should this gap tolerance hold only for *some* figures having gaps? And why should illusory contours occur *across* gaps?

If we consider Fig. 1(a), we note that the illusory lines (and regions of changed brightness) are nearly at right angles to the gaps between the lines, which does not at all fit the triggered line-detector idea. Why should they trigger at right angles to the lines?

An alternative, cognitive account, is to suppose that the perceptual system *postulates a nearer masking object*. This occurs when the gaps are unlikely and when they form the shape of a likely object that would produce gaps by masking. It is important to note that absence of signals, in principle, can provide data for hypotheses; so neural signals should not in any case be equated with data or perception without question. Some evidence for this postulated theory of masking objects has been provided by the experiments of Gregory and Harris (1974), using the visual construction shown in Fig. 4, in which the dots were forced back by stereopsis to lie behind the broken triangle when they could no longer form the corners of a planar masking object. It was found that the illusory contours disappeared or were greatly diminished, though they reappeared at full strength when the dots were brought forward to lie in front of the broken triangle. This depth asymetry confirmed the notion of a postulated masking object. But, of course, as for all experiments, interpretation depends on accepting some assumptions and rejecting others. It was assumed, for example, that there was not this kind of asymetry for depth in the striate feature detectors.

Finally, the class of illusions that have received by far the most discus-

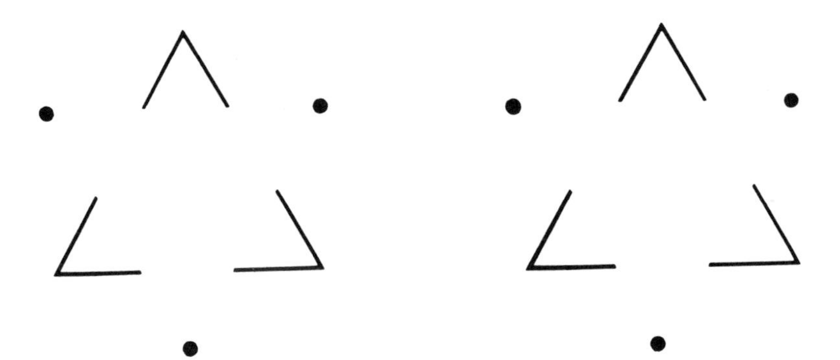

FIG. 4. Stereogram for showing that the illusory object joining the three dots disappears when the dots are forced behind the broken triangle—when it cannot be a masking object.

sion and experimental treatment is that of *distortion* illusions. After a century of research, it is still a question whether they are physiological or cognitive in origin. Many physiological explanations have been put forward, especially that angles produce associated length distortion by interactive neural inhibition (Carpenter & Blakemore, 1973) or by optical defects of the image of the eye. It is now clear, however, that the origin cannot be optical, retinal, or indeed more peripheral than the place of fusion of the two eye's signals normally giving stereoscopy. This was shown by Bela Julesz's demonstration (Julesz, 1971) that a pair of random dot stereograms in which the illusion figure (for example, the pair of Müller–Lyer arrows) was presented to each eye as random dots was seen by neural cross-correlation from the two eyes, which takes place at or after fusion of the signals from the eyes. So any such physiological, signal-channel explanation would have to apply to central mechanisms, and not to the optics of the eyes or the retinas. Signal distortion beyond stereoscopic fusion is not ruled out; but this is less attractive as an explanation, if only because we know less about it and cannot investigate it directly. Also, it is far from clear that retinal image shapes remain mapped in the brain as corresponding shapes—what lateral inhibition would do to produce these distortions cannot be stated.

IV. THE THEORY OF INAPPROPRIATE CONSTANCY SCALING

Perhaps the most discussed cognitive account of distortion illusions is the *theory of inappropriate constancy scaling* (Gregory, 1963). This points to the problem for perception of deriving three-dimensional perceptions from single, two-dimensional retinal images, in which size constancy maintains a nearly constant object size, though retinal images shrink to half their size with each doubling in distance of the object. It is suggested that size constancy is given by active scaling processes, which are set by typical distance cues, especially perspective. So, it is argued, when perspective is presented on the flat plane of a picture, it sets size scaling in much the same way as for the perspective of retinal images normally given by three-dimensional objects in the three-dimensional world. This size scaling should *increase* the perceived size of features that are represented on the flat plane of the figure as *distant*. This is what is observed in illusion figures; features indicated by perspective as *distant* are *expanded* on the picture plane. This could possibly be a mere coincidence. Is it the perspective–depth significance that is producing the distortion (by setting size scaling inappropriately because the figure is in fact flat), or do these angles, converging lines, and so on upset the signals of the visual channel directly, perhaps by interactive physiological effects such as lateral inhibition? It is difficult to decide. The essence of the difficulty is that any display used for a test experiment can be thought of either as a stimulus pattern, that may be distorted neurally in the channel, or as a source of information that may be cognitively misleading. When the display is varied, how can we know whether it is the changed stimulus pattern or the changed information conveyed by the pattern that modifies the illusion? Once we know which features of the stimulus pattern are critical for conveying information (or for setting size scaling) we would know how to change sensory inputs without affecting their accepted information. This consideration is clear by analogy with, for example, different typefaces. These different stimulus patterns convey the same information:

Perceptions are hypotheses
Perceptions are hypotheses

When translated into other languages the patterns would be even more different, though the information conveyed would remain the same.

The converse case is also true. Some very small changes of stimulus pattern have large effects on meaning. This is so when the change is from one to another symbol of the language, or to another sensory cue. A large change in sentence meaning may result from the physically small change of one single letter:

MY DRAIN IS FULL OF GARBAGE
MY BRAIN IS FULL OF GARBAGE

The small physical change from *D* to *B* in the second word changes the meaning of the entire sentence dramatically. In addition, the word *full* means something somewhat different, and the word garbage something very different in the two sentences, though they are physically unchanged. This is surely equivalent to the changes of significance of features of Boring's Young–Old Lady figure. Large changes of meaning generally take place with words that in isolation are clearly ambiguous, much as Necker cubes are ambiguous visually. However, even large physical changes have no effect when there is sufficient redundancy. The writer and the cartoonist base their arts on skilled manipulation of these variables. There are surely implications for neurology in these considerations, for if neural activity itself is a data-related variable, how then can it be interpreted from recorded activity without knowing the context, the coding, and the data-producing procedures in operation in the nervous system? Sorting this out seems like a good task for neuropsychology. This is not, however, our concern here. We are more modestly interested in using such examples to show that it is possible to separate changes in stimulus pattern that produce perceptual changes according to the laws of physics (or physiology) from similar stimulus changes that modify perception by shifts of meaning (according—we may hope—to the cognitive laws of psychology). It follows that we should be able to distinguish between illusions resulting from distorted neural signals (physiological illusions) and illusions resulting from inappropriate uses of information (cognitive illusions). This, however, requires justifiable assumptions of which features are information bearers. But then, as we have said, *any* scientific experiment requires background assumptions, which may be more or less justified.

We have recently reported (Gregory & Harris, 1975) an experiment designed along these lines to test the theory of inappropriate constancy scaling. The prediction is that distortion illusions (such as the Müller–Lyer illusion) should reduce to zero when scaling features of the figure are the same as for the object of which the illusion figure is a representation. By this theory, the Müller–Lyer arrows are accepted by the visual system as a perspective representation of three-dimensional, right-angled corners. If this is so, we should expect zero length distortion in the figure (absence of illusion) when the perspective and other depth-information cues are appropriate for seeing normal corners. The same argument should apply to converging lines on a picture plane, representing parallel lines receding in three-dimensional space (the Ponzo illusion), and so on for any distorting figure representing objects that are normally seen without distortion. To make this last point clear, if three-dimensional corners, such as the lines of the ceiling and walls at the corner of a room—a would-be Müller–Lyer illusion (Fig. 5a,b) or the parallel tracks of a railway—a would-be Ponzo illusion (Fig. 6)—are not distorted when seen

FIG. 5. (a) An inside corner gives essentially the same retinal image as the Müller–Lyer illusion (Fig. 2a). (b) An outside corner gives essentially the same retinal image as the Müller–Lyer illusion (Fig. 2b). The illusion distortion occurs with photographs as with skeleton illusion figures.

normally as three-dimensional objects, then we should predict on this theory that they will not be distorted when represented appropriately on a picture plane. That they are not appropriately represented in the line drawings of the illusion figures as usually presented is clear, because these figures generally appear flat, or in the queer paradoxical depth typical of pictures, although they are perspective line drawings.

FIG. 6. The upper rectangle (signaled by perspective as more distant) line appears longer than its fellow.

There are two reasons for these illusion figures appearing flat or in paradoxical depth. In the first place, they are usually drawn with highly exaggerated perspective. In the second place, the texture of the paper on which they are presented gives the countermanding depth information that the figures are flat. The fact that the perspective is usually drawn with exaggerated angles is clear from comparison with photographs which, like the eye, limit perspective angles to what can be encompassed by the acceptance angle of the lens. The countermanding effect of picture-plane texture is demonstrated by removing the texture (for example, by drawing the figures in luminous paint), after which they do appear in depth, roughly according to their perspective.

In our experiment, we project three-dimensional wire corners (or parallel lines, or whatever) with point-source optics onto a translucent screen. This gives precisely correct perspective when the observer's viewing distance is the same as the distance of the point source from the screen. For any other distance the perspective is inappropriate at the eye. As set up, the perspective was precisely correct with a viewing distance of 40 cm. We then arranged for the perceived depth to be correct by adding a second point source to give stereo projection of the wire models (Gregory 1964). This second source was horizontally separated from the first by the observer's interocular distance. Cross-polarization was used to give each eye its own correct perspective projection. The result was that the flat display appeared in depth, just as though the corners were viewed directly, although the figure remained flat on the screen. With this arrangement (Fig. 7a,b) we could change the perspective, and independently change the perceived depth as given by the stereoscopy, which also removed the countermanding depth cues of the screen texture. The apparent lengths (and so the distortions) were measured with a neutral comparison line of adjustable length given by a CRT, which gave a line of variable length that could also be adjusted in distance. The result was that when the perspective and the depth given by stereoscopy were appropriate to the projected object (the projected wire model), the distortions entirely disappeared. The result for the Müller–Lyer illusion (with three viewing conditions and a range of viewing distances, including the critical appropriate distance) is shown in the graph of Fig. 8).

Since the visual channel is signaling the same angles when the stereo is removed or reversed, and nearly the same angles for the other viewing distances, we take it that it is the *informational* features of the illusion figure that produce the distortion, rather than the stimulus pattern producing distortion of neural signals, since the distortion disappears when the depth scaling is correct for the projected object, though the angles signaled remain the same.

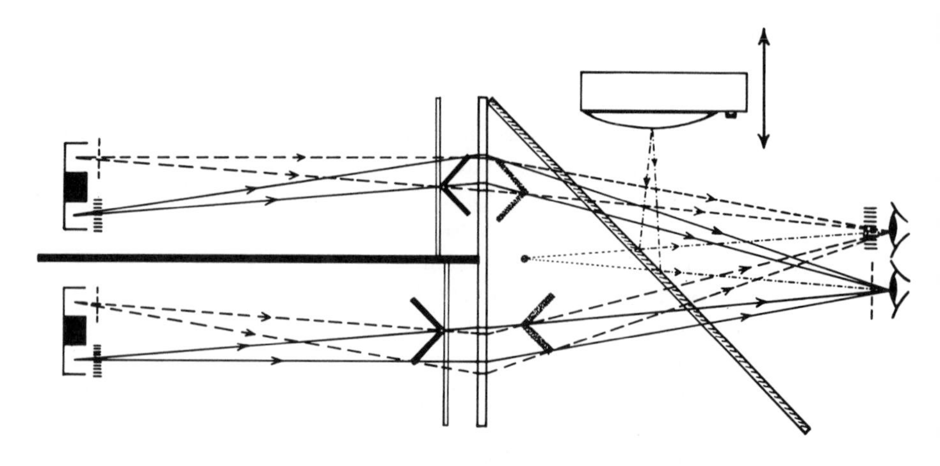

FIG. 7. (a) Plan view of the "illusion destruction" apparatus. (b) Photograph of "illusion destruction" apparatus.

For perspective shapes to be accepted as evidence of depth, or to set size scaling, the visual system must accept assumptions of typical shapes of common objects, such as corners and parallel lines. So we may now make the further prediction that the distortions will reappear in this situation with objects that differ from right angles, or parallel lines, or other generally assumed shapes accepted for "reading" depth from per-

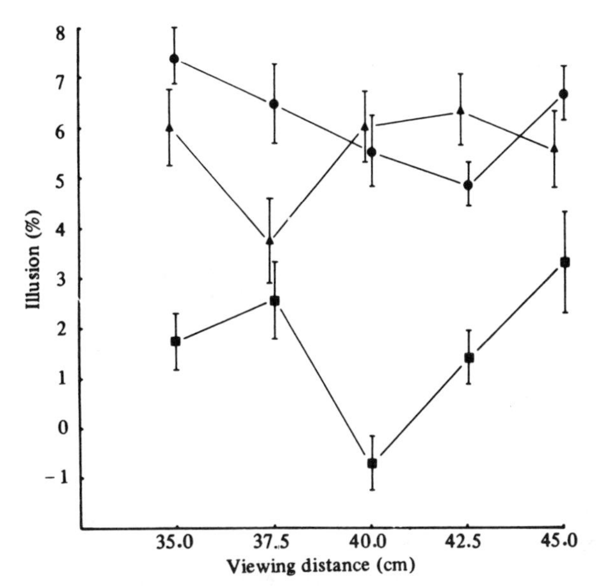

FIG. 8. Results of "illusion destruction" experiment. The lower curve (appropriate stereo projection, ■) shows that the Müller–Lyer illusion disappears completely in the condition (40 cm viewing distance) when perspective and stereo depth are appropriate for the (corner model) object, which is displayed on a screen by stereo shadow projection. The distortion returns with monocular (●) or reversed stereo projection (▲)—when the depth is not perceived correctly. This result shows that the distortion cannot be due to signal distortion in the visual channel. It is cognitive—we believe—given by inappropriate size–depth scaling.

spective. This experiment is at present being carried out in our laboratory. Such experiments could reveal cognitive assumptions that are accepted by perceptual systems for reading stimuli. Which assumptions are accepted and how they are read may depend on early experience, as reflected by measured differences between illusions across races living in different visual environments (Segall *et al.*, 1966; Deregowski, 1973).

We should also expect to find that changes of perceptual assumptions can modify perceptions with no change of stimulus pattern. Changes of assumptions should sometimes convert veridical perceptions into illusions, or vice versa. Such effects of assumptions we do indeed see. Consider perceptual reversal of ambiguous objects such as skeletal wire cubes in depth. Depth reversals follow changes of *assumption* of which face is the front and which the back of the object. This *must* be an assumption, for there is inadequate available sensory information for deciding which is the nearer. Hence the depth ambiguity ceases when sufficient information of depth is provided. It is important to note that the *shape* of the object changes with each depth reversal. Consider its front and back faces: When the physically nearer face switches in depth, to the further face it expands, so the object no longer looks like a cube. It is now a truncated pyramid, with its apparent back face larger than its apparent front. None of the angles is now a right angle. These changes of shape are dramatic and systematic for all observers. They occur with no change in the sensory stimulus pattern and so must be cognitive. This is size scaling following apparent distance. The distortion illusion figures (such as the Müller–Lyer illusion) also set constancy of shape and size, but by their *signaled* distances.

The theory of inappropriate scaling for these illusions asserts that depth cues, such as perspective features, set size independently of *seen* distance. So we have two very different processes operating: (*a*) size set by typically depth-related stimulus features, which serve as *signals* for setting scale, and (*b*) size set by apparent depth, which may be based on *assumptions* of depth rather than from stimuli. This situation may be described as (*a*) scale setting upwards, from stimulus patterns serving as informational signals (which may or may not be appropriate to the physical world); and (*b*) scale setting downwards, from assumptions of distance (which may not be appropriate) contained in the prevailing perceptual hypothesis. Normally there will be little or no conflict between upward and downward scaling. But conflict should be expected when depth cues are presented in flat pictures such as the distortion illusion figures. Precise prediction of what to expect in such situations is almost impossible, for we have no theoretical accounts of how these conflicts are resolved in such bizarre situations as depth represented (inappropriately) on flat planes. We can, however, say that upward scaling operates even when depth is not accepted for perception by the downwards processing.

V. CONCLUDING REMARKS

We find that one illusion can generate further illusions by generating false assumptions that may be acted upon even when the resulting perception is bizarre. We may return to the skeleton wire cube for a particularly clear example of this. When depth-reversed, the skeletal cube swings round *with* movements of the observer, in a most odd way, which is against its physical motion parallax. The direction of rotation is determined by the assumed relative distances of the features of the object. When the assumptions are wrong, the direction of rotation from motion parallax is reversed.

This emphasis on assumptions for perception, and the importance of false assumptions as generators of illusions, is reminiscent of Thomas

Kuhn's (1970) account of the physical sciences depending upon and changing with switches of paradigm assumptions. Curiously, though, Kuhn does not extend his paradigm notion to perception itself. He evidently hopes, like so many philosophers, to find knowledge from perception that is not biased by assumptions. The detailed study of illusions tells us, though, that this is a vain hope. By understanding illusions of perception and errors of science, we can use departures from physical fact to discover which assumptions are accepted for deriving hypotheses from signals. Illusory perceptions are as important for understanding the observer as nonillusory perceptions are important for understanding the world of physics. Departures from physics are the start of journeys to understanding the mind.

References

Carpenter, R. H. S., & Blakemore, C. Interactions between orientations in human vision. *Experimental Brain Research*, 1973, **18,** 287–303.

Deregowski, J. B. Illusion and culture. In *Illusion in nature and art*. R. L. Gregory & E. H. Gombrich (Eds.), London: Duckworth, 1973. Pp. 161–191.

Gombrich, E. H. *Art and illusion*. London: Phaidon, 1960.

Gregory, R. L. Distortion of visual space as inappropriate Constancy Scaling. *Nature (London)*, 1963, **199,** 678–680.

Gregory, R. L. Stereoscopic shadow images. *Nature (London)*, 1964, **203,** 1407–1408.

Gregory, R. L. *The intelligent eye*. London: Weidenfeld and Nicolson, 1970.

Gregory, R. L. *Eye and brain*. London: Weidenfeld and Nicolson, 1972.

Gregory, R. L. The confounded eye. In R. L. Gregory & E. H. Gombrich (Eds.), *Illusion in nature and art*. London: Duckworth, 1973.

Gregory, R. L. Choosing a paradigm for perception. In E. C. Carterette & M. P. Friedman (Eds.), *Handbook of Perception*, Vol. 1. New York: Academic Press, 1974.

Gregory, R. L. Do we need cognitive concepts? In M. Gazzinaga & C. Blakemore (Eds.), *Handbook of Psychobiology*. New York: Academic Press, 1975.

Gregory, R. L. Vision with iso-luminant colour contrast. I. A technique and observations. *Perception*, 1977, *6*, 113–119.

Gregory, R. L., & Harris, J. P. Illusory contours and stereo depth. *Perception & Psychophysics*, 1974, **15**(3), 411–416.

Gregory, R. L., & Harris, J. P. Illusion-destruction by appropriate scaling. *Perception*, 1975, **4,** 203–220.

Hubel, D. H., & Wiesel, T. N. Reception fields, binolular interaction and functional architecture in the cat's visual cortex. *Journal of Physiology (London)* 1962, **160,** 106.

Julesz, B. *Foundations of cyclopean perception*. Chicago: Univ. of Chicago Press, 1971.

Kuhn, T. S. *The structure of scientific revolutions*. Chicago: Univ. of Chicago Press, 1970.

Penrose, L. S., & Penrose, R. Impossible objects: A special type of illusion. *British Journal of Psychology*, 1958, **49,** 31.

Segal, M. H., Campbell, D. T., & Herskovits, M. J. *The influence of culture on visual perception*. Indianapolis: Bobbs-Merrill, 1966.

Stebbing, L. S. Furniture of the earth. In A. Danto & S. Morgenbesser (Eds.), *Philosophy of Science*. Cleveland and New York: World, from L. S. Stebbing (1937) "Philosophy and the Physicist, London: Methuen, 1960.

Chapter 10

DISORDERS OF PERCEPTUAL PROCESSING

FRANCIS J. PIROZZOLO

I. INTRODUCTION

The discipline of human neuropsychology is a new branch of the biobehavioral sciences which has developed rather rapidly in recent years

ISBN 0-12-161909-5

and has contributed to the burgeoning field of cognitive psychology by demonstrating relationships between regional brain systems and higher mental functions. This chapter is an attempt to review in rather broad strokes the experimental and clinical observations that have focused on disorders of perceptual processing.

Perception is a principal factor in virtually every aspect of human activity. Investigations in the neurosciences during the past quarter century have brought about a radical revision of the concept of perception (see Hubel & Wiesel, 1965). Neuropsychological studies of this important area have also contributed significantly to the contemporary psychologist's understanding of human information processing systems. The major sources of data that have shown how injury affects perceptual processing in the human brain come from neuropsychological studies of local brain lesions. A large proportion of this research comes from the detailed analysis of perceptual function in patients who have sustained cerebrovascular accidents, although an impressive amount of data was collected after World War I (Kleist, 1934) and World War II (K. Goldstein, 1948; Luria, 1966; Wepman, 1951; Newcombe, 1969) using patients who survived penetrating missile wounds of the brain. Still other data on the representation of perceptual functions in the brain have been derived from the electrical stimulation of the cortex in epileptics (Penfield & Roberts, 1959) and from studies of the sequelae of neurosurgical procedures, such as commissurotomies (Bogen & Vogel, 1962) for treatment of intractable epilepsy, and from studies of patients who have undergone hemispherectomies (Smith & Burklund, 1966; Smith, 1974; Dennis & Whitaker, 1976; Kohn & Dennis, 1974), for treatment of gliomas, Sturge–Weber–Dimitri syndrome, and other diseases resulting in diffuse damage to one hemisphere.

A. Neuropsychological Assessment

Although the improvement in neurodiagnostic procedures (e.g., computerized tomography) has changed the role of neuropsychological assessment in recent years, the comprehensive mental status examination performed by the neurologist or neuropsychologist is still an important technique in determining the neurologic function of patients with brain dysfunctions. Typically, the patient who presents symptoms of brain damage undergoes a variety of brief screening tests, which are collectively known as the mental status examination. These tests include measures of level of consciousness, attention, emotional behavior, motor coordination, memory, language, perception, and other higher cortical functions (Strub & Black, 1977). Additional neuropsychological testing is indicated in certain situations, and a number of procedures are commonly employed. While the clinical impressions and experimental testing of a neuropsychologist are an integral part of the evaluation because they can uncover discrete or hidden symptoms of brain dysfunction (see, for example, Dennis & Whitaker, 1976; Pirozzolo, Whitaker, Selnes, & Horner, 1977), standardized neuropsychological assessments may be carried out to determine the degree of severity of "organicity." Popular methods for the assessment of higher cortical functions include the Halstead–Reitan Battery (Halstead, 1947; Reitan, 1955; Halstead & Wepman, 1949), Wechsler tests of intelligence and memory (D. Wechsler, 1955; D. Wechsler & Stone, 1973), Bender Motor Gestalt Test (Bender, 1939), Raven Progressive Matrices (Raven, 1956), Boston Diagnostic Aphasia Examination (Goodglass & Kaplan, 1972), and Token Test (DeRenzi & Vignolo, 1962).

Additionally, numerous methods have been modified by neuropsychologists as ancillary tools for assessing the effects of brain damage, such as the dichotic listening, visual hemifield, dichaptic shapes, and finger tapping techniques.

Numerous variables affect the clinical picture of brain disorders, such as diffuse versus focal effects, the nature, extent and progression of the process, age at onset, and premorbid state of the patient (Lezak, 1976). In most neuropsychological research, careful attention is given to these factors that may affect a patient's performance on a given task. Still other problems affect the interpretation of the results of neuropsychological evaluations, and one such problem is whether the neuropsychologically elicited disturbances observed after brain damage are part of a larger clinical picture of cognitive deterioration. This issue is more frequently addressed in the literature on aphasia, especially in discussions of the potential relationship between aphasia and intelligence (see Weisenburg & McBride, 1935; Vygotsky, 1962; Wepman, 1976). Some control over these factors is gained when the neuropsychologist uses subject populations with well-defined, circumscribed brain damage (as opposed to patients with lesions causing mass effects, or patients with progressive cerebral diseases causing generalized intellectual decline).

This chapter is an attempt to describe the major disorders of perceptual processing as discrete symptom complexes. Certain disorders that will be discussed occur fairly frequently, while others are rare and seldom occur as the only predominant sequelae of brain damage. Certain disorders, such as finger agnosia, occur most often in a context with other related disturbances, such as spatial agnosia. Other disorders, including many of the visual agnosias, may occur most often after recovery from sensory defects. It is the task of the science of neuropsychology to determine whether each of these disorders represents a true clinical entity, a part of an interrelated set of symptoms known as a syndrome, or a problem that can be best understood as representing part of a clinical picture of progressive intellectual deterioration.

II. LESIONS OF THE VISUAL PATHWAYS

The ability of the visual system to detect features of the spatial organization of the visual scene is contingent upon the functioning of the primary visual cortex, the areas located bilaterally on the medial aspect of each occipital pole. Lesions involving the optic nerve, the portion of the optic pathway from the retina to the optic chiasm, result in monocular blindness, while damage to the optic tract, the portion of the optic pathway after the decussation, results in severe loss of capacity in correlated regions of the contralateral visual field (Holmes, 1919; Teuber, Battersby, & Bender, 1960). The most extensive early description of the behavior of animals with total lesions of the striate cortex was given by Kluver (1942). Studies of humans (e.g., Teuber *et al.*, 1960) and animals (e.g., Kluver, 1942) show that when visual stimuli are presented in the defective fields nothing is apparently seen (or at least reported), although simple reflexes, such as pupillary constriction in response to sudden increases in illumination and perception of vigorous movement, are retained.

Recently the possibility of residual visual function after visual cortex destruction producing corresponding areas of blindness (scotoma) in the visual fields has been postulated from demonstrations by a number of researchers (Poppel, Held, & Frost, 1973; Botez, 1975). Areas of "blindness," defined by standard methods of perimetry, have been shown to possess much greater visual capacity than previously suspected.

By projecting moving visual stimuli into an area shown to be "blind" by routine clinical or perimetric examinations the experimenters were able to observe that subjects could correctly localize the stimuli within the "blind" region. Accurate performances by these cortically blind subjects implicates another still functional visual system.

The "two visual systems" theory (e.g., Schneider, 1969; Dick, 1976) holds that separate brain mechanisms control localization and discrimination functions. Ablations of the superior colliculi and visual cortex result in two quite different patterns of visual disturbance. Interruption of the fiber systems which course through the colliculi impairs ability to discriminate stimulus velocity and orientation (Keating, 1974).

Although evidence for the concept of "two visual systems" is derived mainly from clinical studies of brain damage and experimentally induced lesions in animals, some recent psychophysical experiments have indicated that peripheral visual acuity does not improve after the correction of refractive errors (Johnson, Leibowitz, Millodot, & Lamont, 1976). Using a method to test the differential effects of blurring on the two separate visual functions, Johnson *et al.* found that there is a characteristic decline in visual acuity with increasing eccentricity but that there is no difference between conditions with corrected refractive error and without correction. Thus, these results suggest that the visual periphery does not play a prominent role in discriminative vision, but that its primary function is in the apprehension of stimuli for the guiding of eye movements that bring the stimuli into foveal vision. This system is mediated through subcortical mechanisms that are sensitive to motion and abrupt stimulus changes. The discriminative system subserves the identification process. The functions of resolution and pattern vision in the system are mainly mediated by the primary visual cortex. Trevarthen (1968) has concluded that the discriminative system is associated with foveal and parafoveal regions while the localization system is represented over the entire visual field.

III. THE CONCEPT OF AGNOSIA

Traditionally, disorders of perceptual recognition have been called *agnosias*, a term which was coined by the neuropsychiatrist Sigmund Freud (1891). A decade earlier, Munk (1881) had demonstrated that ablation of part of the occipital cortex in dogs resulted in *Seelenblindheit* (mindblindness). Animals with these occipital lesions were no longer able to recognize visually presented objects as meaningful elements. The first important clinical observations of this phenomenon were made by Finkelnburg (1870) and Jackson (1876/1958), but not until the classic studies of Goldstein and Gelb (1918, 1920) did these perceptual processing disorders receive great attention. Goldstein and Gelb triggered a debate which has endured for over a half-century (Gardner, 1975), with their publication of the results of psychological studies of a patient who was unable to appreciate the meaning of objects which were presented to him, in spite of the fact that the patient apparently had normal intelligence and sensory acuity. Bay (1951, 1953), however, questioned the existence of visual agnosia, and in a dramatic exposition, demonstrated that the patient examined by Goldstein and Gelb did not have the alleged defects in perceptual processing reported by these investigators. A heated debate ensued in which many of the most eminent scholars in clinical neurology and neuropsychology participated. One position was perhaps best exemplified by Critchley's (1964) argument that the existence of pure agnosic disorders was unwarranted from clinical observations. Clinicians adopting this position suggested that the syndrome which presented as agnosia was accounted for by sensory defects, or perhaps by naming disorders. Numerous investigators argued for the validity of the agnosias and recently several excellent case studies have illustrated that such a disorder does exist (Rubens & Benson, 1971).

Lissauer (1890) contended that three forms of agnosia exist: the apperceptive form, which was a low-level distortion of the sensory impression; the associative form, which was a higher-level defect in the capacity to identify or recognize an object; and a mixed form. Despite the early criticism of the concept of agnosia, it is generally believed that the agnosic syndromes exist (Brown, 1972) and good evidence has been provided for the plausibility of both apperceptive agnosia (which is commonly seen after recovery from visual field defects) and associative agnosia (which has been construed by Geschwind, 1965, among others, as a sensory–verbal disconnection).

In addition to studies pointing to the existence of agnosias in humans, there is an abundance of evidence which suggests that these disturbances can be experimentally induced in animals (Kluver & Bucy, 1938; Pribram, 1962). While the mechanisms for the human agnosias may be somewhat different from those in other primates, these studies have provided excellent analogues for the study of the agnosias in humans.

IV. VISUAL AGNOSIA

A. Prosopagnosia

Prosopagnosia (Bodamer, 1947), a term used to describe disturbances in recognizing familiar faces, occurs most frequently in patients with other visual–spatial disabilities, such as spatial dyscalculia, spatial dyslexia, and directional disorientation (Hecaen & Angelergues, 1963). But it has also been reported to occur in relative isolation and thus has been considered a distinct clinical entity (Hoff & Potzl, 1937; Gloning, Gloning, Hoff, & Tschabitscher, 1966). Clinicopathological findings indicate that prosopagnosia results from lesions involving the posterior areas of the right hemisphere (Joynt & Goldstein, 1975) and from bilateral lesions (Wilbrand, 1887; Benson, Segarra, & Albert, 1974; Hoff & Potzl, 1937). Data from tachistoscopic visual hemifield studies would appear to support the assumption that the right hemisphere is superior in facial recognition (Pirozzolo & Rayner, 1977; Rizzolatti, Umilta, & Berlucchi, 1971). In a recent paper, Cohn, Neumann, and Wood (1977), however, have indicated that in all nine of the cases reported in the neurological literature of prosopagnosia with pathologically verified occlusive lesions, bilateral lesions were found in association with the gnostic deficits. In addition, at least one of the fusiform gyri was destroyed, which led these authors to suggest that the fusiform gyrus is the anatomical substrate for facial recognition.

B. Balint's Syndrome

Balint (1909) found that a patient with bilateral lesions involving the anterior occipital region suffered from a functional constriction of the visual fields. The psychological deficits in Balint's syndrome, more recently known as simultaneous agnosia (or simultanagnosia), have been extensively studied by Holmes (1919), Luria (1959), and Kinsbourne and Warrington (1962). While patients with this syndrome have full visual fields on perimetric examination, they are unable to attend to more than one visual element. If an object (e.g., a letter) is presented in the fovea and another is presented simultaneously in parafoveal vision subjects will not report the parafoveal stimulus. Using a tachistoscope, Goldstein was among the first to demonstrate this defect. In an elegant series of studies, Luria and his colleagues (Luria, 1959; Luria, Pravdina-Vinarskaya, & Yarbus, 1963) have attempted to elucidate the causal mechanisms involved in this syndrome. Believing that lesions in the anterior occipital regions prohibited excitation of sufficient portions of visual cortex, Luria (1959) improved visual perception in these patients by caffeine injections. The accompanying "fixation ataxia" may also account for the behavioral deficits in Balint's syndrome. Luria *et al.* (1963) also found that the exploratory eye movements of a patient with simultanagnosia were severely disorganized but that the fixation ataxia disappeared with caffeine injections.

C. Apractagnosia

Apractagnosia is a visual–spatial disorder involving the misperception of extrapersonal space. Deficits in this disorder include directional and topographical disorientation and the related problems of dressing apraxia and visuo-constructive defects. The disorder was described as early as 1912 by Kleist as "optic apraxia," but only the more recent reviews of Critchley (1953) and Brown (1972) contain adequate descriptions and definitions of the syndrome. Kleist (1934) later settled on the term "constructional apraxia" to describe the syndrome, which unfortunately has only served to confuse the issue by its omission of the visual–spatial component.

Excellent studies of topographical ability have been reported by Meyer (1900), Hécaen (1962), and Semmes, Weinstein, Ghent, and Teuber (1955), but there is little agreement over the localization of this function. Semmes *et al.* found patients with left posterior lesions to be most impaired on a topographical task, while McFie and Zangwill (1960) found a more extensive disability in patients with right hemisphere lesions.

1. Constructional Apraxia as a Spatial Agnosia

The test battery of almost every clinical psychologist, neurologist, or neuropsychologist includes tasks requiring the patient to draw geometric figures or to perform some other visual constructive task. Brain-damaged patients with either left- or right-sided lesions involving the parietal lobe frequently cannot copy figures accurately. There is now some strong evidence that addresses the question of qualitative differences between the right hemisphere patients' disability and the left hemisphere patients' disability (Warrington, James, & Kinsbourne, 1966). The results of these studies suggest that the left hemisphere patients manifest motor executional (apraxic) difficulties, while the right hemisphere patients show deficits in handling the spatial relationships of the task. It can be concluded, therefore, that the latter disorder is of an agnosic or perceptual nature, while the former is of an apraxic nature.

D. Autotopagnosia

Autotopagnosia is the disturbed perception of one's body image and body parts, best typified by the classic case of Pick's (1922), who, when asked to point to her ear, looked around a table top and replied that she must have lost it. The disability usually includes not only a disorder of one's body scheme, but also the inability to name body parts on command (whether of the patient or the examiner).

Finger agnosia and finger naming difficulties are the best known examples of specific autotopagnosias. Finger agnosia is the predominant characteristic of the Gerstmann syndrome (Gerstmann, 1924), which also includes dysgraphia, directional disorientation, and dyscalculia. While it is unclear as to whether the Gerstmann syndrome is a definite clinical entity (Benton, 1977), it is generally agreed that when all of the four symptoms which constitute the so-called Gerstmann syndrome occur in a single patient that it strongly suggests a lesion located (see Fig. 1) in the left inferior parietal region (Strub & Geschwind, 1974).

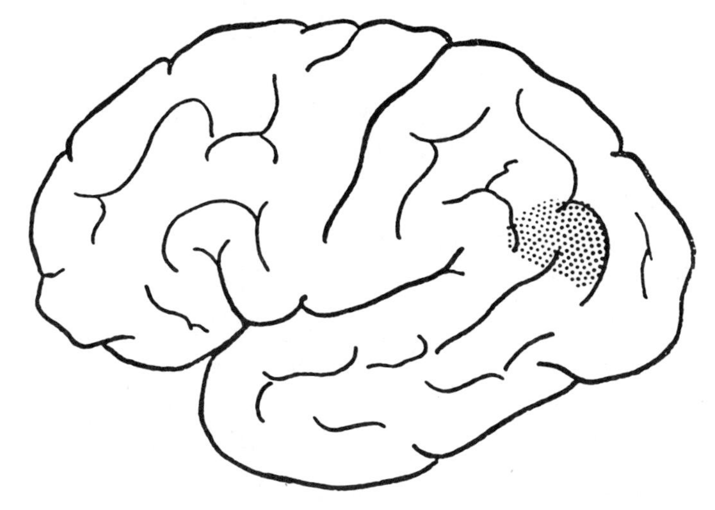

FIG. 1. The left angular gyrus. Site of lesions most commonly associated with the Gerstmann syndrome.

E. Color Agnosia

Wilbrand (1887) described a patient with intact color perception but with a marked disability in color naming. He gave the disorder the name "amnestic color blindness," and considered it a *dysnomia*. Conversely, Sittig (1921) studied a similar case with a left temporo-occipital lesion and was the first to emphasize the agnosic quality of the disorder. These two early reports characterize the difficulty clinicians have had in relating the symptom of defective color recognition to a class of disorders, either the aphasias or the agnosias. While the clinical picture is variable in color agnosia, several excellent case studies have demonstrated that the syndrome should not be considered a symptom of aphasia. The studies of Kleist (1934), Stengel (1948), and Kinsbourne and Warrington (1964) have shown that the syndrome of color agnosia is not characterized solely by a color-naming difficulty, but also by a defect in nonverbal recognition (e.g., sorting) of colors. An interesting feature of some cases reported in the literature (e.g., Kleist, 1934) is a relative sparing of recognition for white, gray, and black, which suggests that color perception is not intact in amnestic color blindness. The lesions associated with this disorder are generally in the left temporo-occipital region (Gloning, Gloning, & Hoff, 1968).

F. Agnosic Alexia

Benson (1977) has enumerated three forms of alexia, acquired reading disturbances: *agnosic alexia* (pure word blindness), *aphasic alexia*, and *frontal alexia*. Goldstein (1948) suggested that the alexias could be divided into two categories depending upon the absence or presence of aphasic disturbances. Primary (agnosic) alexia, or the syndrome of alexia without agraphia, occurs without significant aphasic symptoms and therefore is considered a "pure alexia" (Gloning, Gloning, Seidelberger, & Tschabitsher, 1955). Since agraphia and other aphasic symptomatology, such as anomia, occur in aphasic alexia and a motor aphasia is the predominant symptom in frontal alexia, these two alexias are considered secondary to disruption of the central language system.

Déjèrine (1892) was the first to provide evidence that two lesions are necessary to produce agnosic alexia, one in the splenium of the corpus callosum, which prohibits transfer of visual images from the right hemisphere to the language zones in the left hemisphere, and one in the angular gyrus, which disconnects the left visual cortex from the language zones.

The syndrome of agnosic alexia is particularly interesting because of the relative preservation of the ability to write. While spontaneous writing by agnosic alexics is usually less than perfect, writing to dictation is generally adequate. It is assumed that writing is intact because the auditory–kinesthetic associations are left intact by the lesions which have produced the visual–verbal disconnection.

V. THE SOMATOSENSORY SYSTEM AND TACTILE AGNOSIA

The somatic sensations are generally considered to be those of touch, pain, temperature, position sense, and unconscious proprioceptive sense. Neuropsychology has been largely concerned only with the tactile and nociceptive senses and therefore discussion shall be limited to these. The basic organization of the somatosensory system can be characterized as follows: First order cells outside the central nervous system receive sensory impressions and transmit impulses to the spinal cord (via the dorsal root ganglia or in the Gasserian ganglion) or brainstem. Some of the fibers project directly to the thalamus while others project to the reticular formation. The termination of this projection system (see Fig. 2) lies in post-central cerebral cortex (SSI and SSII). Lesions in the thalamus, spinothalamic tract, or spinal cord cause a variety of bilateral and unilat-

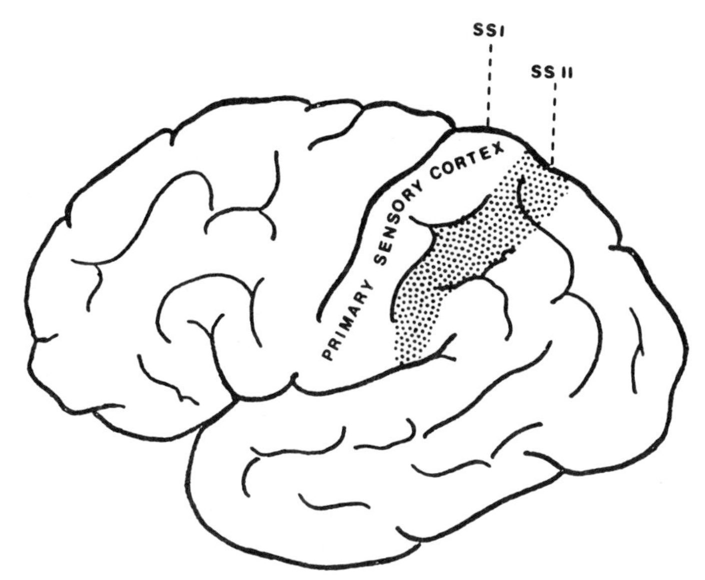

FIG. 2. Somatosensory cortex.

eral sensory losses while lesions in somatosensory cortex can result in contralateral or bilateral disturbances.

Lesions that involve the secondary zones of the parietal lobe often result in "tactile asymbolia" (Wernicke, 1894), or "parietal tactile agnosia" (Neilsen, 1946). The basic defect in this disorder is not one of sensory loss, but of a higher-level symbolic process involving the inability to recognize an object by touch. This disturbance, also known as astereognosia, has been regarded as an indication of a pathological narrowing of tactile perception (Luria, 1973), and also as a sensory–verbal disconnection (Geschwind, 1965). Perhaps the most compelling evidence for the mechanisms involved in astereognosis has come from the split-brain studies of Sperry, Gazzaniga, and Bogen (e.g., 1969). Tactile recognition of objects palpated with the left hand is severely disturbed in commissurotomized patients, presumably because the language zones of the left hemisphere are disconnected from the sensory representation of the object in the right hemisphere.

VI. NOCICEPTION AND "ASYMBOLIA FOR PAIN"

Interest in pain has increased dramatically during the past decade, but there have been surprisingly few attempts to assess the effects of cerebral lesions on pain perception in laboratory animals (Liebeskind & Paul, 1977). A typical finding of many studies has been a failure to attenuate pain perception when lesions are inflicted at supraspinal levels. Nevertheless, there is some behavioral evidence that an agnosia for nociception exists. Schilder and Stengel (1931) gave the first detailed description of the syndrome of *asymbolia for pain* in humans, and associated it with lesions involving the supramarginal gyrus of the left hemisphere (see Fig. 3). Patients with this syndrome can correctly distinguish certain sensations, but do not respond to the aversive nature of painful stimuli. Rubins and Friedman (1948) have presented cases which confirm the clinicopathological impressions of Schilder and Stengel, showing destruction of the secondary somatosensory area in the left parietal lobe. One possible explanation for this disorder has been suggested by Geschwind (1965), who postulated that a sensory-rhinencephalic disconnexion is responsible for the indifference to noxious sensory stimuli.

VII. LESIONS OF THE AUDITORY PATHWAYS

Although the auditory system has been the subject of much research, knowledge of the process of sensory transduction, the exact mechanism by which the organ of Corti converts mechanical energy into neural activity, and the brain–behavior relationships in the defective auditory

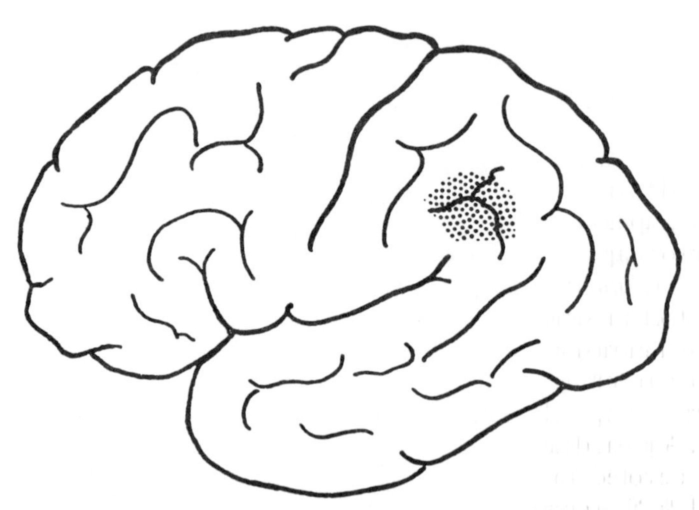

FIG. 3. The left supramarginal gyrus. Site of lesions responsible for the syndrome of "asymbolia for pain."

system, is very incomplete. The subthalamic relay nuclei in the auditory system are numerous and have complex interrelationships. It is not within the scope of this chapter to include a description of this intricate system, but it should be pointed out that lesions involving these nuclei or their pathways have a variety of unilateral and bilateral effects, although, in general, the system is very forgiving, no doubt because of the neural duplicity and decussation at several levels.

The primary projection zone of the auditory cortex is in the transverse gyri of Heschl and neurology textbooks have traditionally suggested that bilateral lesions cause deafness, but that exceptions to the rule exist (Peele, 1977). Unilateral lesions, unlike those in the visual system, seem to be well compensated by the second, intact gyrus and do not produce a total loss of hearing or a reduction in acuity for ordinary sounds, but simply cause the threshold for auditory sensations to be increased.

While clinical studies would generally appear to support these observations, recent evidence from animal experimentation (e.g., Heffner, 1977) suggests that bilateral lesions of the auditory cortex result only in transient disturbances of function. Heffner has reported, for instance, that bilateral ablation of both primary and secondary auditory cortex results only in a temporary inability to classify previously learned sounds in a two-choice experimental situation. Nevertheless, this gnostic disability diminishes within a few months and the animals perform at normal levels. It may be that other cortical auditory areas (such as Area 42 and Area 22) which receive afferents from the medial geniculate nucleus can compensate for destruction of Area 41.

VIII. SENSORY APHASIA

Perhaps the most distinctive disorder of auditory perceptual processing is the deterioration of "phonemic hearing," which is the hallmark of sensory aphasia, first recognized by Carl Wernicke (1874). The syndrome of sensory aphasia is more or less synonymous with the terms *Wernicke's, posterior, fluent*, and *receptive aphasia*. While articulation and fluency are intact in sensory aphasics, expressive speech is unintelligible with frequent intrusions of paraphasias, neologisms, and sound transpositions. Lesions resulting in sensory aphasia involve the posterior portion of the superior temporal gyrus (see Fig. 4) in the left hemisphere (Wernicke's area). A great deal of neurolinguistic and neuropsychological research has been devoted to sensory aphasia in recent years (e.g., Boller, 1978; Tallal & Newcombe, 1978; Buckingham & Kertesz, 1974), and in most large medical centers there is no lack of available patients with this

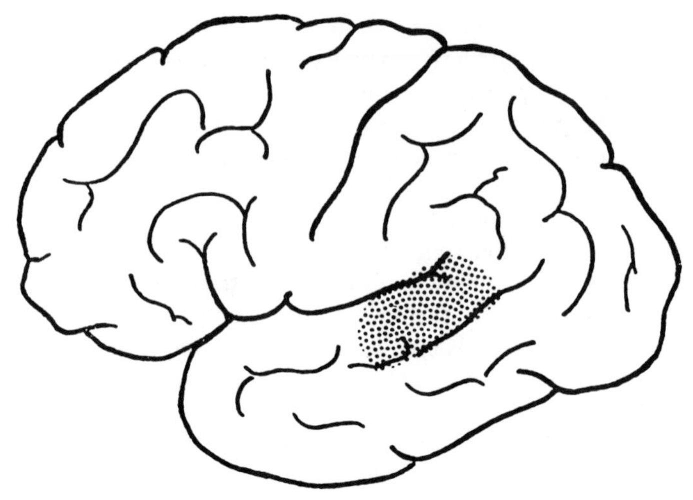

FIG. 4. Wernicke's area. Site of lesions most commonly associated with sensory aphasia.

disturbance. Considerably less is known about other auditory perceptual processing disorders which will be discussed later, perhaps owing to their relatively infrequent occurrence.

IX. AUDITORY AGNOSIA

Auditory agnosia, a term introduced by Freud (1891), is a rare disturbance of auditory perceptual processing including both "pure word deafness" and an agnosia for nonlinguistic material.

A. Pure Word Deafness

The syndrome of "pure word deafness" or subcortical sensory aphasia was introduced in the German literature by Kussmaul in 1877 (*reinen Wortthaubheit*) and in the French literature by Serieux in 1893 (*surdite verbale pure*). It is a disorder in which there is a selective loss of auditory–linguistic comprehension, while speech production, reading, and writing are intact (Goldstein, 1974). Hugo Liepmann (1898) provided the first postmortem evidence that a subcortical lesion disconnecting Wernicke's area from the transverse gyri of Heschl (primary auditory cortex) in the left and right hemispheres was responsible for the disorder. This clinicopathological correlation showing sparing of the auditory association cortex as well as primary auditory cortex, but with a grey matter lesion disconnecting these zones has stood up for over three-quarters of a century. Ulrich's analysis (1978) of 29 cases of auditory agnosia reported in the literature reveals that recurrent lesions involving both temporal lobes are usually responsible, and that the left hemisphere is usually the site of the first insult.

B. Sound Agnosia

The question as to the existence of *sound agnosia*, an agnosia that is selective for nonlinguistic sounds, and that does not include *sensory amusia* (agnosia for musical sounds), was raised as early as 1919 by S. E. Henschen, who could find no documentation of its existence in the literature. Fifteen years later, Kleist (1934) was also unsuccessful in finding a case with selective loss of nonlinguistic sound recognition. Spreen, Benton, and Fincham (1965) and Neilsen and Sult (1939), however, were able to document cases of sound agnosia with relative preservation of speech perception. It is assumed that many other cases exist but that the symptoms are mistaken for a hearing loss.

C. Auditory Affective Agnosia

A subtype of auditory agnosia that occurs almost exclusively after right temporoparietal lesions is the syndrome of *auditory affective agnosia* (Heilman, Scholes, & Watson, 1975), a disorder in recognizing the cues which may communicate the affective tone of speech. Although the mechanism of auditory affective agnosia is still far from clear, several studies have demonstrated that patients with right-sided lesions are markedly inferior to normals and even aphasics in the ability to comprehend and recall affectively charged stimuli (Gainotti, 1972; A. F. Wechsler, 1972).

D. Sensory Amusia

Milner (1962) studied the effects of partial left and right temporal lobectomies and determined that verbal materials were perceived with greater difficulty by patients who had undergone left temporal lobectomies, while patients who had right temporal lobectomies performed more poorly on musical perception tasks. Milner's results have been misinterpreted to suggest that verbal functions are carried out by the left hemisphere while musical functions are carried out by the right hemisphere. Unfortunately, the few documented cases of *sensory amusia* do not generate such a clear picture of the cerebral localization of disorders of musical perception (in addition, there is not a single case of sensory amusia in the literature in which the disorder occurs in isolation). Generally, the expressive amusias are more frequently associated with left temporal lesions (Wertheim, 1969) and there is no agreement as to the typical lesion site responsible for the receptive disorder. Wertheim (1969) has concluded that sensory amusia is usually associated with left temporal pole lesions, in agreement with a suggestion made by Henschen (1925), while Pick (1931) and Ustvedt (1937) each have argued for areas in the left auditory association cortex.

X. RELATED DISORDERS

A. Anosognosia

Von Monakow (1885) was the first clinician to provide evidence that neurological patients with certain posterior cerebral lesions were apparently unaware of their disability. Babinski (1914) later applied the term *anosognosia* to this disorder and further speculated that the site of the lesion was more frequently in the right rather than the left hemisphere. Still later, Gerstmann (1942) differentiated three forms of the disorder: the unawareness of hemiplegia, the denial of affected body parts, and the distortion of sensory impressions from the affected limbs.

B. The Neglect Syndrome

The perceptual processing disorders that result from lesions involving the posterior association cortex of the right hemisphere are predominantly those relating to the *syndrome of neglect* (Heilman & Watson, 1977). Although there are a variety of terms that refer to this syndrome, such as inattention, extinction to simultaneous stimulation, amorphosynthesis, disregard, and hemispatial agnosia, the presenting behavior of patients with neglect is remarkably uniform. The syndrome is characterized by inattention to sensory stimuli coming from one hemifield in the absence of a field defect. The disorder has been demonstrated in the visual (Holmes, 1919), auditory (Diamond & Bender, 1965), and tactile modalities (Critchley, 1949), with occasional reports of patients having neglect in all three perceptual modalities (Denny-Brown *et al.*, 1952). The excellent work of Heilman and his associates (Heilman & Watson, 1977; Heilman & Valenstein, 1972, etc.) has implicated the role of a unilaterally impaired orienting response involving a cortico–subcortical loop in the neglect syndrome.

XI. THE CHEMICAL SENSES

Very little is known about the physiology of the two chemical senses, the mechanisms of sensory transduction, and the clinical disorders that affect the functioning of these systems. The only exception to this statement comes from the large amount of data which is accruing from taste aversion studies (e.g., Nachman & Hartley, 1975; Garcia & Ervin, 1968). Loss of gustatory sensation can result from damage to the cranial nerves which carry sensory information from the tongue (the facial nerve and the glossopharyngeal nerve), as well as brainstem lesions involving the Nucleus or Tractus Solitarius, thalamic and opercular–insular cortical lesions. Gustatory agnosia is a purely theoretical notion, but recent studies of hypogeusia, as well as taste aversion, have established that strong associations between gustatory and other meaningful stimuli can be formed and destroyed.

Olfactory perception plays a relatively minor role in human behavior. Anosmia, or defective olfactory sensation, can often go undetected. The only significant behavioral impairment, of course, is on the appreciation of the taste of food. Most anosmia is accounted for by allergic rhinitis and not by a neurological disorder. The most common neurological cause of olfactory disorders are lesions affecting the olfactory bulb, although lesions affecting the prepyriform and periamygdaloid cortex produce anosmia as do olfactory groove meningiomas. Less is known about olfactory perception than any other perceptual modality. There are great system-limiting constraints on research involving this lower-level sensory system, such as the technical problems of single unit recording from olfactory fibers.

XII. DEVELOPMENTAL DISORDERS OF PERCEPTUAL PROCESSING

Many of the disorders of higher cortical function discussed in this chapter occur in children not as a result of verified brain damage, but probably as a result of late or incomplete maturation of the brain regions underlying a given function. While investigators have been singularly unsuccessful in discovering the pathophysiology involved in these disorders, the search is among the most active areas of research in neuropsychology.

Developmental dyslexia is the syndrome of delayed reading acquisition in children with normal intelligence, sensory acuity, and emotional adjustment and who have had adequate instructional and sociocultural opportunities to learn to read. Interestingly, the first clinical observations of the syndrome were made by Morgan (1896), whose interest was aroused by Hinshelwood's (1895) studies of acquired dyslexia. Thus, the historical roots of research on developmental dyslexia are in the disciplines of behavioral neurology and neuropsychology. Morgan suggested at the time that the angular gyrus, the brain region known to be structurally altered in acquired dyslexia, was underdeveloped in developmental dyslexics. Fisher (1910), and more recently, Geschwind (1965) and Pirozzolo (1977a) have concurred with this etiologic hypothesis by showing that other functions that are subserved by this region have also been found to be delayed in children with dyslexia.

Recently several investigators have been able to demonstrate that developmental dyslexia is not a single homogeneous clinical entity and that there are, in fact, two or more independent neuropsychological syndromes subsumed under the heading of developmental dyslexia (Kinsbourne & Warrington, 1963a; Mattis, French, & Rapin, 1975; Pirozzolo, 1977a, 1978a; Pirozzolo & Rayner, 1978a). One group of dyslexics is characterized by visual–spatial gnostic and oculomotor difficulties (Pirozzolo & Rayner, 1978b), while a second group has symptoms of a mild auditory–linguistic disorder (Pirozzolo, Rayner, & Whitaker, 1977).

Another possible clue to mechanisms related to developmental dyslexia has been reported by Hier, LeMay, Rosenberger, and Perlo (1978). These investigators found a reversal of cerebral asymmetry in subjects with developmental dyslexia. Geschwind and Levitsky (1968) were the first to discover that most right-handed adults had a larger left than right planum temporale region, thus indicating that the larger size of this left area may provide a more favorable anatomical substrate for the development of language. Hier and his colleagues found the reverse in certain dyslexics, that is, the width of the right hemisphere is greater than that of the left hemisphere. This is the first correlative anatomical evidence that supports Orton's (1937) speculation that dyslexics have anomalous cerebral dominance.

Numerous other disorders of perceptual processing have been found in children who have no other evidence of neurological dysfunction. A "developmental Gerstmann syndrome," consisting of the symptoms of left–right disorientation, spatial dysgraphia, finger agnosia, and dyscalculia, has been described by Kinsbourne and Warrington (1963b), Benson and Geschwind (1970), and Pirozzolo and Rayner (1978b). Pontius (1976) has identified a group of children who appear to suffer from a developmental prosopagnosia. Rapin, Mattis, Rowan, & Golden (1977) have recently identified a subsample of children with developmental language disabilities that closely correspond to the deficits observed in adults with pure word deafness. The prospects for future research in the areas of childhood language and perceptual disorders are outstanding and, with research on remediation and rehabilitation, are one of the foremost challenges of clinical neuropsychology.

XIII. CONCLUSIONS

This chapter was intended to review the functional disorders which affect perceptual processing in the human brain. A review of some of the operating characteristics for each of the perceptual systems has been presented, in addition to brief discussions of the major disorders of perceptual processing. One of the assumptions in this review, and indeed in all of neuropsychological research, is that the study of these disorders will have implications for the normal perceptual process through the investigation of the patterns of impairment to which these processes are subject.

While great progress has been made in recent years toward understanding the normal and abnormal functioning perceptual systems, many issues remain to be solved. Sensory neurophysiologists have made major achievements toward understanding how single neurons reconstruct sensory impressions, but knowledge of the perceptual function of groups of neurons has not advanced so dramatically. In addition to elucidating the mechanisms of sensory transduction, a more precise knowledge of cell specificity, plasticity, and restoration of function is needed. It also seems important to specify the role of efferent systems on perception, such as the role of eye movements in visual perception. Luria (1966) has suggested, for instance, that visual perceptual difficulties typically occur after damage to the frontal motor association cortex. Preliminary evidence about the role of disordered eye movements in visual perceptual processing suggests a strong relationship between oculomotor scanning defects and subsequent visual disorientation (Pirozzolo, 1978b; Pirozzolo & Rayner, 1978a,b).

A problem that has not been addressed in this chapter, but which is a popular topic for research in neuropsychology, concerns the manner in which the cerebral hemispheres differ in their abilities to process certain stimuli. Split-brain research (e.g., Sperry *et al.*, 1969), visual hemifield research (e.g., Pirozzolo & Rayner, 1977; Pirozzolo, 1977b), and dichotic listening studies (e.g., Kimura, 1967) have provided some important data on how information processing strategies differ between the hemispheres, but solutions to the problem of defining the operating characteristics of the hemispheres remain to be found.

References

Babinski, J. Contribution a l'etude des troubles mentaux dan l'hemiplegie organique cerebrale (anosognosie). *Revue de Neurologie*, 1914, **27,** 845–848.

Balint, R. Seelenlahmung des Schauens, optische Ataxie, raumliche Storung der Aufmerksamkeit. *Monatsschrift für Psychiatrie und Neurologie,* 1909, **25,** 51–81.

Bay, E. Uber den Begriff der Agnosie. *Nervenarzt*, 1951, **22,** 179–187.

Bay, E. Disturbances of visual perception and their examination. *Brain*, 1953, **76** 515–551.

Bender, L. *A visual motor Gestalt test and its clinical use*. New York: The American Orthopsychiatric Association, 1939.

Benson, D. F. The third alexia. *Archives of Neurology*, 1977, **34,** 327–331.

Benson, D. F., & Geschwind, N. The developmental Gerstmann syndrome. *Neurology*, 1970, **20,** 203–208.

Benson, D. F., Segarra, J., & Albert, M. Visual agnosia-prosopagnosia. *Archives of Neurology*, 1974, **30,** 307–310.

Benton, A. Reflections on the Gerstmann syndrome. *Brain and Language*, 1977, **4,** 44–61.

Bodamer, J. Die Prosop-Agnosie (die Agnosie des Physiognomieerkeenens). *Archiv für Psychiatrie und Nervenkrankheiten*, 1947, **179,** 6–53.

Bogen, J., & Vogel, P. J. Cerebral commissurotomy: A case report. *Bulletin of the Los Angeles Neurological Society*, 1962, **27,** 169–172.

Boller, F. Comprehension disorders in aphasia: A historical review. *Brain and Language*, 1978, **5,** 149–165.

Botez, M. I. Two visual systems in clinical neurology: Readaptive role of the primitive system in visual agnostic patients. *European Neurology*, 1975, **13,** 101–122.

Brown, J. W. *Aphasia, apraxia and agnosia*. Springfield, Illinois: Thomas, 1972.

Buckingham, H., & Kertesz, A. A linguistic analysis of fluent aphasia. *Brain and Language*, 1974, **1,** 43–62.

Cohn, R., Neumann, M., & Wood, D. H. Prosopagnosia: A clinicopathological study. *Annals of Neurology*, 1977, **1,** 177–182.

Critchley, M. The phenomenon on tactile inattention with special reference to parietal lesions. *Brain*, 1949, **72,** 538–561.

Critchley, M. *The parietal lobes*. London: Edward Arnold, 1953.

Critchley, M. The problem of visual agnosia. *Journal of Neurological Science*, 1964, **1,** 274–290.

Déjèrine, J. Des differentes varietes de cecite verbale. *Memoires de la Societe de Biologie*, 1892, **Feb. 27,** 1–30.

Dennis, M., & Whitaker, H. Language acquisition following hemidecortication: Linguistic superiority of the left over the right hemisphere. *Brain and Language*, 1976, **3,** 404–433.

Denny-Brown, D., Meyer, J. S., & Horenstein, S. The significance of perceptual rivalry resulting from parietal lesion. *Brain*, 1952, **75,** 433–471.

DeRenzi, E., & Vignolo, T. The token test: A sensitive test to detect receptive disturbances in aphasia. *Brain*, 1962, **85,** 665–678.

Diamond, S., & Bender, M. On auditory extinction and allorcusis. *Transactions of the American Neurological Association*, 1965, **90,** 154–157.

Dick, A. O. Spatial abilities. In H. A. Whitaker & H. A. Whitaker (Eds.), *Studies in neurolinguistics* (Vol. 2). New York: Academic Press, 1976.

Finkelnburg, F. C. Niederrheinische Gesellschaft in Bonn. Medicinische Section. *Berlin Klinische Wochenscrift*, **7,** 1870, 449–450; 460–461.

Fisher, J. H. Congenital word-blindness (inability to learn to read). *Transactions of the Ophthalmological Society of the United Kingdom*, 1910, **30,** 216–225.

Freud, S. *Zur Auffassung der Aphasien*. Wien: F. Duetiche, 1891.

Gainotti, G. Emotional behavior and hemispheric side of the lesion. *Cortex*, 1972, **8,** 41–55.

Garcia, J., & Ervin, F. R. Gustatory–visceral and telerceptor–cutaneous conditioning: Adaptation in external and internal milieus. *Communications in Behavior Biology*, 1968, **1,** 389–415.

Gardner, H. *The shattered mind*. New York: Random House, 1975.

Gerstmann, J. Fingeragnosie. *Weinische Klinische Wochenschrift*, 1924, **40,** 1010–1012.

Gerstmann, J. Problem of perception of disease and of impaired body territories with organic lesions. *Archives of Neurology and Psychiatry*, 1942, **48,** 890–913.

Geschwind, N. Disconnexion syndromes in animals and man. *Brain*, 1965, **88,** 237–294; 585–644.

Geschwind, N., & Levitsky, W. Human brain: Left–right asymmetries in temporal speech region. *Science*, 1968, **161,** 186–188.

Gloning, I., Gloning, K., & Hoff, H. *Neuropsychological syndromes in lesions of the occipital lobes and the adjacent areas*. Paris: Gauthiers-Villars, 1968.

Gloning, I., Gloning, K., Seidelberger, F., & Tschabitscher, H. Ein Fall von Wortblindheit mit Obduktionsbefund. *Wein. Z. Nervenheilkd,* 1955, **12,** 194–215.

Gloning, I., Gloning, K., Hoff, H., & Tschabitscher, H. Zur Prosopagnosie. *Neuropsychologia*, 1966, **4,** 113–132.

Goldstein, K., & Gelb, A. Psychologische analysen hirnpathologischer Falle auf Grund von Untersuchungen Hirnvertetzter. *Zeitschrift für die Gesamte Neurologie und Psychiatrie*, 1918, **41,** 1–142.

Goldstein, K., & Gelb, A. *Psychologische Analyse hirnpatologischer Falle*. Leipzig: Barth, 1920.

Goldstein, K. *Language and language disturbances*. New York: Grune & Stratton, 1948.

Goldstein, M. Auditory agnosia for speech. *Brain and Language*, 1974, **1,** 195–204.

Goodglass, H., & Kaplan, E. *The assessment of aphasia and related disorders*. Philadelphia: Lea & Febiger, 1972.

Halstead, W. C. *Brain and intelligence*. Chicago: Univ. of Chicago Press, 1947.

Halstead, W. C., & Wepman, J. M. The Halstead–Wepman screening test. *Journal of Speech and Hearing Disorders*, 1949, **14,** 9–15.

Hecaen, H. Clinical symptomatology in right and left hemisphere lesions. In V. Mountcastle (Ed.), *Interhemispheric relations and cerebral dominance*. Baltimore: Johns Hopkins Press, 1962.

Hecaen, H., & Angelergues, R. *La cecite psychique: Etude critique de la notion d'agnosie*. Paris: Masson et Cie, 1963.

Heffner, H. E. Effect of auditory cortex ablation on the perception of meaningful sounds. *Neuroscience Abstracts*, 1977, **3,** 6.

Heilman, K. M., Scholes, R., & Watson, R. T. Auditory affective agnosia. *Journal of Neurology, Neurosurgery, and Psychiatry,* 1975, **38,** 69–72.

Heilman, K. M., & Valenstein, E. Auditory neglect in man. *Archives of Neurology*, 1972, **26,** 32–35.

Heilman, K. M., & Watson, R. T. The neglect syndrome—A unilateral defect of the orienting response. In S. Harnad, R. W. Doty, L. Goldstein, J. Jaynes, & G. Krauthamer (Eds.), *Lateralization in the nervous system*. New York: Academic Press, 1977.

Henschen, S. Clinical and anatomical contributions to brain pathology. *Archives of Neurology and Psychiatry*, 1925, **13,** 226–249.

Henschen, S. E. On the hearing sphere. *Acta Oto-laryngologica*, 1919, **1,** 423–486.

Hier, D. B., LeMay, M., Rosenberger, P., & Perlo, V. P. Developmental dyslexia. *Archives of Neurology*, 1978, **35,** 90–92.

Hinschelwood, J. Wordblindness and visual memory. *Lancet*, 1895, **2,** 1564–1570.

Hoff, H., & Potzl, O. Experimentelle Nachbildung von Anosognsie. *Zeitschrift für Neurologie und Psychiatrie*, 1937, **137,** 722–734.

Holmes, G. Disturbances of visual space perception. *British Medical Journal*, 1919, **2,** 230–233.

Hubel, D., & Wiesel, T. Receptive fields and functional architecture in two non-striate visual areas (18 and 19) of the cat. *Journal of Neurophysiology*, 1965, **28,** 229–289.

Jackson, J. H. Case of large cerebral tumour without optic neuritis and with left hemiplegia and imperception. In J. Taylor (Ed.), *Selected writings of John Hughlings Jackson*, (2 vols.). New York: Basic Books, 1958. (Originally published, 1876).

Johnson, C. A., Leibowitz, H. W., Millodot, M., & Lamont, A. Peripheral visual acuity and refractive error: Evidence for two visual systems? *Perception and Psychophysics*, 1976, **20,** 460–462.

Joynt, R., & Goldstein, M. Minor cerebral hemisphere. In W. J. Friedlander (Ed.), *Advances in Neurology* (Vol. 7). New York: Raven Press, 1975.

Keating, E. G. Impaired orientation after primate tectal lesions. *Brain Research*, 1974, **67,** 538–541.

Kimura, D. Functional asymmetry of the brain in dichotic listening. *Cortex*, 1967, **3,** 163–178.

Kinsbourne, M., & Warrington, E. K. A disorder of simultaneous form perception. *Brain*, 1962, **85,** 461–486.

Kinsbourne, M., & Warrington, E. K. Developmental factors in reading and writing backwardness. *British Journal of Psychology*, 1963, **54,** 145–156. (a)

Kinsbourne, M., & Warrington, E. K. The developmental Gerstmann syndrome. *Archives of Neurology*, 1963, **8,** 490–501. (b)

Kinsbourne, M., & Warrington, E. K. Observations on colour agnosia. *Journal of Neurology, Neurosurgery, and Psychiatry*, 1964, **27,** 296–299.

Kleist, K. Der Gan und der Gegenwurte Stand der Apraxie-forschung. *Ergebaisse Neurologie und Psychiatrie*, 1912, **1,** 342–452.

Kleist, K. *Gehirnpathologie*. Leipzig: Barth, 1934.

Kluver, H. Functional significance of the geniculostriate system. *Biological Symposium*, 1942, **7,** 253–299.

Kluver, H., & Bucy, P. C. An analysis of certain effects of bilateral temporal lobectomy in the rhesus monkey, with special reference to "psychic blindness." *Journal of Psychology*, 1938, **5,** 33–54.

Kohn, B., & Dennis, M. Hemispheric specialization after hemidecortication. In M. Kinsbourne & W. L. Smith (Eds.), *Hemispheric disconnection and cerebral function*. Springfield, Illinois: Thomas, 1974.

Kussmaul, A. Disturbances of speech. In H. von Ziemssen (Ed.), *Cyclopedia of the Practice of Medicine* (Vol. 14). New York: William Wood, 1877.

Lezak, M. D. *Neuropsychological assessment*. New York: Oxford Univ. Press, 1976.

Liebeskind, J. C., & Paul, L. Psychological and physiological mechanisms of pain. *Annual Review of Psychology*, 1977, **28,** 41–60.

Liepmann, H. Ein Fall von reiner Sprachtaubheit. In C. Wernicke (Ed.), *Psychiatrische Abhandlunger*, Breslau: Schletter, 1898.

Lissauer, H. Ein Fall von Seelenblindheit nebst einem Beitrag zur Theorie derselben. *Archiv für Psychiatrie und Nervenkrankheit*, 1890, **21,** 222–270.
Luria, A. R. Disorders of simultaneous perception in a case of bilateral occipitoparietal brain injury. *Brain*, 1959, **82,** 437–449.
Luria, A. R. *Higher cortical functions in man*. New York: Basic Books, 1966.
Luria, A. R. *The working brain*. New York: Basic Books, 1973.
Luria, A. R., Pravdina-Vinarskaya, E. N., & Yarbus, A. L. Disorders of ocular movement in a case of simultanagnosia. *Brain*, 1963, **86,** 219–228.
Mattis, S., French, J., & Rapin, E. Dyslexia in children and young adults: Three independent neuropsychological syndromes. *Developmental Medicine and Child Neurology*, 1975, **17,** 150–163.
McFie, J., & Zangwill, O. Visual-constructive disabilities associated with lesions of the left cerebral hemisphere. *Brain*, 1960, **83,** 243–260.
Meyer, O. Ein-und Doppelseitige Homonyme Hemianopsie mit orientierungsstorungen. *Monatsschrift für Psychiatrie und Neurologie*, 1900, **8,** 440.
Milner, B. Laterality effects in audition. In V. B. Mountcastle (Ed.), *Interhemispheric relations and cerebral dominance*. Baltimore: Johns Hopkins Press, 1962.
Morgan, W. P. A case of congenital word blindness. *British Medical Journal*, 1896, **2,** 1378.
Munk, H. Ueber die Functionen der Grosshirnrinde. *Gesammelte Mittheilunge aus den Jahren 1877–1880*. Berlin: Hirschwald, 1881.
Nachman, M., & Hartley, P. L. Role of illness in producing learned taste aversions in rats: A comparison of several rodenticides. *Journal of Comparative and Physiological Psychology*, 1975, **89,** 1010–1018.
Neilson, J. M. *Agnosia, apraxia and aphasia*. New York: Hoeber, 1946.
Neilsen, J. M., & Sult, C. W. Agnosias and the body scheme. *Bulletin of the Los Angeles Neurological Society*, 1939, **4,** 69–76.
Newcombe, F. *Penetrating missile wounds of the brain*. Oxford: Oxford Univ. Press, 1969.
Orton, S. T. *Reading, writing and speech problems in children*: New York: Norton, 1937.
Peele, T. *The neuroanatomic basis of clinical neurology*. New York: McGraw-Hill, 1977.
Penfield, W., & Roberts, L. *Speech and brain mechanisms*. Princeton, New Jersey: Princeton Univ. Press, 1959.
Pick, A. Storung der Orientierung am eigenen Korper. *Psychologische Forschung*, 1922, **1,** 303–318.
Pick, A. Aphasie. In A. Bethe & G. V. Bergmann, *Handbuch der Normalen und Pathologischen Physiologie* (Vol. 15). Berlin: 1931.
Pirozzolo, F. J. *The neuropsychology of developmental reading disorders*. New York: Praeger, 1977. (a)
Pirozzolo, F. J. Lateral asymmetries in visual perception: A review of tachistoscopic visual half-field research. *Perceptual and Motor Skills*, 1977, **45,** 695–701. (b)
Pirozzolo, F. J. Cerebral asymmetries and reading acquisition. *Academic Therapy*, 1978, **13,** 261–266. (a)
Pirozzolo, F. J. Slow saccades. *Archives of Neurology*, 1978, **35,** in press. (b)
Pirozzolo, F. J., & Rayner, K. Hemispheric specialization in reading and word recognition. *Brain and Language*, 1977, **4,** 248–261.
Pirozzolo, F. J., & Rayner, K. The neural control of eye movement in acquired and developmental reading disorders. In H. A. Whitaker & H. A. Whitaker (Eds.), *Studies in Neurolinguistics* (Vol. 4). New York: Academic Press, 1978. (a)
Pirozzolo, F. J., & Rayner, K. Disorders of oculomotor scanning and graphic orientation in developmental Gerstmann syndrome. *Brain and Language*, 1978, **5,** 119–126. (b)
Pirozzolo, F. J., Rayner, K., & Whitaker, H. A. *Left hemisphere mechanisms in dyslexia*.

Paper presented to the Fifth Annual Convention of the International Neuropsychological Society, Sante Fe, New Mexico, 1977.

Pirozzolo, F. J., Whitaker, H. A., Selnes, O. A., & Horner, F. Linguistic specialization of the left hemisphere. *Neuroscience Abstracts*, 1977, **3,** 748.

Pontius, A. A. Dyslexia and specifically distorted drawings of the face: A new subgroup with prosopagnosia-like signs. *Experientia*, 1976, **32,** 1432–1435.

Poppel, E., Held, R., & Frost, D. Residual visual function after brain wounds involving the central visual pathways in man. *Nature*, 1973, **243,** 295–296.

Pribram, K. Discussion. In V. B. Mountcastle (Ed.), *Interhemispheric relations and cerebral dominance*. Baltimore: Johns Hopkins Press, 1962.

Rapin, I., Mattis, S., Rowan, A. J., & Golden, G. G. Verbal auditory agnosia in children. *Developmental Medicine and Child Neurology*, 1977, **19,** 192–207.

Raven, J. C. *Raven progressive matrices*. New York: The Psychological Corporation, 1956.

Reitan, R. Investigation of the validity of Halstead's measures of biological intelligence. *Archives of Neurology and Psychiatry*, 1955, **48,** 474–477.

Rizzolatti, C., Umilta, C., & Berlucchi, G. Opposite superiorities of the right and left cerebral hemispheres in discriminative reaction time to physiognomical and alphabetical material. *Brain*, 1971, **94,** 431–442.

Rubens, A., & Benson, D. F. Associative visual agnosia. *Archives of Neurology*, 1971, **24,** 305–316.

Rubins, J. L., & Friedman, E. D. Asymbolia for pain. *Archives of Neurology and Psychiatry*, 1948, **60,** 554.

Schilder, P., & Stengel, E. Asymbolia for pain. Report of a case. *Archives of Neurology and Psychiatry*, 1931, **25,** 598–600.

Schneider, G. E. Two visual systems. *Science*, 1969, **163,** 895–902.

Semmes, J., Weinstein, S., Ghent, L., & Teuber, H-L. Spatial orientation in man after cerebral injury. I. Analyses by locus of lesion. *Journal of Psychology*, 1955, **39,** 227–244.

Serieux, P. Sur un cas de surdite verbale pure. *Revue de Medicine*, 1893, **13,** 733.

Sittig, O. Storungen im Verhalten gegenuber Farben bei Aphasischen. *Monatsschift fur Psychiatrie und Neurologie*, 1921, **49,** 63–88; 169–187.

Smith, A. Dominant and nondominant hemispherectomy. In M. Kinsbourne & W. L. Smith (Eds.), *Hemispheric disconnection and cerebral function*. Springfield, Illinois: Thomas, 1974.

Smith, A., & Burkland, C. W. Dominant hemispherectomy. *Science*, 1966, **153,** 1280–1282.

Sperry, R. W., Gazzaniga, M. S., & Bogen, J. E. Interhemispheric relationships: The neocortical commissures; syndromes of hemisphere disconnections. In P. S. Vinken & G. Bruyn (Eds.), *Handbook of clinical neurology* (Vol. 4). Amsterdam: North Holland, 1969.

Spreen, O., Benton, A., & Fincham, R. Auditory agnosia without aphasia. *Archives of Neurology*, 1965, **13,** 84–92.

Stengel, E. The syndrome of visual alexia with colour agnosia. *Journal of Mental Science*, 1948, **94,** 46–58.

Strub, R., & Geschwind, N. Gerstmann syndrome without aphasia. *Cortex*, 1974, **10,** 378–387.

Strub, R. L., & Black, F. W. *The mental status examination in neurology*. Philadelphia: F. A. Davis, 1977.

Tallal, P. & Newcombe, F. Impairment of auditory perception and language comprehension in dysphasia. *Brain and Language*, 1978, **5,** 13–24.

Teuber, H-L., Battersby, W. A., & Bender, M. B. *Visual field defects after penetrating wounds of the brain*. Cambridge: Harvard Univ., 1960.

AUTHOR INDEX

Numbers in italics refer to the pages on which the complete references are listed.

G

H

I

J

M

R

S

SUBJECT INDEX

W

HANDBOOK OF PERCEPTION

EDITORS: *Edward C. Carterette and Morton P. Friedman*

Department of Psychology
University of California, Los Angeles
Los Angeles, California

Volume I: Historical and Philosophical Roots of Perception. 1974

Volume II: Psychophysical Judgment and Measurement. 1974

Volume III: Biology of Perceptual Systems. 1973

Volume IV: Hearing. 1978

Volume V: Seeing. 1975

Volume VIA: Tasting and Smelling. 1978

Volume VIB: Feeling and Hurting. 1978

Volume VII: Language and Speech. 1976

Volume VIII: Perceptual Coding. 1978

Volume IX: Perceptual Processing. 1978

Volume X: Perceptual Ecology. 1978

. . . . .

CONTENTS OF OTHER VOLUMES

Volume I: Historical and Philosophical Roots of Perception

CONTENTS OF OTHER VOLUMES

Volume IV: Hearing

Volume V: Seeing

Volume VII: Language and Speech

CONTENTS OF OTHER VOLUMES

Volume X: Perceptual Ecology

A 8
B 9
C 0
D 1
E 2
F 3
G 4
H 5